U.S. ARMY RANGER HANDBOOK

U.S. ARMY RANGER HANDBOOK

Revised and Updated Edition

U.S. Department of Defense

Skyhorse Publishing

Skyhorse Publishing books may be purchased in bulk at special discounts for sales promotion, corporate gifts, fund-raising, or educational purposes. Special editions can also be created to specifications. For details, contact the Special Sales Department, Skyhorse Publishing, 307 West 36th Street, 11th Floor, New York, NY 10018 or info@skyhorsepublishing.com.

Skyhorse® and Skyhorse Publishing® are registered trademarks of Skyhorse Publishing, Inc.®, a Delaware corporation.

Visit our website at www.skyhorsepublishing.com.

10 9 8 7 6 5 4

Library of Congress Cataloging-in-Publication Data is available on file.

ISBN: 978-1-61608-877-4

Printed in the United States of America

TABLE OF CONTENTS

RANGER CREED xi
STANDING ORDERS ROGER'S RANGERS xi
RANGER HISTORY xii
PREFACE xiii

CHAPTER 1 LEADERSHIP
- PRINCIPLES 1-1
- DUTIES, RESPONSIBILITIES, AND ACTIONS 1-2
- ASSUMPTION OF COMMAND 1-8

CHAPTER 2 OPERATIONS
- TROOP-LEADING PROCEDURES 2-1
- COMBAT INTELLIGENCE 2-5
- WARNING ORDER 2-6
- OPERATION ORDER 2-10
- FRAGMENTARY ORDER 2-14
- ANNEXES 2-17
- COORDINATION CHECKLISTS 2-25
- TASK, PURPOSE, OPERATION 2-29
- TERRAIN MODEL 2-30

CHAPTER 3 FIRE SUPPORT
- BASIC FIRE SUPPORT TASKS, TARGETING, AND INTERDICTION 3-1
- CAPABILITIES 3-2
- RISK ESTIMATE DISTANCES 3-2
- TARGET OVERLAYS 3-3
- CALL FOR FIRE 3-5
- CLOSE AIR SUPPORT 3-8
- CLOSE COMBAT ATTACK AVIATION 3-10

CHAPTER 4 COMMUNICATIONS

EQUIPMENT
- MILITARY RADIOS 4-1
- MAN-PACK RADIO ASSEMBLY (AN/PRC-119F) 4-4
- AUTOMATED NET CONTROL DEVICE 4-5
- BASIC TROUBLESHOOTING 4-6

ANTENNAS
- REPAIRS 4-6
- CONSTRUCTION AND ADJUSTMENT 4-7
- FIELD EXPEDIENT (FE) OMNI DIRECTIONAL ANTENNAS 4-8

ANTENNA LENGTH PLANNING CONSIDERATIONS 4-12

CHAPTER 5 DEMOLITIONS
INITIATING (PRIMING) SYSTEMS 5-3
DETONATION (FIRING) SYSTEMS 5-4
SAFETY 5-4
EXPEDIENT EXPLOSIVES - IMPROVISED SHAPED CHARGE 5-4
EXPEDIENT EXPLOSIVES - PLATTER CHARGE 5-5
EXPEDIENT EXPLOSIVES - GRAPESHOT CHARGE 5-6
DEMOLITION KNOTS 5-7
MINIMUM SAFE DISTANCES 5-8
BREACHING CHARGES 5-8
TIMBER CUTTING CHARGES 5-11

CHAPTER 6 MOVEMENT
FORMATIONS 6-1
MOVEMENT TECHNIQUES 6-1
STANDARDS 6-4
FUNDAMENTALS 6-4
TACTICAL MARCHES 6-5
MOVEMENT DURING LIMITED VISIBILITY CONDITIONS 6-6
DANGER AREAS 6-7

CHAPTER 7 PATROLS
PRINCIPLES
PLANNING 7-1
RECONNAISSANCE 7-1
SECURITY 7-1
CONTROL 7-1
COMMON SENSE 7-1

PLANNING
TASK ORGANIZATION 7-1
INITIAL PLANNING AND COORDINATION 7-3
COMPLETION OF PLAN 7-3

RECONNAISSANCE PATROLS
FUNDAMENTALS OF RECONNAISSANCE 7-5
TASK STANDARDS 7-5
ACTIONS ON THE OBJECTIVE, AREA RECONNAISSANCE 7-5
ACTIONS ON THE OBJECTIVE, ZONE RECONNAISSANCE 7-8

COMBAT PATROLS

PLANNING CONSIDERATIONS 7-9
AMBUSH 7-10
HASTY AMBUSH 7-11
DELIBERATE (POINT/AREA) AMBUSH 7-12
PERFORM RAID 7-15

SUPPORTING TASKS
LINKUP 7-18
DEBRIEF 7-18
OBJECTIVE RALLY POINT 7-19
PATROL BASE 7-20

MOVEMENT TO CONTACT
TECHNIQUES 7-23
TASK STANDARDS 7-24

CHAPTER 8 **BATTLE DRILLS**
REACT TO CONTACT (VISUAL, IED, DIRECT FIRE [RPG]) (07-3-D9501) 8-1
BREAK CONTACT (07-3-D9505) 8-6
REACT TO AMBUSH (FAR) (07-3-D9503) 8-9
REACT TO AMBUSH (NEAR) (07-3-D9502) 8-12
KNOCK OUT BUNKER (07-3-D9406) 8-15
ENTER AND CLEAR A ROOM (07-4-D9509) 8-18
ENTER A TRENCH TO SECURE A FOOTHOLD (07-3-D9410) 8-21
BREACH A MINED WIRE OBSTACLE (07-3-D9412) 8-25
REACT TO INDIRECT FIRE (07-3-D9504) 8-28

CHAPTER 9 **MILITARY MOUNTAINEERING**
TRAINING 9-1
DISMOUNTED MOBILITY 9-1
TASK ORGANIZATION 9-1
RESCUE EQUIPMENT 9-2
MOUNTAINEERING EQUIPMENT 9-3
ANCHORS 9-5
KNOTS 9-8
BELAYS 9-13
CLIMBING COMMANDS 9-15
ROPE INSTALLATIONS 9-15
RAPPELLING 9-22

CHAPTER 10 **MACHINE GUN EMPLOYMENT**
SPECIFICATIONS 10-1
DEFINITIONS 10-2

CLASSES OF AUTOMATIC WEAPONS FIRE 10-3
OFFENSE 10-8
DEFENSE 10-9
CONTROL OF MACHINE GUNS 10-10
AMMUNITION PLANNING 10-11

CHAPTER 11 CONVOY OPERATIONS
PLANNING 11-1
FIVE PHASES OF TRUCK MOVEMENT 11-1

CHAPTER 12 URBAN OPERATIONS
AN URBAN PERSPECTIVE 12-1
STRATEGIC IMPORTANCE OF URBAN AREAS 12-1
MODERN ARMY URBAN OPERATIONS 12-1
TASK ORGANIZATION 12-1
FULL SPECTRUM OPERATIONS 12-1
PREPARATIONS FOR FUTURE URBAN OPERATIONS 12-2
CONDUCT OF LIVE, VIRTUAL, AND CONSTRUCTIVE TRAINING 12-3
RANGERS – URBAN WARRIORS 12-3
PRINCIPLES 12-4
METT-TC 12-4
CLOSE QUARTERS COMBAT 12-6
REHEARSALS 12-6
TIPS FOR MARKING BUILDINGS AND ROOMS 12-8

CHAPTER 13 WATERBORNE OPERATIONS
ROPE BRIDGE 13-1
PONCHO RAFT 13-4
OTHER WATERCRAFT 13-5

CHAPTER 14 EVASION AND SURVIVAL
EVASION
PLANNING CONSIDERATIONS 14-1
INITIAL EVASION POINT 14-1
EVASION MOVEMENT 14-1
ROUTES 14-1
COMMUNICATIONS 14-2
HIDE SITE 14-2
HOLE-UP AREA 14-2
CAMOUFLAGE 14-2

SURVIVAL
MEMORY AID 14-3
SURVIVAL KITS 14-3
NAVIGATION 14-3
TRAPS AND SNARES 14-11
PROCESSING OF FISH OR GAME 14-15
SHELTERS 14-19
FIRES 14-21
METHODS 14-23

CHAPTER 15 **AVIATION**
REVERSE PLANNING SEQUENCE 15-1
SELECTION AND MARKING OF PICKUP AND LANDING ZONES 15-1
AIR ASSAULT FORMATIONS 15-2
PICKUP ZONE OPERATIONS 15-6
SAFETY 15-9
REQUIREMENTS 15-9
DESERT 15-11
MOUNTAINS 15-12
OBSERVATION HELICOPTERS 15-13
ATTACK HELICOPTERS 15-15
UTILITY HELICOPTERS 15-17
CARGO HELICOPTERS 15-19

CHAPTER 16 **FIRST AID**
LIFESAVING STEPS 16-1
CARE UNDER FIRE 16-1
PRIMARY SURVEY 16-1
AIRWAY MANAGEMENT 16-1
BREATHING 16-2
BLEEDING 16-2
SHOCK 16-2
EXTREMITY INJURIES 16-2
ABDOMINAL INJURIES 16-2
BURNS 16-3
HOT WEATHER (HEAT) INJURIES 16-4
POISONOUS PLANT IDENTIFICATION 16-7
FOOT CARE 16-7
LITTER 16-7
HYDRATION AND ACCLIMATIZATION 16-9
WORK, REST, AND WATER CONSUMPTION 16-9

APPENDIX A **RESOURCES**
REACT TO INDIRECT FIRE
REACT TO CONTACT
REACT TO A NEAR AMBUSH
REACT TO A FAR AMBUSH
BREAK CONTACT
FORMATIONS AND ORDER OF MOVEMENT
LINKUP
LINEAR DANGER AREA
LARGE OPEN DANGER AREA
CROSSING A SMALL OPEN AREA
SQUAD ATTACK
RAID BOARDS (LEFT)/ GENERAL INFO
RAID BOARDS (MIDDLE) ACTIONS ON OBJECTIVE
RAID BOARDS (MIDDLE) TASK ORGANIZATION
RAID BOARDS (RIGHT) SOP
AMBUSH SOP (LEFT)
AMBUSH BOARDS (MIDDLE)
LEADER'S RECONNAISSANCE
AMBUSH BOARDS RIGHT, SOP

APPENDIX B **QUICK REFERENCE CARDS**
CASUALTY FEEDER CARD
GTA
MEDEVAC AND AIRCRAFT REQUESTS
IED/UXO
RANGE CARD - PDF
RANGE CARD - FPL
STANDARD RANGE CARD
OBSERVED FIRE REFERENCE CARD AND RULER (INSIDE BACK COVER)
9 LINE MEDEVAC

GLOSSARY

INDEX

RANGER CREED

Recognizing that I volunteered as a Ranger, fully knowing the hazards of my chosen profession, I will always endeavor to uphold the prestige, honor, and high esprit de corps of the Rangers.

Acknowledging the fact that a Ranger is a more elite Soldier who arrives at the cutting edge of battle by land, sea, or air, I accept the fact that as a Ranger my country expects me to move further, faster, and fight harder than any other Soldier.

Never shall I fail my comrades I will always keep myself mentally alert, physically strong, and morally straight and I will shoulder more than my share of the task whatever it may be, one hundred percent and then some.

Gallantly will I show the world that I am a specially selected and well trained Soldier. My courtesy to superior officers, neatness of dress, and care of equipment shall set the example for others to follow.

Energetically will I meet the enemies of my country. I shall defeat them on the field of battle for I am better trained and will fight with all my might. Surrender is not a Ranger word. I will never leave a fallen comrade to fall into the hands of the enemy and under no circumstances will I ever embarrass my country.

Readily will I display the intestinal fortitude required to fight on to the Ranger objective and complete the mission, though I be the lone survivor.

STANDING ORDERS, ROGERS' RANGERS

MAJOR ROBERT ROGERS, 1759

1. Don't forget nothing.
2. Have your musket clean as a whistle, hatchet scoured, sixty rounds powder and ball, and be ready to march at a minute's warning.
3. When you're on the march, act the way you would if you was sneaking up on a deer. See the enemy first.
4. Tell the truth about what you see and what you do. There is an army depending on us for correct information. You can lie all you please when you tell other folks about the Rangers, but don't never lie to a Ranger or officer.
5. Don't never take a chance you don't have to.
6. When we're on the march we march single file, far enough apart so one shot can't go through two men.
7. If we strike swamps, or soft ground, we spread out abreast, so it's hard to track us.
8. When we march, we keep moving till dark, so as to give the enemy the least possible chance at us.
9. When we camp, half the party stays awake while the other half sleeps.
10. If we take prisoners, we keep' em separate till we have had time to examine them, so they can't cook up a story between' em.
11. Don't ever march home the same way. Take a different route so you won't be ambushed.
12. No matter whether we travel in big parties or little ones, each party has to keep a scout 20 yards ahead, 20 yards on each flank, and 20 yards in the rear so the main body can't be surprised and wiped out.
13. Every night you'll be told where to meet if surrounded by a superior force.
14. Don't sit down to eat without posting sentries.
15. Don't sleep beyond dawn. Dawn's when the French and Indians attack.
16. Don't cross a river by a regular ford.
17. If somebody's trailing you, make a circle, come back onto your own tracks, and ambush the folks that aim to ambush you.
18. Don't stand up when the enemy's coming against you. Kneel down, lie down, hide behind a tree.
19. Let the enemy come till he's almost close enough to touch, then let him have it and jump out and finish him up with your hatchet.

RANGER HISTORY

The history of the American Ranger is a long and colorful saga of courage, daring, and outstanding leadership. It is a story of men whose skills in the art of fighting have seldom been surpassed. Only the highlights of their numerous exploits are told here.

Rangers mainly performed defensive missions until, during King Phillip's War in 1675, Benjamin Church's Company of Independent Rangers (from Plymouth Colony) conducted successful raids on hostile Indians. In 1756, Major Robert Rogers, of New Hampshire, recruited nine companies of American colonists to fight for the British during the French and Indian War. Ranger techniques and methods of operation inherently characterized the American frontiersmen. Major Rogers was the first to capitalize on them and incorporate them into the fighting doctrine of a permanently organized fighting force.

The method of fighting used by the first Rangers was further developed during the Revolutionary War by Colonel Daniel Morgan, who organized a unit known as "Morgan's Riflemen." According to General Burgoyne, Morgan's men were "....the most famous corps of the Continental Army, all of them crack shots."

Francis Marion, the "Swamp Fox," organized another famous Revolutionary War Ranger element known as "Marion's Partisans." Marion's Partisans, numbering anywhere from a handful to several hundred, operated both with and independent of other elements of General Washington's Army. Operating out of the Carolina swamps, they disrupted British communications and prevented the organization of loyalists to support the British cause, substantially contributing to the American victory.

The American Civil War was again the occasion for the creation of special units such as Rangers. John S. Mosby, a master of the prompt and skillful use of cavalry, was one of the most outstanding Confederate Rangers. He believed that by resorting to aggressive action he could compel his enemies to guard a hundred points. He would then attack one of the weakest points and be assured numerical superiority.

With America's entry into the Second World War, Rangers came forth to add to the pages of history. Major William O. Darby organized and activated the 1st Ranger Battalion on June19, 1942 at Carrickfergus, North Ireland. The members were all hand picked volunteers; 50 participated in the gallant Dieppe Raid on the northern coast of France with British and Canadian commandos. The 1st, 3rd, and 4th Ranger Battalions participated with distinction in the North African, Sicilian and Italian campaigns. Darby's Ranger Battalions spearheaded the Seventh Army landing at Gela and Licata during the Sicilian invasion and played a key role in the subsequent campaign, which ended in the capture of Messina. They infiltrated German lines and mounted an attack against Cisterna, where they virtually annihilated an entire German parachute regiment during close in, night, bayonet, and hand to hand fighting.

The 2nd and 5th Ranger Battalions participated in the D Day landings at Omaha Beach, Normandy. It was during the bitter fighting along the beach that the Rangers gained their official motto. As the situation became critical on Omaha Beach, the division commander of the 29th Infantry Division stated that the entire force must clear the beach and advance inland. He then turned to Lieutenant Colonel Max Schneider, Commander of the 5th Ranger Battalion, and said, "Rangers, lead the way." The 5th Ranger Battalion spearheaded the breakthrough. This enabled the Allies to drive inland, away from the invasion beaches.

The 6th Ranger Battalion, operating in the Pacific, conducted Ranger-type missions behind enemy lines. These missions involved reconnaissance and hard hitting, long-range raids. These Rangers were the first American group to return to the Philippines, destroying key coastal installations prior to the invasion. A reinforced company from the 6th Ranger Battalion formed the rescue force that liberated American and Allied POWs from the Japanese prison camp at Cabanatuan.

Another Ranger type unit was the 5307th Composite Unit (Provisional), organized and trained as a long range penetration unit for employment behind enemy lines in Japanese occupied Burma. The unit commander was Brigadier General (later Major General) Frank D. Merrill. Its 2,997 officers and men became popularly known as "Merrill's Marauders."

The men of Merrill's Marauders were volunteers from the 5th, 154th, and 33rd Infantry Regiments and from other Infantry regiments engaged in combat in the Southwest and South Pacific. These men responded to a call from Chief of Staff, General George C. Marshall, for volunteers for a hazardous mission. These volunteers were to have a high state of physical ruggedness and stamina and were to come from jungle trained and jungle tested units.

Before joining the Northern Burma Campaign, Merrill's Marauders trained in India under British Major General Orde C. Wingate. From February to June 1943, they learned long range penetration tactics and techniques like those developed and first employed by General Wingate. The operations of the Marauders were closely coordinated with those of the Chinese 22nd and 38th Divisions in a drive to recover northern Burma and clear the way for the construction of Ledo Road, which was to link the Indian railhead at Ledo with the old Burma Road to China. The Marauders marched and fought through jungle and over mountains from Hukwang Valley in Northwest Burma, to Myitkyina and the Irrawaddy River. In 5 major and 30 minor engagements, they met and defeated the veteran soldiers of the Japanese 18th Division. Operating in the rear of the main force of the Japanese, they prepared the way for the Southward advances of the Chinese by disorganizing supply lines and communications. The climax of the Marauder's operations was the capture of Myitkyina Airfield, the only all weather strip in northern Burma. This was the final victory of "Merrill's Marauders," which disbanded in August 1944. Remaining personnel merged into the 475th Infantry Regiment, which fought its last battle on February 3 and 4, 1945, at Loi Kang Ridge, China. This Infantry Regiment is the father of the 75th Ranger Regiment.

Soon after the Korean War started in June 1950, the 8th Army Ranger Company was formed of volunteers from American units in Japan. The Company was trained in Korea and distinguished itself in combat during the drive to the Yalu River, performing task force and spearhead operations. During the massive Chinese intervention of November 1950, this small, vastly outnumbered unit withstood five enemy assaults on its position.

In September 1950, a D.A. message called for volunteers to train as Airborne Rangers. Five thousand regular Army paratroopers from the 82nd Airborne Division volunteered. Nine hundred were chosen to form the first eight Airborne Ranger companies. Nine more companies were formed from regular Army and National Guard Infantry division volunteers. These seventeen Airborne Ranger companies were activated and trained at Fort Benning, Georgia. Most received more training in the Colorado mountains.

In 1950 and 1951, some 700 men of the 1st, 2nd, 3rd, 4th, 5th, and 8th Airborne Ranger companies fought to the front of every American Infantry Division in Korea. Attacking by land, water, and air, these six Ranger companies raided, penetrated, and ambushed North Korean and Chinese forces. They were the first Rangers to make combat jumps. After the Chinese intervention, these Rangers were the first Americans to re cross the 38th parallel. The 2nd Airborne Ranger Company was the only African American Ranger unit in the history of the American Army. The men of the six Ranger companies who fought in Korea paid the bloody price of freedom. One in nine of this gallant brotherhood died on the battlefields of Korea.

Other Airborne Ranger companies led the way while serving with Infantry divisions in the U.S., Germany, and Japan. These volunteers fought as members of line Infantry units in Korea. They volunteered for the Army, the Airborne, the Rangers, and for combat. The first men to earn and wear the coveted Ranger Tab, these men are the original Airborne Rangers. One Ranger, Donn Porter, received the Medal of Honor posthumously. Fourteen Korean War Rangers rose to general officer. Dozens more became colonels, senior NCOs, and civilian leaders.

In October 1951, the Army Chief of Staff, General J. Lawton Collins, directed that Ranger training extend to all Army combat units. He directed the Commandant of the Infantry School to establish a Ranger Department. This new department would develop and conduct a Ranger course of instruction. His goal was to raise the standard of training in all combat units. The program built on lessons learned from World War II and the Korean conflict.

During the Vietnam Conflict, fourteen Ranger companies consisting of highly motivated volunteers served with distinction from the Mekong Delta to the DMZ. Assigned to separate brigade, division, and field force units, they conducted long range reconnaissance and exploitation operations into enemy held areas. They provided valuable combat intelligence. Initially designated as long-range reconnaissance patrol (LRRP), then long-range patrol (LRP) companies, these units were later designated as C through P (there is no Juliet Company) Rangers, 75th Infantry.

After Vietnam, the Army Chief of Staff, General Abrams, recognized the need for a highly trained and highly mobile reaction force. He activated the first battalion sized Ranger units since World War II, the 1st and 2nd Battalions (Ranger), 75th Infantry.

The 1st Battalion trained at Fort Benning, Georgia and was activated February 8, 1974 at Fort Stewart, Georgia. The 2nd Battalion was activated on October 3, 1974. The 1st Battalion is now based at Hunter Army Airfield, Georgia; the 2nd Battalion is based at Fort Lewis, Washington.

General Abrams' farsighted decision and the combat effectiveness of the Ranger battalions were proven in the U.S. invasion of Grenada, Operation ***"Urgent Fury,"*** October 1983. The mission was to protect American citizens and restore democracy. The Ranger battalions "led the way" with a daring, low level airborne assault (from 500 feet) to seize the airfield at Point Salines. They continued operations for several days, eliminating pockets of resistance and rescuing American medical students. Due to this success, in 1984, D.A. increased the strength of Ranger units to their highest levels in 40 years. To do this, it activated another Ranger battalion as well as a Ranger Regimental Headquarters. After these units, the 3rd Battalion (Ranger), 75th Infantry, and Headquarters Company (Ranger), 75th Infantry, were activated, there were over 2,000 Soldiers assigned to Ranger units. On February 3, 1986, the 75th Infantry was renamed the 75th Ranger Regiment.

On December 20, 1989, the 75th Ranger Regiment was again called to show its effectiveness in combat. For the first time since reorganizing in 1984, the Regimental Headquarters and all three Ranger battalions deployed together. During Operation ***"Just Cause"*** in Panama, the 75th Ranger Regiment spearheaded the assault into Panama by conducting airborne assaults on the Torrijos/Tocumen Airport and Rio Hato Airfield. Their mission: to facilitate the restoration of democracy in Panama and to protect the lives of American citizens. Between December 20, 1989 and January 7, 1990, the regiment performed many follow on missions in Panama.

Early in 1991, elements of the 75th Ranger Regiment deployed to Saudi Arabia in support of Operation ***"Desert Storm."***

In August 1993, elements of the 75th Ranger Regiment deployed to Somalia in support of Operation ***"Restore Hope,"*** and returned November 1993.

In 1994, elements of the 75th Ranger Regiment deployed to Haiti in support of Operation ***"Uphold Democracy."***

In 2000 – 2001, elements of the 75th Ranger Regiment deployed to Kosovo in support of Operation ***"Joint Guardian."***

Since September 11, 2001, the 75th Ranger Regiment has led the way in the Global War on Terrorism. In October 2001, elements of the 75th Ranger Regiment deployed to Afghanistan in support of Operation ***"Enduring Freedom."*** In March 2003, elements of the Regiment deployed in support of Operation ***"Iraqi Freedom."***

The performance of the Rangers significantly contributed to the overall success of these operations and upheld the Ranger tradition. As in the past, the Regiment stands ready to execute its mission to conduct special operations in support of the United States' policies and objectives.

RANGER MEDAL OF HONOR RECIPIENTS

Millett, Lewis L. Sr	Captain	Feb 7 1951	Co. E 2 / 27th Infantry
Porter, Donn F.*	Sergeant	Sept 7 1952	Co. G 2 / 14th Infantry
Mize, Ola L.	Sergeant	June 10-11 1953	Co. K 3 / 15th Infantry
Dolby, David C.	Staff Sergeant	May 21 1966	Co. B 1 / 8th (ABN) Cavalry
Foley, Robert F.	Captain	Nov 5 1966	Co. A 2 / 27th Infantry
Zabitosky, Fred M.	Staff Sergeant	Feb 19 1968	5th Special Forces
Bucha, Paul W.	Captain	May 16-19 1968	Co. D 3 / 187 Infantry
Rabel, Laszlo*	Staff Sergeant	Nov 13 1968	74th Infantry (LRRP)
Howard, Robert L.	Sergeant First Class	Dec 30 1968	5th Special Forces
Law, Robert D. *	Specialist 4	Feb 22 1969	Co. I 75th Infantry (Ranger)
Kerrey, J. Robert	Lieutenant	Mar 14 1969	Seal Team 1
Doane, Stephen H.*	1st Lieutenant	Mar 25 1969	Co. B 1 / 5th Infantry
Pruden, Robert J.*	Staff Sergeant	Nov 22 1969	Co. G 75th Infantry (Ranger)
Littrell, Gary L.	Sergeant First Class	April 4-8 1970	Advisory Team 21 (Ranger)
Lucas, Andre C.*	Lt Colonel	Jul 1-23 1970	HHC 2 / 506 Infantry
Gordon, Gary I. *	Master Sergeant	Oct 3 1993	Task Force Ranger
Shughart, Randall D. *	Sergeant First Class	Oct 3 1993	Task Force Ranger

*Awarded posthumously

PREFACE

The MCOE SH 21-76 (Ranger Handbook) is mainly written for U.S. Army Rangers and other light Infantry units, however it should also serve as a handy reference for other U.S. military units. It covers how Infantry squad- and platoon-sized elements conduct combat operations in varied terrains. It cites other Army resources to ensure continuity.

This handbook provides squad and platoon leaders with the roles, tactics, knowledge, and operational requirements to employ combat multipliers in a combat environment.

The proponent of this publication is the U.S. Army Maneuver Center of Excellence (MCoE). The preparing agency is the U.S. Army Ranger School.

Send comments, recommendations, and other correspondence related to this manual to the following address:

E-Mail	john.edmunds@conus.army.mil
Office/Fax	(706) 544-6448 / - 6421 (DSN 834)
US Mail	Commander, Ranger Training Brigade ATTN: ATSH-RB / Edmunds 10850 Schneider Rd, Bldg 5024 Ft Benning, GA 31905

Chapter 1
LEADERSHIP

Leadership, the most essential element of combat power, gives purpose, direction, and motivation in combat. The leader balances and maximizes maneuver, firepower, and protection against the enemy. This chapter discusses how he does this by exploring the principles of leadership (Be, Know, Do); the duties, responsibilities, and actions of an effective leader; and the leader's assumption of command.

1-1.1. PRINCIPLES. (Figure 1-1).

Figure 1-1. BE, KNOW, DO–THE PRINCIPLES OF LEADERSHIP

BE

- Technically and tactically proficient
- Able to accomplish to standard all tasks required for the wartime mission.
- Courageous, committed, and candid.
- A leader with integrity.

KNOW

- The four major factors of leadership and how they affect each other are—
  - –Led
  - –Leader
  - –Situation
  - –Communications
- Yourself, and the strengths and weaknesses in your character, knowledge, and skills. Seek continual self-improvement, that is, develop your strengths and work to overcome your weaknesses.
- Your Rangers, and look out for their well being by training them for the rigors of combat, taking care of their physical and safety needs, and disciplining and rewarding them.

DO

- Seek responsibility and take responsibility for your actions; exercise initiative; demonstrate resourcefulness; and take advantage of opportunities on the battlefield that will lead to you to victory; accept fair criticism, and take corrective actions for your mistakes.
- Assess situations rapidly, make sound and timely decisions, gather essential information, announce decisions in time for Rangers to react, and consider the short- and long-term effects of your decision.
- Set the example by serving as a role model for your Rangers. Set high but attainable standards; be willing do what you require of your Rangers; and share dangers and hardships with them.
- Keep your subordinates informed to help them make decisions and execute plans within your intent, encourage initiative, improve teamwork, and enhance morale.
- Develop a sense of responsibility in subordinates by teaching, challenging, and developing them. Delegate to show you trust them. This makes them want more responsibility.
- Ensure the Rangers understand the task; supervise them, and ensure they accomplish it. Rangers need to know what you expect, when and what you want them to do, and to what standard.
- Build the team by training and cross-training your Rangers until they are confident in their technical and tactical abilities. Develop a team spirit that motivates them to go willingly and confidently into combat.
- Know your unit's capabilities and limitations, and employ them accordingly.

1-1.2. **DUTIES, RESPONSIBILITIES, AND ACTIONS.** To complete all assigned tasks, every Ranger in the patrol must do his job. Each must accomplish his specific duties and responsibilities and be a part of the team **(Figure 1-2).**

Figure 1-2. DUTIES, RESPONSIBILITIES, AND ACTIONS

PLATOON LEADER

- Is responsible for what the patrol does or fails to do. This includes tactical employment, training, administration, personnel management, and logistics. He does this by planning, making timely decisions, issuing orders, assigning tasks, and supervising patrol activities. He must know his Rangers and how to employ the patrol's weapons. He is responsible for positioning and employing all assigned or attached crew-served weapons and employment of supporting weapons.
- Establishes time schedule using backwards planning. Considers time for execution, movement to the objective, and the planning and preparation phase of the operation.
- Takes the initiative to accomplish the mission in the absence of orders. Keeps higher informed by using periodic situation reports (SITREP).
- Plans with the help of the platoon sergeant (PSG), squad leaders, and other key personnel (team leaders, FO, attachment leaders).
- Stays abreast of the situation through coordination with adjacent patrols and higher HQ; supervises, issues FRAGOs, and accomplishes the mission.
- If needed to perform the mission, requests more support for his patrol from higher headquarters.
- Directs and assists the platoon sergeant in planning and coordinating the patrol's sustainment effort and casualty evacuation (CASEVAC) plan.
- During planning, receives on-hand status reports from the platoon sergeant and squad leaders.
- Reviews patrol requirements based on the tactical plan.
- Ensures that all-round security is maintained at all times.
- Supervises and spot-checks all assigned tasks, and corrects unsatisfactory actions.
- During execution, positions himself where he can influence the most critical task for mission accomplishment; usually with the main effort, to ensure that his platoon achieves its decisive point
- Is responsible for positioning and employing all assigned and attached crew-served weapons.
- Commands through his squad leaders IAW the intent of the two levels higher commanders.
- Conducts rehearsals.

PLATOON SERGEANT (PSG)

- The PSG is the senior NCO in the patrol and second in succession of command. He helps and advises the patrol leader, and leads the patrol in the leader's absence. He supervises the patrol's administration, logistics, and maintenance, and he prepares and issues paragraph 4 of the patrol OPORD.

DUTIES

- Organizes and controls the patrol CP IAW the unit SOP, patrol leader's guidance, and METT-TC factors.
- Receives squad leader's requests for rations, water, and ammunition. Work with the company first sergeant or XO to request resupply. Directs the routing of supplies and mail.
- Supervises and directs the patrol medic and patrol aid-litter teams in moving casualties to the rear.
- Maintains patrol status of personnel, weapons, and equipment; consolidates and forwards the patrol's casualty reports (DA Forms 1155 and 1156); and receives and orients replacements.

- Monitors the morale, discipline, and health of patrol members.
- Supervises task-organized elements of patrol:
  - — Quartering parties.
  - — Security forces during withdrawals.
  - — Support elements during raids or attacks.
  - — Security patrols during night attacks.
- Coordinates and supervises patrol resupply operations.
- Ensures that supplies are distributed IAW the patrol leader's guidance and direction.
- Ensures that ammunition, supplies, and loads are properly and evenly distributed (a critical task during consolidation and reorganization).
- Ensures the casualty evacuation plan is complete and executed properly.
- Ensures that the patrol adheres to the platoon leader's time schedule.
- Assists the platoon leader in supervising and spot-checking all assigned tasks, and corrects unsatisfactory actions.

ACTIONS DURING MOVEMENT AND HALTS

- Takes actions necessary to facilitate movement.
- Supervises rear security during movement.
- Establishes, supervises, and maintains security during halts.
- Knows unit location.
- Performs additional tasks as required by the patrol leader and assists in every way possible. Focuses on security and control of patrol.

ACTIONS AT DANGER AREAS

- Directs positioning of near-side security (usually conducted by the trail squad or team).
- Maintains accountability of personnel.

ACTIONS IN THE OBJECTIVE AREA

- Assists with ORP occupation.
- Supervises, establishes, and maintains security at the ORP.
- Supervises the final preparation of men, weapons, and equipment in the ORP IAW the patrol leader's guidance.
- Assists the patrol leader in control and security.
- Supervises the consolidation and reorganization of ammunition and equipment.
- Establishes, marks, and supervises the planned casualty collection point (CCP), and ensures that the personnel status (to include WIA/KIA) is accurately reported to higher.
- Performs additional tasks assigned by the patrol leader and reports status to platoon leader.

ACTIONS IN THE PATROL BASE

- Assists in patrol base occupation.
- Assist in establishing and adjusting perimeter.
- Enforces security in the patrol base.
- Keeps movement and noise to a minimum.
- Supervises and enforces camouflage.

- Assigns sectors of fire.
- Ensures designated personnel remain alert and equipment is maintained to a high state of readiness.
- Requisitions supplies, water, and ammunition, and supervises their distribution.
- Supervises the priority of work and ensures its accomplishment.
  - ***Security plan***.
    - Ensures crew-served weapons have interlocking sectors of fire.
    - Ensures claymore mines are emplaced to cover dead space.
    - Ensures range cards and sector sketch are complete.
  - ***Alert plan***.
  - ***Evacuation plan***.
  - ***Withdrawal plan***.
  - ***Alternate patrol base***.
  - ***Maintenance plan***.
  - ***Hygiene plan***.
  - ***Messing plan***.
  - ***Water plan***.
  - ***Rest plan***.
- Performs additional tasks assigned by the patrol leader and assists him in every way possible.

SQUAD LEADER (SL)

- Is responsible for what the squad does or fails to do. He is a tactical leader who leads by example.

DUTIES

- Completes casualty feeder reports and reviews the casualty reports completed by squad members.
- Directs the maintenance of the squad's weapons and equipment.
- Inspects the condition of Rangers' weapons, clothing, and equipment.
- Keeps the PL and PSG informed on status of squad.
- Submits ACE report to PSG.

ACTIONS THROUGHOUT THE MISSION

- Obtains status report from team leaders and submits reports to the PL and PSG.
- Makes a recommendation to the PL/PSG when problems are observed.
- Delegates priority task to team leaders, and supervises their accomplishment IAW squad leader's guidance.
- Uses initiative in the absence of orders.
- Follows the PL's plan and makes recommendations.

ACTIONS DURING MOVEMENT AND HALTS

- Ensures heavy equipment is rotated among members and difficult duties are shared.
- Notifies PL of the status of the squad.
- Maintains proper movement techniques while monitoring route, pace, and azimuth.
- Ensures the squad maintains security throughout the movement and at halts.

- Prevents breaks in contact.
- Ensures subordinate leaders are disseminating information, assigning sectors of fire, and checking personnel.

ACTION IN THE OBJECTIVE AREA

- Ensures special equipment has been prepared for actions at the objective.
- Maintains positive control of squad during the execution of the mission.
- Positions key weapons systems during and after assault on the objective.
- Obtains status reports from team leaders and ensures ammunition is redistributed and reports status to PL.

ACTIONS IN THE PATROL BASE

- Ensures patrol base is occupied according to the plan.
- Ensures that his sector of the patrol base is covered by interlocking fires; makes final adjustments, if necessary.
- Sends out LP or OPs in front of assigned sector. (METT-TC dependent).
- Ensures priorities of work are being accomplished, and reports accomplished priorities to the PL and PSG.
- Adheres to time schedule.
- Ensures personnel know alert and evacuation plans and locations of key leaders, OPs, and the alternate patrol base.

WEAPONS SQUAD LEADER

- Is responsible for all that the weapons squad does or fails to do. His duties are the same as those of the squad leader. Also, he controls the machine guns in support of the patrol's mission. He advises the PL on employment of his squad.

DUTIES

- Supervises machine gun teams to ensure they follow priorities of work.
- Inspects machine gun teams for correct range cards, fighting positions, and understanding of fire plan.
- Supervises maintenance of machine guns, that is, ensures that maintenance is performed correctly, that deficiencies are corrected and reported, and that the performance of maintenance does not violate the security plan.
- Assists PL in planning.
- Positions at halts and danger areas and according to the patrol SOP any machine guns not attached to squads.
- Rotates loads. Machine gunners normally get tired first.
- Submits ACE report to PSG.
- Designates sectors of fire, principal direction of fire (PDF), and secondary sectors of fire for all guns.
- Gives fire commands to achieve maximum effectiveness of firepower:
  - — Shifts fires.
  - — Corrects windage or elevation to increase accuracy.
  - — Alternates firing guns.
  - — Controls rates of fire and fire distribution.
- Knows locations of assault and security elements, and prevents fratricide.
- Reports to the PL.

TEAM LEADER (TL)

- Controls the movement of his fire team and the rate and placement of fire. To do this, leads from the front and uses the proper commands and signals. Maintains accountability of his Rangers, weapons, and equipment. Ensures his Rangers maintain unit standards in all areas, and are knowledgeable of their tasks and the operation. The following checklist outlines specific duties and responsibilities of team leaders during mission planning and execution. The TL leads by example:

ACTIONS DURING PLANNING AND PREPARATION

- *Warning Order*
  - — Assists in control of the squad
  - — Monitors squad during issuance of the order
- *OPORD Preparation*
  - — Posts changes to schedule
  - — Posts and updates team duties on warning order board
  - — Submits ammunition and supply requests
  - — Picks up ammunition and supplies
  - — Distributes ammunition and special equipment
  - — Performs all tasks given in the SL's special instructions paragraph
- *OPORD Issuance and Rehearsal*
  - — Monitors squad during issuance of the order
  - — Assists SL during rehearsals
  - — Takes actions necessary to facilitate movement
  - — Enforces rear security
  - — Establishes, supervises, and maintains security at all times
  - — Performs other tasks as SL requires, and helps him in every way possible, particularly in control and security

ACTIONS IN THE ORP

- Assists in the occupation of the ORP.
- Helps supervise, establish, and maintain security.
- Supervises the final preparation of Rangers, weapons, and equipment in the ORP IAW the SL's guidance.
- Assists in control of personnel departing and entering the ORP.
- Reorganizes perimeter after the reconnaissance party departs.
- Maintains communication with higher headquarters.
- Upon return of reconnaissance party, helps reorganize personnel and redistribute ammunition and equipment; ensures accountability of all personnel and equipment is maintained.
- Disseminates PIR to his team.
- Performs additional tasks assigned by the SL.

ACTIONS IN THE PATROL BASE

- Inspects the perimeter to ensure team has interlocking sectors of fire; prepares team sector sketch.

- Enforces the priority of work and ensures it is properly accomplished.
- Performs additional tasks assigned by the SL and assists him in every way possible.

MEDIC

- Assists the PSG in directing aid and litter teams; monitors the health and hygiene of the platoon.

DUTIES

- Treats casualties, conducts triage, and assists in CASEVAC under the control of the PSG.
- Aids the PL or PSG in field hygiene matters. Personally checks the health and physical condition of platoon members.
- Requests Class VIII (medical) supplies through the PSG.
- Provides technical expertise to and supervision of combat lifesavers.
- Ensures casualty feeder reports are correct and attached to each evacuated casualty.
- Carries out other tasks assigned by the PL or PSG.

RADIO OPERATOR

- Is responsible for establishing and maintaining communications with higher headquarters and within the patrol.

DUTIES DURING PLANNING

- Enters the net at the specified time.
- Ensures that all frequencies, COMSEC fills, and net IDs, are preset in squad/platoon radios.
- Informs SL and PL of changes to call signs, frequencies, challenge and password, and number combination based on the appropriate time in the ANCD.
- Ensures the proper function of all radios and troubleshoots and reports deficiencies to higher.
- Weatherproofs all communications equipment.

DUTIES DURING EXECUTION

- Serves as en route recorder during all phases of the mission.
- Tracks time after the initiation of the assault.
- Records all enemy contact and reports it to higher in a SALUTE format.
- Reports all OPSKEDs to higher.
- Consolidates and records all PIR.

FORWARD OBSERVER (FO)

- Works for the PL. Serves as the eyes of the FA and mortars. Is mainly responsible for locating targets, and for calling for and adjusting indirect fire support. Knows the terrain where the platoon is operating; knows the tactical situation. Knows the mission, the concept, and the unit's scheme of maneuver and priority of fires.

DUTIES DURING PLANNING

- Selects targets to support the platoon's mission based on the company OPORD, platoon leader's guidance, and analysis of METT-TC factors.
- Prepares and uses situation maps, overlays, and terrain sketches.

DUTIES DURING EXECUTION

- Informs the FIST headquarters of platoon activities and of the fire support situation
- Selects new targets to support the platoon's mission based on the company OPORD, the platoon leader's guidance, and an analysis of METT-TC factors.
- Calls for and adjusts fire support.
- Operates as a team with the radio operator.
- Selects OPs.
- Maintains communications as prescribed by the FSO.
- Maintains the current 8-digit coordinate of his location at all times.

1-1.3. ASSUMPTION OF COMMAND. Any platoon/squad member might have to take command of his element in an emergency, so every Ranger must be prepared to do so. During an assumption of command, situation permitting, the Ranger assuming command accomplishes the tasks (not necessarily in this order) based on METT-TC shown in **Figure 1-3**.

Figure 1-3. TASKS FOR ASSUMPTION OF COMMAND

• *INFORMS*	The unit's subordinate leaders of the command and notifies higher.
• *CHECKS*	Security.
• *CHECKS*	Crew-served weapons.
• *PINPOINTS*	Location.
• *COORDINATES* and *CHECKS*	Equipment.
• *CHECKS*	Personnel status.
• *ISSUES*	FRAGO (if required).
• *REORGANIZES*	As needed, maintaining unit integrity when possible.
• *MAINTAINS*	Noise and light discipline.
• *CONTINUES*	Patrol base activities, especially security, if assuming command in a patrol base.
• *RECONNOITERS*	Or, at the least, conducts a map reconnaissance.
• *FINALIZES*	Plan.
• *EXECUTES*	The mission.

Chapter 2
OPERATIONS

This chapter provides techniques and procedures used by Infantry platoons and squads throughout the planning and execution phases of tactical operations. Specifically, it discusses the troop leading procedures, combat intelligence, combat orders, and planning techniques and tools needed to prepare a platoon to fight. These topics are time sensitive and apply to all combat operations. When they have time, leaders can plan and prepare in depth. If they have less time, they must rely on previously rehearsed actions, battle drills, and standing operating procedures (SOPs).

2-1. TROOP LEADING PROCEDURES. Figure 2-1 shows the steps in the troop leading procedures (TLPs). These steps are what a leader does to prepare his unit to accomplish a tactical mission. The TLP starts when the leader is alerted for a mission or receives a change or new mission. He can perform Steps 3 through 8 in any order, or at the same time. He can also use the tools of the tactician shown in **Figure 2-2**:

Figure 2-1. STEPS IN THE TROOP LEADING PROCEDURE

1. Receive the mission.	5. Reconnoiter.
2. Issue a warning order.	6. Complete the plan.
3. Make a tentative plan.	7. Issue the complete order.
4. Initiate movement.	8. Supervise.

Figure 2-2. TOOLS OF THE TACTICIAN RELATIONSHIP

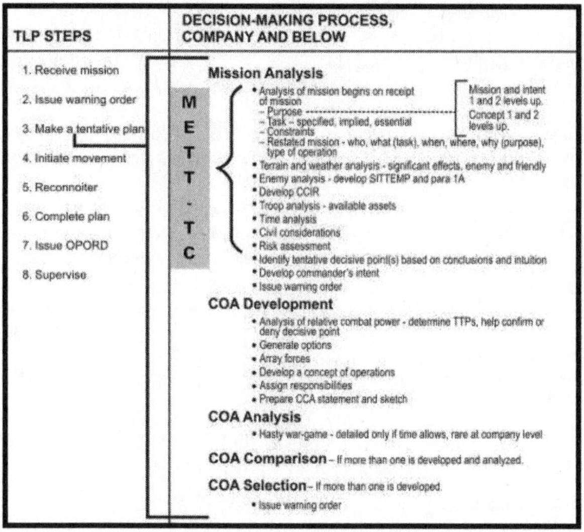

a. **STEP 1–RECEIVE THE MISSION.** The leader may receive the mission in a warning order, an operation order (OPORD), or a fragmentary order (FRAGO). He should use no more than one third of the available time for his own planning and for issuing his OPORD. The remaining two thirds is for subordinates to plan and prepare for the operation. Leaders should also consider other factors such as available daylight and travel time to and from orders and rehearsals.

b. **STEP 2–ISSUE A WARNING ORDER.** The leader provides initial instructions in a warning order. The warning order contains enough information to begin preparation as soon as possible. The warning order mirrors the five paragraph OPORD format. A warning order may include–

- The mission or nature of the operation (***mission statement***).
- Time and place for issuance of the operation order (***coordinating instructions***).
- Who is participating in the operation (***coordinating instructions***).
- Time of the operation (***timeline***).

c. **STEP 3–MAKE A TENTATIVE PLAN.** The leader develops an estimate of the situation to use as the basis for his tentative plan. This is the leader's mission analysis. He will use METT-TC when developing his tentative plan.

(1) ***Conduct a Detailed Mission Analysis***.

(a) *Concept and Intent*. Higher commanders' concepts and intents two levels up. This information is found in paragraph 1b for two levels up and in paragraphs 2 and 3 for one higher.

(b) *Unit Tasks*. Tasks that are clearly stated in the order (***Specified Tasks***) or tasks that become apparent as the OPORD is analyzed (***Implied Tasks***).

EXAMPLES OF SPECIFIED AND IMPLIED TASKS

SPECIFIED TASKS
- Retain Hill 545 to prevent envelopment of B Co.
- Provide one squad to the 81-mm platoon to carry ammo.
- Establish an OP vic GA124325 NLT 301500 Nov 89.

IMPLIED TASKS
- Provide security during movement.
- Conduct resupply operations.
- Coordinate with adjacent units.

(c) *Unit Constraints*. The leader identifies any constraints placed on his unit. Constraints can take the form of a requirement (for example, maintain a reserve of one squad) or a prohibition on action (for example, no movement beyond phase line orange prior to H hour).

(d) *Mission Essential Task(s)*. After reviewing all the factors shown in previous paragraphs, the leader identifies the mission essential task(s). Failure to accomplish a mission essential task equals failure to accomplish the mission. The mission essential task should be in the maneuver paragraph.

(e) *Restated Mission*. The restated mission focuses the remainder of the estimate process. It clearly, concisely states the mission (purpose to be achieved) and the mission essential task(s) required to achieve it. It identifies WHO, WHAT (the task), WHEN (the critical time), WHERE (usually a grid coordinate), and WHY (the purpose the unit must achieve).

EXAMPLES OF RESTATED MISSIONS

(*Who?*) 1st Platoon attacks (*What?*) to seize (*Where?*) Hill 482 vic NB 457371 (OBJ Blue) (*When?*) NLT 090500Z Dec 92 L 482 (*Why?*) to enable the company's main effort to destroy enemy command bunker.

(*Who?*) 1st Platoon, C Company defends (*What?*) to destroy from (*Where?*) AB163456 to AB163486 to AB123486 to AB123456 (*When?*) NLT 281530Z Oct 97 (*Why?*) to prevent enemy forces from enveloping B Company, 1-66 Infantry (L) from the South.

(2) **Analyze the situation and develop a course of action.** Each COA must be:

Feasible: It accomplishes the mission and supports the commander's concept.

Reasonable: The unit remains an effective force after completing the mission.

Distinguishable: It is not just a minor variation of another COA.

(3) Upon developing a COA, the unit leader will assign **C2 headquarters**, complete generic task organization assigning all organic and attached elements, and prepare COA statement and sketch.

(4) With the restated mission from Step 1 to provide focus, the leader continues the **estimate process** using the remaining factors of METT-TC:

(a) What is known about the enemy (**Figure 2-3**)?

(b) How will *terrain* and weather affect the operation? Analyze terrain using OACOK.

Figure 2-3. ENEMY

Composition	• This is an analysis of the forces and weapons that the enemy can bring to bear. Determine what weapons systems they have available, and what additional weapons and units are supporting him.
Disposition	• The enemy's disposition is how he is arrayed on the terrain, such as in defensive positions, in an assembly area, or moving in march formation.
Strength	• Percentage strength, and number of PAX
Recent Activities	• Identify recent and significant enemy activities that may indicate future intentions.
Reinforcement Capabilities	• Determine positions for reserves and estimated time to counterattack or reinforce.
Possible COAs	• Determine the enemy's possible COA. Analyzing these COAs may ensure that the friendly unit is not surprised during execution.

Observation and Fields of Fire. Determine locations that provide the best observation and fields of fire along the approaches, near the objective, or on key terrain. The analysis of fields of fire is mainly concerned with the ability to cover the terrain with direct fire.

Avenues of Approach. Avenues of approach are developed next and identified one level down. Aerial and subterranean avenues must also be considered. Use **Figure 2-4** for offensive considerations to avenues of approach.

Figure 2-4. OFFENSIVE CONSIDERATIONS

Offensive Considerations (Friendly)	• How can these avenues support my movement? • What are the advantages / disadvantages of each? (Consider enemy, speed, cover, and concealment.) • What are the likely enemy counterattack routes?
Offensive Considerations (Enemy)	• How can the enemy use these approaches? • Which avenue is most dangerous? Least? (Prioritize each approach.) • Which avenues would support a counterattack?

Cover and Concealment. The analysis of cover and concealment is often inseparable from the fields of fires and observation. Weapon positions must have both to be effective and to be survivable. Infantry units are capable of improving poor cover and concealment by digging in and camouflaging their positions. When moving, the terrain is used to provide cover and concealment.

Obstacles. Identify the existing and reinforcing obstacles and hindering terrain that will affect mobility.

Key Terrain. Key terrain is any location or area that the seizure, retention, or control of affords a marked advantage to either combatant. Using the map and information already gathered, look for key terrain that dominates avenues of approach or the objective area. Next, look for decisive terrain that if held or controlled will have an extraordinary impact on the mission.

(5) ***Analyze Courses of Action (War Game)***. This analysis is conducted by war gaming the friendly courses of action against the enemy's most probable courses of action.

(6) ***Compare Courses of Action***. The leader compares the COAs and selects the one that is most likely to accomplish the assigned mission. He considers the advantages and disadvantages for each COA. He also considers how the critical events impact on COAs.

(7) ***Make a Decision***. The leader selects the COA that he believes has the best chance of accomplishing the mission.

d. **STEP 4–START NECESSARY MOVEMENT.** The unit may need to begin movement while the leader is still planning or forward reconnoitering. This step may occur anytime during the TLP.

e. **STEP 5–RECONNOITER.** If time allows, the leader makes a personal reconnaissance. When time does not allow, the leader must make a map reconnaissance. Sometimes the leader must rely on others (for example, scouts) to conduct the reconnaissance.

f. **STEP 6–COMPLETE THE PLAN.** The leader completes his plan based on the reconnaissance and any changes in the situation.

g. **STEP 7–ISSUE THE COMPLETE ORDER.** Platoon and squad leaders normally issue oral operation orders to aid subordinates in understanding the concept for the mission. If possible, leaders should issue the order with one or both of the following aids: within sight of the objective, on the defensive terrain, or on a terrain model or sketch. Leaders may require subordinates to repeat all or part of the order or demonstrate on the model or sketch their understanding of the operation. They should also quiz their Rangers to ensure that all Rangers understand the mission.

h. **STEP 8–SUPERVISE AND REFINE.** The leader supervises the unit's preparation for combat by conducting rehearsals and inspections.

(1) ***Rehearsals.*** Rehearsals include the practice of having squad leaders brief their planned actions in execution sequence to the platoon leader. The leader should conduct rehearsals on terrain that resembles the actual ground and in similar light conditions.

(a) *Purpose.* The leader uses rehearsals to:

- Practice essential tasks (improve performance).
- Reveal weaknesses or problems in the plan.
- Coordinate the actions of subordinate elements.
- Improve Ranger understanding of the concept of the operation (foster confidence in Rangers).

(b) *Times and Tasks.* The platoon may begin rehearsals of battle drills and other SOP items before the receipt of the operation order. Once the order has been issued, it can rehearse mission specific tasks. Some important tasks to rehearse include—

- Actions on the objective.
- Assaulting a trench, bunker, or building.
- Actions at the assault position.
- Breaching obstacles (mine and wire).
- Using special weapons or demolitions.
- Actions on unexpected enemy contact.

(c) *Types.*

Backbrief.
- Key leaders sequentially brief the actions required during operation.
- Patrol leader controls.
    - Conducted twice: right after FRAGO (confirmation brief) and again after subordinates develop their own plan.

Reduced force.
- Conducted when time is key constraint.
    - Conducted when security must be maintained.
- Key leaders normally attend.
- Mock ups, sand tables, and small scale replicas used.

Full force.
- Most effective type.
- First executed in daylight and open terrain.
- Secondly conduct in same conditions as operation.
- All Rangers participate.
- May use force on force.

Techniques.
- Force on force.
- Map (limited value and limited number of attendees).
- Radio (cannot mass leaders; confirms communications).
- Sand table or terrain model (key leaders; includes all control measures).
- Rehearsal of Concept (ROC) drill (similar to sand table / terrain model; subordinates actually move themselves).

(d) *Inspections.* Squad leaders should conduct initial inspections shortly after receipt of the warning order. The platoon sergeant spot checks throughout the unit's preparation for combat. The platoon leader and platoon sergeant make a final inspection. They should inspect—
- Weapons and ammunition.
- Uniforms and equipment.
- Mission essential equipment.
- Soldier's understanding of the mission and individual responsibilities.
- Communications.
- Rations and water.
- Camouflage.
- Deficiencies noted during earlier inspections.

2-2. COMBAT INTELLIGENCE. Gathering information is one of the most important aspects of conducting a patrolling operation. This paragraph details what information to collect and how to report it:

a. **Reports.** All information must be quickly, completely, and accurately reported. Use the SALUTE report format **(Figure 2-5)** for reporting and recording information.

Figure 2-5. SALUTE REPORT FORMAT

SIZE – Seven enemy personnel
ACTIVITY– Traveling SW
LOCATION – GA123456
UNIT / UNIFORM – OD uniforms with red six-point star on left shoulder
TIME – 210200JAN10
EQUIPMENT – Carry one machine gun and one rocket launcher

b. **Field Sketches.** Try to include a sketch with each report. Include only any aspects of military importance such as targets, objectives, obstacles, sector limits, or troop dispositions and locations (use symbols from **FM 1-02**). Use NOTES to explain the drawing, but they should not clutter the sketch. Leave off personnel, weapons, and equipment; these items go on the SALUTE report, not on this one.

c. **Captured Documents.** The leader collects documents and turns them in with his reports. He marks each document with the time and place of capture.

d. **Prisoners.** If prisoners are captured during a patrolling operation, they should be treated IAW the Geneva Convention and handled by the 5 S & T rule:

(1) ***Search***
(2) ***Silence***
(3) ***Segregate***
(4) ***Safeguard***
(5) ***Speed to rear***
(6) ***Tag***

e. **Debrief.** Immediately upon return from a mission, the unit is debriefed using the standard NATO report format.

2-3. WARNING ORDER. A warning order **(WARNO)** gives subordinates advance notice of an upcoming operation. This gives them time to prepare. A warning order is brief but complete. **Figure 2-6** shows an example format; **Figure 2-7** shows an example warning order.

NOTE: A warning order only authorizes execution when it clearly says so.

Figure 2-6. WARNING ORDER FORMAT

WARNING ORDER ________

*Roll call, pencil/pen/paper, RHB, map, protractor, leader's monitor, hold all questions till the end.
References: Refer to higher headquarters' OPORD, and identify map sheet for operation.
Time Zone Used throughout the Order: (Optional)
Task Organization: Optional; see paragraph 1c.

1. **SITUATION**. Find this in higher's OPORD para 1a(1-3).

a. **Area of Interest**. Outline the area of interest on the map.
(1) Orient relative to each point on the compass (N, S, E, W)
(2) Box in the entire AO with grid lines

b. **Area of Operations.** Outline the area of operation on the map. Point out the objective and current location of your unit.
(1) Trace your Zone using boundaries
(2) Familiarize by identifying natural (terrain) and man-made features in the zone your unit is operating.

c. **Enemy Forces**. Include significant changes in enemy composition, dispositions, and courses of action. Information not available for inclusion in the initial WARNO can be included in subsequent warning orders (WHO, WHAT, WHERE).

d. **Friendly Forces**. Optional; address only if essential to the WARNO.
(1) Give higher commander's mission (WHO, WHAT, WHEN, WHERE, WHY).
(2) State higher commander's intent. (Higher's [go to mapboard] OPORD para 1b[2]), give task and purpose.
(3) Point out friendly locations on the map board

e. **Attachments and Detachments**. Give initial task organization, only address major unit changes, and then go to the map board.

2. **MISSION**. State mission twice (WHO, WHAT, WHEN, WHERE, WHY).

3. **EXECUTION**.

a. **Concept of Operations**. Provide as much information as available. The concept should describe the employment of maneuver elements. Give general direction, distance, time of travel, mode of travel, and major tasks to be conducted. Use grids and terrain features. Cover all movements. Specify points where the ground tactical plan starts and stops.

b. **Tasks to Subordinate Units**. Provide specified tasks to subordinate units. Focus on non-tactical instructions for planning and preparation of the operations order. Leaders should also include tactical instructions for executing the mission using control, movement, AOO for each element in task organization. Planning guidance consists of tasks assigned to elements in the form and order of teams, special teams, and key individuals.

c. **Coordinating Instructions**. Include any information available at that time. If you know it, then at least cover the following items:

- Uniform and equipment common to all.
- Consider the factors of METT-TC and tailor the load for each Ranger.
- Timeline. (State when, what, where, who and all specified times. Reverse plan. Use 1/3 - 2/3 rule).
- Give specific priorities in order of completion.
- Give information about coordination meetings.
- Time of OPORD.
- Rehearsals / Inspections by priority.
- Earliest movement time.

4. **SUSTAINMENT.** Include any known logistics preparation for the operation.

a. Logistics.

(1) Maintenance. Include weapons and equipment DX time and location.

(2) Transportation. State method and mode of transportation for infil/exfil. Identify any coordination needed for external assets. Task subordinate leader (if needed) to generate load plan, number of lifts/serials, and bump plan.

(3) Supply. Only include classes of supply that require coordination or special instructions (rations, fuel, ammo etc).

b. Personnel Services Support. State any pertinent services for Soldiers (religious services etc).

c. Army Health System Support. Identify any medical equipment, support, or preventative medicine that needs to be coordinated.

5. **COMMAND AND CONTROL**.

a. Command. Succession of Command. State the succession of command if not covered in the unit's SOP.

b. Control.

(1) Command Posts. Describe the employment of command posts (CPs), including the location of each CP and its time of opening and closing, as appropriate. Typically at platoon level the only reference to command posts will be the company CP.

(2) Reports. List reports not covered in SOPs.

c. Signal. Describe the concept of signal support, including current SOI edition or refer to higher OPORD. Give subordinates guidance on tasks to complete for preparation of the OPORD and the mission. Give time, place, and uniform for the OPORD. Give a time hack and ask for questions.

Figure 2-7. EXAMPLE WARNING ORDER

WARNING ORDER (SQUAD)

Roll call, Camp Darby Special (CDS) 1:50,000 map, pen, paper, pencil, protractor, Ranger Handbook (RHB), hold all questions, and TL's monitor task organization. **ATM-SEC/BRM-ASLT/HQ SUPPORT**

1. **SITUATION**. (BRIEF).

a. **Area of Interest.** Orient the Map (N,S,E,W) Our squad's area of interest is boxed in by the 86 grid line to the north, the 18 grid line to the east, the 77 grid line to the south, and the 13 grid line to the west.

b. **Area of Operation.** We will be operating in Zone C. Trace Zone C with boundaries. Familiarize Zone C with three natural and man-made features. Our objective is located here (point on map) at GA 152 796, and our current location is here (point on map) at GA 196 790.

c. **Enemy.** 3 W's Para 1c(1-3). Describe enemy recent locations and activities.
WHO? The Aragon Liberation Front (ALF).
WHAT? Ambushed an ACDC patrol
WHERE? GA 156 804

d. **Friendly.** 4 W's Para 2. Mission and intent one and two levels up. Task and purpose of adjacent patrols. Provide the big picture concept.

Higher's Mission and Intent

Mission

WHO? 1st PLT, B CO.

WHAT? (***Task***) Conduct area ambushes to destroy enemy forces.

WHERE? On OBJ Black NLT 302300NOV2010

WHY? (***Purpose***) To prevent the enemy from maintaining control of OBJ Black.

Intent

-Find, fix, and finish enemy forces in Zone C.

-Enemy personnel and equipment are destroyed

-ALF resupply denied

e. **Attachments / Detachments.** MG TM 300530NOV2010.

2. **MISSION**. Clear and Concise, 5 W's, PARA 3, X2, Task and Purpose

1st SQD, 1st PLT, B CO (DO) conducts a point ambush to **destroy (TASK)** enemy personnel and equipment on OBJ Red (GA 152 793) NLT 302300NOV10 in order to **prevent (PURPOSE)** the enemy from maintaining control of OBJ Red.

3. **EXECUTION**.

a. **Concept of Operations:** (Orient Rangers to sketch or terrain model). We are currently located at Camp Darby, GA 1962 7902. We will depart Camp Darby moving generally northwest for 6,000 meters. The movement should take approximately 20 minutes. We will be travelling by truck to our insertion point, GA 176 812, where we will dismount the trucks. Our ground tactical plan will begin when we will move generally southwest for 3,000 meters. The movement should take 3 hours as we will be travelling by foot to our tentative ORP at Grid GA 154 795. Here will we finalize the preparing of M, W, and E. We will then move generally southwest for 400 meters to our objective at Grid GA 152 793. It will take us 30 minutes to an hour to complete our movement by foot. The reason for this is stealth on the objective while we occupy our positions. From our objective we will travel generally northwest for 3,000 meters, to our link-up site, GA 152 819. It will take four hours to move by foot since we will be travelling during limited visibility. Once complete with link-up, we will move generally southeast for 8,000 meters. The movement should take approximately 30 minutes as we will be travelling by truck back to Camp Darby. Once back on Camp Darby we will debrief and prepare to conduct follow-on operations.

b. **Tasks to Subordinate Units**. NON-TACTICAL AND TACTICAL INSTRUCTIONS METT-TC

Planning Guidance- Teams / / Special Teams / / Key Individuals (Control—Movement—Objective)

HQs: 2nd in the OOM / / M240 will provide supporting fires into the kill zone during AOO / / RTO will be the recorder en route and during actions on the OBJ. You will write Para 5 of the SQD OPORD, ensure all radios are operational with proper frequencies loaded, also ensure we enter the net on time.

ATM: ATM will be 1st OOM / Responsible for land navigation / ATM is flank security for AOO / / 1-2 Ranger EPW TM / 1-2 Ranger aid and litter TM / 1-2 Ranger DEMO TM / 1-2 Ranger ORP clearing TM / 2-2 Ranger flank security TM for AOO / 1-2 Ranger Linkup Security TM / / 1 SAW gunner to ASSLT element for AOO / 1 compass man / 1 pace man / ATL will be the security TM LDR for AOO. You are responsible for writing Para 1 and Linkup Annex of SQD OPORD, draw all sketches (FOOM, DAs, BDs, Linkup, Truck, AOO), Terrain Model, Routes, Fire Support Overlay (sterile and non-sterile).

BTM: BTM 3rd OOM / BTM is assault for AOO / / 1-2 man EPW TM / 1-2 Ranger aid and litter TM / 1-2 Ranger DEMO TM / 1-2 Ranger S/O TM / / 1 GDR to Security Team for AOO / 1 compass man / 1 pace man / BTL is the assault TL for AOO. You are 2nd in the chain of command and in charge at all times during my absence. You must write Para 4 and the Truck Annex of the SQD OPORD, prepare Supply, DX, and AMMO Lists, draw and issue all items. Ensure that everyone does a test fire and that all equipment is tied down IAW 4th RTB SOP. Update the squad status card and hand receipt.

TL: You will update the WARNO Board with all the correct information. As a task is accomplished, you will line it out. Post your COC, DP, and JD (Special Teams and Key Individuals). Come see me for further guidance at the conclusion of this Warning Order.

c. **Coordinating Instructions**. (METT-TC) Tailor the load, per Ranger, no SOPs.

(1) ***Packing List***. IAW RTB's seasonal packing list. Write on note cards or paper and read off by item

WORN

RUCK

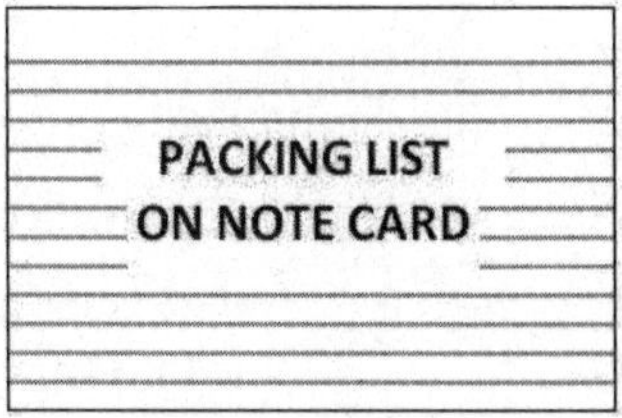

(2) ***Time Schedule.***

WHEN	WHAT	WHERE	WHO	F
*0630	Warning Order	Bay Area	All	3D1
0700	Initial Inspection	Bay Area	All	
*0730	REQ-Ammo / Supply	CO TOC	BTL / RB	4B1
*0745	P / U-Ammo / Supply	CO TOC	BTL / Detail	4B1
0745	Adj Unit Coord	Sqd Bays	ATL / CM	3D1
0750	Test Fire	T / F Area	All	
*0800	S-2 / S-3 / Fires Coord	Plt Bay	SL / RTO	3D1
*0815	Enter Net	Bay Area	RTO	
0930	Squad OPORD	Bay Area	All	
1300	Rehearsal	Bay Area	All	
1330	Final Inspection	Bay Area	All	
*1400	Truck Linkup	Co Toc	All	3D1
1500	Depart Darby	Co Toc	All	
*1500	Insertion Complete	TBD	All	3D1
1800	In ORP	TBD	All	
*2000	In Position	GA 152 793	All	
*2300	Mission Complete	TBD	All	
*0200	Linkup Complete	TBD	All	
0500	S-2 Debrief	Bn TOC	All	
Specified Times. Use 1/3 – 2/3 rule and reverse planning.				

4. **SUSTAINMENT.**

a. Logistics.

(1) Maintenance. Weapons and equipment DX will be at 0700 in the company CP.

(2) Transportation. Method of transportation for infill will be UH-60, and foot for exfil. Bravo team leader will generate the load plan, bump plan, and number of chalks/lifts.

(3) Supply.

(a) CLI. Each man will have two MREs and six quarts of water for the operation.
(b) CLV. Bravo team leader will draw enough ammunition for each man to carry a basic load IAW the squad SOP.

(Example Squad Ammunition SOP)

A. INDIVIDUAL WEAPONS (PER WPN SYSTEM)

WPN	QTY	REMARKS
M16 / M4	210	5.56mm Ball
M249	625	5.56mm Ball, Link (4:1 mix)
M240B	825	7.62mm Ball, Link (4:1 mix)
M203	12	40mm HEDP
M203	3	40mm Shot
M203	4	40mm Illumination
M203	2	40mm TP

B. SUPPLEMENTAL AMMO (TOTAL)

TYPE	QTY	REMARKS
CLAYMORE	3	M18A1
DEMO KIT	2	C4 / M81 / M14
M67 FRAGS	24	2 Per Man
5.56mm Tracer	126	42 Per Leader
AT-4	2	1 Per Team
HC Smoke	6	
Red Smoke	2	
Yellow Smoke	4	

b. Personnel Services Support. Religious services will be held at the 0800 in the chapel.
c. Army Health System Support. Alpha team leader, coordinate for one additional CLS bag.

5. **COMMAND AND CONTROL.**
a. Command.
(1) Location of Commander/Patrol Leader. The patrol leader will be located in the squad bay during phase I (Mission Preparation). Location of patrol leader for all other phases will be briefed in the OPORD.
(2) Succession of Command. State the succession of command if not covered in the unit's SOP. SL, BTL, ATL, RTO.
b. Control.
(1) Command Posts. The platoon CP is located at GA 166 807, and the company CP is located at GA 196 790.
(2) Reports. Pertinent reports will be covered in the OPORD.
c. Signal.
(1) The battalion will be operating on 37.950 SC/PT, Call sign "Darby 74"
(2) Our squad frequency is 77.000 SC/PT, Call sign "Bravo One One"
(3) All other signals such as frequencies, call signs, challenges and passwords will be given during the OPORD.

*Additional Guidance:
1. Give subordinates additional guidance on tasks to complete for preparation of the OPORD and the mission
2. Give time, place and uniform of the OPORD
3. Give a time hack and ask for questions

2-4. OPERATION ORDER. An operation order (OPORD) is a directive issued by a leader to his subordinates in order to effect the coordinated execution of a specific operation. A five paragraph format (example shown in **Figure 2-8**) is used to organize the briefing, to ensure completeness, and to help subordinate leaders understand and follow the order. Use a terrain model or sketch along with a map to explain the order. When possible, such as in the defense, give the order while observing the objective. The platoon/ squad leader briefs his OPORD orally off NOTES that follow the five paragraph format. Before the issuance of the OPORD, the leader ensures that the following resources are in place: pencil, pen, paper, RHB, map, protractor. Leader's monitor subordinates. Then he calls roll and says, "Please hold all questions till the end."

Figure 2-8. EXAMPLE SQUAD OPORD FORMAT

OPERATION ORDER

[Plans and orders normally contain a code name and are numbered consecutively within a calendar year.]
References: The heading of the plan or order lists maps, charts, data, or other documents the unit will need to understand the plan or order. The user need not reference the SOP, but may refer to the SOP in the body of the plan or order. He refers to a map by map series number (and country or geographic area, if required), sheet number and name, edition, and scale, if required. "Datum" refers to the mathematical model of the earth that applies to the coordinates on a particular map. It is used to determine coordinates. Different nations use different datum for printing coordinates on their maps. The datum is usually referenced in the marginal information of each map.

Time zone used throughout the order: If the operation will take place in one time zone, use that time zone throughout the order (including annexes and appendixes). If the operation spans several time zones, use Zulu time.

Task organization: Describe the allocation of forces to support the commander's concept. You may show task organization in one of two places: ***just above paragraph 1, or in an annex***, if the task organization is long or complex.

- Go to the map.
- Apply the **Orient, Box, Trace,** and **Familiarize** technique to (only) the areas the unit is moving through. (Get this info from the platoon OPORD.)
- Determine the effects of seasonal vegetation within the AO.

1. **SITUATION**.

a. Area of Interest. Describe the area of interest or areas outside of your area of operation that can influence your area of operation.

b. Area of Operations. Describe the area of operations. Refer to the appropriate map and use overlays as needed.

(1) **Terrain:** Using the OAKOC format, state how the terrain will affect both friendly and enemy forces in the AO. Use the OAKOC from higher's OPORD. Refine it based on your analysis of the terrain in the AO. Follow these steps to brief terrain.

(2) **Weather.** Describe the aspects of weather that impact operations. Consider the five military aspects of weather to drive your analysis (V,W,T,C,P- Visibility, Winds, Temperature/Humidity, Cloud Cover, Precipitation)

Temp High	Sunrise	Moonrise
Temp Low	Sunset	Moonset
Wind Speed	BMNT	Moonphase
Wind Direction	EENT	Percent Illumination
** This is the information the squad leader received from the platoon OPORD.*		

c. **Enemy Forces**. The enemy situation in higher headquarters' OPORD (paragraph 1c) forms the basis for this. Refine it by adding the detail your subordinates require. • *Point out* on the map the location of recent enemy activity known and suspected.

(1) State the enemy's ***composition, disposition, and strength***.

(2) Describe his ***recent activities***.

(3) Describe his known or suspected ***locations and capabilities***.

(4) Describe the enemy's most likely and most dangerous ***course of action***.

d. **Friendly Forces**. Get this information from paragraphs 1d, 2, and 3 of the higher headquarters' OPORD.

(1) Higher Headquarters' Mission and Intent

(a) Higher Headquarters Two Levels Up

1 Mission- State the *mission* of the Higher Unit (*2 levels up*).

2 Intent- State intent 2 levels up.

(b) Higher Headquarters One Level Up

1 Mission- State the *mission* of the Higher Unit (*1 level up*).

2 Intent- State intent 1 levels up.

(2) Mission of Adjacent Units. State *locations of units* to the left, right, front, and rear. State those units' *tasks and purposes*; and say how those units will *influence* yours, particularly adjacent unit patrols.

a. Show *other units' locations* on map board.

b. Include statements about the *influence each of the above patrols* will have on your mission, if any.

c. Obtain this information from higher's OPORD. It gives each leader an idea of what other units are doing and where they are going. This information is in paragraph 3b(1) (Execution, Concept of the Operation, Scheme of Movement and Maneuver).

d. Also include any information obtained when the leader conducts adjacent unit coordination.

e. **Attachments and Detachments**. Avoid repeating information already listed in Task Organization. Try to put all information in the Task Organization. However, when not in the Task Organization, list units that are attached or detached to the headquarters that issues the order. State when attachment or detachment will be in effect, if that differs from when the OPORD is in effect such as on order or on commitment of the reserve. Use the term "remains attached" when units will be or have been attached for some time.

2. **MISSION**. State the mission derived during the planning process. A mission statement has no subparagraphs. Answer the 5 W's: Who? What (task)? Where? When? and Why (purpose)?

- State the **mission** clearly and concisely. Read it twice.
- Go to map and point out the exact **location of the OBJ** and the **unit's present location**

3. **EXECUTION**

a. **Commander's Intent.** State the commander's intent which is his clear, concise statement of what the force must do and the conditions the force must establish with respect to the enemy, terrain, and civil considerations that represent the desired end state.

b. **Concept of Operations**. Write a clear, concise concept statement. Describe how the unit will accomplish its mission from start to finish. Base the number of subparagraphs, if any, on what the leader considers appropriate, the level of leadership, and the complexity of the operation. The following subparagraphs from **FM 5-0** show what might be required within the concept of the operation. Ensure that you state the purpose of the war fighting functions within the concept of the operation (Figure 1).

Figure 1. WARFIGHTING FUNCTIONS

Fire support
Movement and Maneuver
Protection
Command and Control
Intelligence
Sustainment (formerly called "CSS")

c. **Scheme of Movement and Maneuver**. Describe the employment of maneuver units in accordance with the concept of operations. Address subordinate units and attachments by name. State each one's mission as a task and purpose. Ensure that the subordinate units' missions support that of the main effort. Focus on actions on the objective. Include a detailed plan and criteria for engagement / disengagement, an alternate plan in case of compromise or unplanned enemy force movement, and a withdrawal plan. The brief is to be sequential, taking you from start to finish, covering all aspects of the operation.

- Brief from the start of your operation, to mission complete.
- Cover all routes, primary and alternate, from insertion, through AOO, to link-up, until mission complete.
- Brief your plan for crossing known danger areas.
- Brief your plan for reacting to enemy contact.
- Brief any approved targets/CCPs as you brief your routes.

d. Scheme of ***Fires***. State scheme of fires to support the overall concept and state who (which maneuver unit) has priority of fire. You can use the PLOT-CR format (purpose, location, observer, trigger, communication method, resources) to plan fires. Refer to the target list worksheet and overlay here, if applicable. Discuss specific targets and point them out on the terrain model (Chapter 3, Fire Support).

e. ***Casualty Evacuation***. Provide a detailed CASEVAC plan during each phase of the operation. Include CCP locations, tentative extraction points, and methods of extraction.

f. **Tasks to Subordinate Units**. Clearly state the missions or tasks for each subordinate unit that reports directly to the headquarters issuing the order. List the units in the task organization, including reserves. Use a separate subparagraph for each subordinate unit. State only the tasks needed for comprehension, clarity, and emphasis. Place tactical tasks that affect two or more units in Coordinating Instructions (subparagraph 3h). Platoon leaders may task their subordinate squads to provide any of the following special teams: reconnaissance and security, assault, support, aid and litter, EPW and search, clearing, and demolitions. You may also include detailed instructions for the platoon sergeant, RTO, compass-man, and pace-man.

h. **Coordinating Instructions**. This is always the last subparagraph under paragraph 3. List only the instructions that apply to two or more units, and which are seldom covered in unit SOPs. Refer the user to an annex for more complex instructions. The information listed below is required.

(1) **Time Schedule**. State time, place, uniform, and priority of rehearsals, backbriefs, inspections, and movement.

(2) **Commander's Critical Information Requirements**. Include PIR and FFIR

(a) **Priority intelligence requirements**. PIR includes all intelligence that the commander must have for planning and decision making.

(b) **Friendly force information requirements**. FFIR include what the commander needs to know about friendly forces available for the operation. It can include personnel status, ammunition status, and leadership capabilities.

(3) **Essential elements of friendly information**. EEFI are critical aspects of friendly operations that, if known by the enemy, would compromise, lead to failure, or limit success of the operation.

(4) **Risk-Reduction Control Measures**. These are measures unique to the operation. They supplement the unit SOP and can include mission-oriented protective posture, operational exposure guidance, vehicle recognition signals, and fratricide prevention measures.

(5) **Rules of Engagement** (ROE).

(6) **Environmental Considerations**.

(7) **Force Protection**.

4. SUSTAINMENT. Describe the concept of sustainment to include logistics, personnel, and medical.

a. Logistics.

(1) Sustainment Overlay. Include current and proposed company trains locations, CCPs (include marking method), equipment collection points, HLZs, AXPs, and any friendly sustainment locations (FOBs, COPs etc).

(2) Maintenance. Include weapons and equipment DX time and location.

(3) Transportation. State method and mode of transportation for infil/exfil, load plan, number of lifts/serials, bump plan, recovery assets, recovery plan.

(4) Supply.

Class I--Rations plan.
Class III--Petroleum.
Class V--Ammunition.
Class VII--Major end items.
Class VIII--Medical.
Class IX--Repair parts.
Distribution Methods.

(5) Field Services. Include any services provided or required (laundry, showers etc).

b. Personnel Services Support.

(1) Method of marking and handling EPWs.

(2) Religious Services.

c. Army Health System Support.

(1) Medical Command and Control. Include location of medics, identify medical leadership, personnel controlling medics, and method of marking patients.

(2) Medical Treatment. State how wounded or injured Soldiers will be treated (self aid, buddy aid, CLS, EMT etc).

(3) Medical Evacuation. Describe how dead or wounded, friendly and enemy personnel will be evacuated and identify aid and litter teams. Include special equipment needed for evacuation.

(4) Preventive Medicine. Identify any preventive medicine Soldiers may need for the mission (sun block, chap stick, insect repellant, in-country specific medicine).

5. COMMAND AND CONTROL. *State where command and control facilities and key leaders are located during the operation.*

a. Command.

(1) Location of Commander/Patrol Leader. State where the commander intends to be during the operation, by phase if the operation is phased.

(2) Succession of Command. State the succession of command if not covered in the unit's SOP.

b. Control.

(1) Command Posts. Describe the employment of command posts (CPs), including the location of each CP and its time of opening and closing, as appropriate. Typically at platoon level the only reference to command posts will be the company CP.

(2) Reports. List reports not covered in SOPs.

c. Signal. Describe the concept of signal support, including current SOI edition or refer to higher OPORD.

(1) Identify the ***SOI index*** that is in effect

(2) Identify ***methods of communication by priority***

(3) Describe ***pyrotechnics and signals***, to include arm and hand signals (demonstrate)

(4) Give ***code words*** such as OPSKEDs

(5) Give ***challenge and password*** (use behind friendly lines)

(6) Give ***number combination*** (use forward of friendly lines)

(7) Give ***running password***

(8) Give ***recognition signals*** (near/ far and day/ night)

***Actions after Issuance of OPORD:**

-Issue annexes

-Highlight next hard time

-Give time hack

-ASK for questions

2-5. FRAGMENTARY ORDER. A FRAGO is an abbreviated form of an operation order, usually issued daily, which eliminates the need for restating portions of the OPORD. It is issued after an OPORD to change or modify that order or to execute a branch or sequel to that order. **Figure 2-10** shows an annotated FRAGO format.

Figure 2-10. ANNOTATED FRAGO FORMAT

FRAGMENTARY ORDER____________

Time Zone referenced throughout order:

Task Organization:

1. SITUATION [*Brief changes from base OPORD specific to this day's operation*]

a. Area of Interest. State any changes to the area of interest.

b. Area of Operations. State any changes to the area of operations.

(1)Terrain [*Note any changes that will effect operation in new area of operations*]:Observation/Fields of Fire, Cover and Concealment, Obstacles, Key Terrain, and Avenues of Approach

(2) *Weather and Light Data:*

High:	BMNT:	Moonrise:
Low:	Sunrise:	Moonset:
Wind Speed:	Sunset:	% Illum:
Wind Direction:		EENT:
Forecast:		

c. **Enemy.**
(1) Composition, disposition and strength.
(2) Capabilities.
(3) Recent activities.
(4) Most likely COA.

d. **Friendly.**
(1) Higher mission.
(2) Adjacent patrols task/purpose.
(3) Adjacent patrol objective/route (if known).

2. MISSION (Who, what [task], when, where, why [purpose]—from higher HQ maneuver paragraph).

3. EXECUTION

a. **Commander's Intent.** Include any changes or state "No Change".

b. **Concept of Operations**. Include any changes or state "No Change".

c. **Scheme of Movement and Maneuver**. Include any changes or state "No Change".

d. **Scheme of Fires.** Include any changes or state "No Change".

e. **Casualty Evacuation**. Include any changes or state "No Change".

f. **Tasks to Subordinate Units**. Include any changes or state "No Change".

h. **Coordinating Instructions**. Include any changes or state "No Change".
(1) **Time Schedule**.
(2) **Commander's Critical Information Requirements**.
(a) **Priority intelligence requirements**.
(b) **Friendly force information requirements**.
(3) **Essential elements of friendly information**.
(4) **Risk-Reduction Control Measures**.
(5) **Rules of Engagement** (ROE).
(6) **Environmental Considerations**.
(7) **Force Protection**.

4. SUSTAINMENT. Only cover changes from base order – use standard format and items that have not changed should be briefed "no change."

a. Logistics.
(1) Sustainment Overlay.
(2) Maintenance.
(3) Transportation.
(4) Supply.
Class I:

Class III:
Class V:
Class VII:
Class VIII:
Class IX:
Distribution Methods:

(5) Field Services.

b. Personnel Services Support.

(1) Method of marking and handling EPWs.

(2) Religious Services.

c. Army Health System Support.

(1) Medical Command and Control.

(2) Medical Treatment.

(3) Medical Evacuation.

(4) Preventive Medicine.

5. COMMAND AND CONTROL. Only brief changes to base order. If there are changes state where command and control facilities and key leaders are located during the operation.

a. Command.

(1) Location of Commander/Patrol Leader. State where the commander intends to be during the operation, by phase if the operation is phased.

(2) Succession of Command. State the succession of command if not covered in the unit's SOP.

b. Control.

(1) Command Posts. Describe the employment of command posts (CPs), including the location of each CP and its time of opening and closing, as appropriate. Typically at platoon level the only reference to command posts will be the company CP.

(2) Reports. List reports not covered in SOPs.

c. Signal. Describe the concept of signal support, including current SOI edition or refer to higher OPORD.

(1) Identify the ***SOI index*** that is in effect

(2) Identify ***methods of communication by priority***

(3) Describe ***pyrotechnics and signals***, to include arm and hand signals (demonstrate)

(4) Give ***code words*** such as OPSKEDs

(5) Give ***challenge and password*** (use behind friendly lines)

(6) Give ***number combination*** (use forward of friendly lines)

(7) Give ***running password***

(8) Give ***recognition signals*** (near/ far and day/ night)

*Field FRAGO Guidance:

1. The field FRAGO should take no more than 40 minutes to issue, with 30 minutes for the target. The proposed planning guide is as follows:

a. Paragraphs 1 and 2:	5 minutes
b. Paragraph 3:	20 to 30 minutes
c. Paragraphs 4 and 5:	5 minutes

2. The FRAGO should focus on actions on the objective. The PL may use subordinates to prepare para 1, 4, 5 and routes and fires for the FRAGO. It is acceptable for subordinates to brief the portions of the FRAGO they prepare.
3. Use of sketches and a terrain model are critical to allow rapid understanding of the operation/FRAGO.
4. Rehearsals are critical as elements of the constrained planning model. The FRAGO used with effective rehearsals reduces preparation time and allows the PL more time for movement and recon.
5. Planning in a field environment will necessarily reduce the amount of time leaders have for in-depth mission planning. The TLP give leaders a framework to plan missions and produce orders when time is short.

2-6. ANNEXES. Operation order annexes are issued after an OPORD only if more information is needed about truck movement, air assault, patrol bases, small boats, linkups, or stream crossings, for example. Brevity is standard. Annexes are always issued after the operation order. **Figure 2-11** shows example formats for some types of annexes.

Figure 2-11. EXAMPLE ANNEX FORMATS

AIR MOVEMENT ANNEX

1. **SITUATION**.
   a. **Enemy**.
      (1) Enemy air capability.
      (2) Enemy ADA capability.
      (3) Include in Weather: % Illum, Illum angle, NVG Window, Ceiling and Visibility.
2. **MISSION**.
3. **EXECUTION**.
   a. **Concept of Operations**.
   b. **Tasks to Subordinate Units**.
   c. **Coordinating instructions**.
      (1) Pickup Zone.
         (a) Name/Number.
         (b) Coordinates.
         (c) Load Time.
         (d) Takeoff Time.
         (e) Markings.
         (f) Control.
         (g) Landing Formation.
         (h) Approach/Departure Direction.
         (i) Alternate PZ Name/Number.
         (j) Penetration Points.
         (k) Extraction Points.
      (2) Landing Zone.
         (a) Name/Number.
         (b) Coordinates.
         (c) H-Hour.
         (d) Markings.
         (e) Control.
         (f) Landing Formation/Direction.
         (g) Alt LZ Name/Number.
         (h) Deception Plan.
         (i) Extraction LZ.
      (3) Laager Site.
         (a) Communications.
         (b) Security Force.
      (4) Flight Routes and Alternates.
      (5) Abort Criteria.
      (6) Down Aircraft/Crew (Designated Area of Recovery (DAR).
      (7) Special Instructions.
      (8) Cross-FLOT Considerations.
      (9) Aircraft Speed.
      (10) Aircraft Altitude.
      (11) Aircraft Crank Time.
      (12) Rehearsal Schedule/Plan.

(13) Actions on Enemy Contact (Enroute and on the Ground).

4. SUSTAINMENT.

a. Logistics.

(1) Sustainment Overlay. Include forward area refuel/rearm points.

(2) Maintenance. Specific to aircraft.

(3) Transportation.

(4) Supply.

Class I:

Class III:

Class V:

Class VII:

Class VIII:

Class IX:

Distribution Methods:

5. **COMMAND AND CONTROL**.

a. Command.

(1) Location of Commander/Patrol Leader. State where the commander intends to be during the operation, by phase if the operation is phased.

(2) Succession of Command. State the succession of command if not covered in the unit's SOP.

b. Control.

(2) Reports. List reports not covered in SOPs.

c. Signal. Describe the concept of signal support, including current SOI edition or refer to higher OPORD.

(1) Air/ground call signs and frequencies.

(2) Air/ground emergency code.

(3) Passwords/number combinations.

(4) Fire net/Quickfire net.

(5) Time zone.

(6) Time hack.

PATROL BASE ANNEX

1. **SITUATION**

a. Enemy Forces

b. Friendly Forces

c. Attachments and Detachments

2. **MISSION**.

3. **EXECUTION**.

a. **Concept of Operations**.

b. Scheme of Movement and Maneuver.

c. Scheme of Fires.

d. **Tasks to Subordinate Units**.

(1) ***Teams***.

- Security.
- Recon.
- Surveillance.
- LP/OPs.

(2) ***Individuals***.

e. **Coordinating instructions.**
(1) Occupation plan.
(2) Operations plan.
• Security Plan.
• Alert Plan.
• Priority of work.
• Evacuation plan.
• Alternate patrol base (used when primary is unsuitable or compromised).

4. SUSTAINMENT. Only brief specifics not covered in base order.
a. Logistics.
(1) Sustainment Overlay. Include water plan, maintenance plan, hygiene plan, rations plan, and rest plan.
(2) Maintenance.
(3) Transportation.
(4) Supply.
Class I:
Class III:
Class V:
Class VII:
Class VIII:
Class IX:
Distribution Methods:
(5) Field Services.
b. Personnel Services Support.
(1) Method of marking and handling EPWs.
(2) Religious Services.
c. Army Health System Support.
(1) Medical Command and Control.
(2) Medical Treatment.
(3) Medical Evacuation.
(4) Preventive Medicine.

5. **COMMAND AND CONTROL.**
a. Command.
(1) Location of Commander/Patrol Leader. State where the commander intends to be during the operation, by phase if the operation is phased.
(2) Succession of Command. State the succession of command if not covered in the unit's SOP.
b. Control.
(1) Command Posts. Describe the employment of command posts (CPs), including the location of each CP and its time of opening and closing, as appropriate. Typically at platoon level the only reference to command posts will be the company CP.
(2) Reports. List reports not covered in SOPs.
c. Signal. Describe the concept of signal support, including current SOI edition or refer to higher OPORD.
(1) Identify the ***SOI index*** that is in effect
(2) Identify ***methods of communication by priority***
(3) Describe ***pyrotechnics and signals***, to include arm and hand signals (demonstrate)
(4) Give ***code words*** such as OPSKEDs
(5) Give ***challenge and password*** (use behind friendly lines)
(6) Give ***number combination*** (use forward of friendly lines)
(7) Give ***running password***
(8) Give ***recognition signals*** (near/ far and day/ night)

SMALL BOAT ANNEX

1. **SITUATION**.

   a. **Area of Operations.**

      (1) Terrain.

         (a) River width.
         (b) River depth and water temperature.
         (c) Current.
         (d) Vegetation.

      (2)Weather.

         (a) Tide.
         (b) Surf.
         (c) Wind.

   b. **Enemy Forces**. State any changes/additions to identification, location, activity, and strength.

   c. **Friendly Forces (unit furnishing support)**.

   d. **Attachments and Detachments**.

   e. **Organization for Movement**.

2. **MISSION**.

3. **EXECUTION**.

   a. Concept of Operations.

   b. Scheme of Movement and Maneuver.

   c. Scheme of Fires.

   d. Tasks to Subordinate Units.

      (1) Security.
      (2) Tie-down teams.

         (a) Load equipment.
         (b) Secure equipment.

      (3) Designation of coxswains and boat commanders.
      (4) Selection of navigator(s) and observer(s).

   e. Coordinating Instructions.

      (1) Formations and order of movement.
      (2) Route and alternate route.
      (3) Method of navigation.
      (4) Actions on enemy contact.
      (5) Rally points.
      (6) Embarkation plan.
      (7) Debarkation plan.
      (8) Rehearsals.
      (9) Time schedule.

4. SUSTAINMENT. Only brief specifics not covered in base order.

   a. Logistics.

      (1) Sustainment Overlay.
      (2) Maintenance.
      (3) Transportation. Include disposition of boats, paddles, and life jackets upon debarkation.
      (4) Supply.

         Class I:
         Class III:
         Class V:

Class VII:
Class VIII:
Class IX:
Distribution Methods: Include method of distribution of paddles and life jackets.

5. **COMMAND AND CONTROL**.

a. Command.

(1) Location of Commander/Patrol Leader. State where the commander intends to be during the operation, by phase if the operation is phased.

(2) Succession of Command. State the succession of command if not covered in the unit's SOP.

b. Control.

(1) Command Posts. Describe the employment of command posts (CPs), including the location of each CP and its time of opening and closing, as appropriate. Typically at platoon level the only reference to command posts will be the company CP.

(2) Reports. List reports not covered in SOPs.

c. Signal. Describe the concept of signal support, including current SOI edition or refer to higher OPORD.

(1) Identify the ***SOI index*** that is in effect
(2) Identify ***methods of communication by priority***
(3) Describe ***pyrotechnics and signals***, to include arm and hand signals (demonstrate)
(4) Give ***code words*** such as OPSKEDs
(5) Give ***challenge and password*** (use behind friendly lines)
(6) Give ***number combination*** (use forward of friendly lines)
(7) Give ***running password***
(8) Give ***recognition signals*** (near/ far and day/ night)

STREAM CROSSING ANNEX

1. **SITUATION**.

a. Area of Operations.

(1) Terrain.

(a) River width.
(b) River depth and water temperature.
(c) Current.
(d) Vegetation.
(e) Obstacles.

(2) Weather.

b. Enemy Forces. Enemy (location, identification, activity).
c. Friendly Forces.
d. Attachments and Detachments.

2. **MISSION.**

3. **EXECUTION.**

a. Concept of Operations.
b. Scheme of Movement and Maneuver.
c. Scheme of Fires.
d. Tasks to Subordinate Units.

(1) Elements.
(2) Teams.
(3) Individuals.

e. Coordinating Instructions.

(1) Crossing procedure/techniques.

(2) Security.
(3) Order of crossing.
(4) Actions on enemy contact.
(5) Alternate plan.
(6) Rallying points.
(7) Rehearsal plan.
(8) Time schedule.

4. SUSTAINMENT. Only brief specifics not covered in base order.

a. Logistics.

(1) Sustainment Overlay.
(2) Maintenance.
(3) Transportation.
(4) Supply.

Class I:
Class III:
Class V:
Class VII:
Class VIII:
Class IX:
Distribution Methods:

5. **COMMAND AND CONTROL.**

a. Command.

(1) Location of Commander/Patrol Leader. State where the commander intends to be during the operation, by phase if the operation is phased.

(2) Succession of Command. State the succession of command if not covered in the unit's SOP.

b. Control.

(1) Command Posts. Describe the employment of command posts (CPs), including the location of each CP and its time of opening and closing, as appropriate. Typically at platoon level the only reference to command posts will be the company CP.

(2) Reports. List reports not covered in SOPs.

c. Signal. Describe the concept of signal support, including current SOI edition or refer to higher OPORD.

(1) Identify the ***SOI index*** that is in effect
(2) Identify ***methods of communication by priority***
(3) Describe ***pyrotechnics and signals***, to include arm and hand signals (demonstrate)
(4) Give ***code words*** such as OPSKEDs
(5) Give ***challenge and password*** (use behind friendly lines)
(6) Give ***number combination*** (use forward of friendly lines)
(7) Give ***running password***
(8) Give ***recognition signals*** (near/ far and day/ night)

TRUCK ANNEX

1. **SITUATION.**

a. Enemy.
b. Friendly.
c. Attachments and Detachments.

2. **MISSION.**

3. **EXECUTION.**

a. Concept of Operations.
b. Scheme of Movement and Maneuver.

c. Scheme of Fires.
d. Tasks to Subordinate Units.
e. Coordinating Instructions.
(1) Times of departure and return.
(2) Loading plan and order of movement.
(3) Route (primary and alternate).
(4) Air guards.
(5) Actions on enemy contact (vehicle ambush) during movement, loading, and downloading.
(6) Actions at the detrucking point.
(7) Rehearsals.
(8) Vehicle speed, separation, and recovery plan.
(9) Broken vehicle instructions.

4. SUSTAINMENT. Only brief specifics not covered in base order.
a. Logistics.
(1) Sustainment Overlay.
(2) Maintenance.
(3) Transportation.
(4) Supply.
Class I:
Class III:
Class V:
Class VII:
Class VIII:
Class IX:
Distribution Methods:
(5) Field Services.
b. Personnel Services Support.
(1) Method of marking and handling EPWs.
(2) Religious Services.
c. Army Health System Support.
(1) Medical Command and Control.
(2) Medical Treatment.
(3) Medical Evacuation.
(4) Preventive Medicine.

5. **COMMAND AND CONTROL**.
a. Command.
(1) Location of Commander/Patrol Leader. State where the commander intends to be during the operation, by phase if the operation is phased.
(2) Succession of Command. State the succession of command if not covered in the unit's SOP.
b. Control.
(1) Command Posts. Describe the employment of command posts (CPs), including the location of each CP and its time of opening and closing, as appropriate. Typically at platoon level the only reference to command posts will be the company CP.
(2) Reports. List reports not covered in SOPs.
c. Signal. Describe the concept of signal support, including current SOI edition or refer to higher OPORD.
(1) Identify the ***SOI index*** that is in effect
(2) Identify ***methods of communication by priority***
(3) Describe ***pyrotechnics and signals***, to include arm and hand signals (demonstrate)
(4) Give ***code words*** such as OPSKEDs

(5) Give ***challenge and password*** (use behind friendly lines)
(6) Give ***number combination*** (use forward of friendly lines)
(7) Give ***running password***
(8) Give ***recognition signals*** (near/ far and day/ night)

2-7. COORDINATION CHECKLISTS. The checklists shown in **Figure 2-12** include items that a platoon/squad leader must check when planning for a combat operation. In some cases, he coordinates directly with the appropriate staff section. In most cases, the company commander/ platoon leader provides this information. The platoon/ squad leader can carry copies of these checklists to keep him from overlooking anything that may be vital to the mission.

Figure 2-12. COORDINATION CHECKLISTS

INTELLIGENCE COORDINATION CHECKLIST

The unit one level higher constantly updates intelligence. This ensures that the platoon leader's plan reflects the most recent enemy activity.

1. Identification of enemy unit.
2. Weather and light data.
3. Terrain update.
    a. Aerial photos.
    b. Trails and obstacles not on map.
4. Known or suspected enemy locations.
5. Weapons.
6. Probable course of action.
7. Recent enemy activities.
8. Reaction time of reaction forces.
9. Civilians on the battlefield.
10. Update to CCIR.

OPERATIONS COORDINATION CHECKLIST

The platoon/ squad leader coordinates with the company commander/ platoon leader to confirm the mission and operational plan, receive last minute changes, and either update subordinates in person or issue a FRAGO:

1. Mission backbrief.
2. Identification of friendly units.
3. Changes in the friendly situation.
4. Route selection, LZ/ PZ/ DZ selection.
5. Linkup procedures.
    a. Contingencies
    b. QRF
    c. QRF Frequency
6. Transportation/movement plan.
7. Resupply (with S-4).
8. Signal plan.
9. Departure and reentry of forward units.
10. Special equipment requirements.
11. Adjacent units in the area of operations.
12. Rehearsal areas.
13. Method of insertion/extraction.

FIRE SUPPORT COORDINATION CHECKLIST

The platoon/squad leader coordinates the following with the forward observer (FO):

1. Mission backbrief.
2. Identification of supporting unit.
3. Mission and objective.
4. Route to and from the objective (include alternate routes).
5. Time of departure and expected time of return.
6. Unit target list (from fire plan).
7. Type of available support (artillery, mortar, naval gunfire and aerial support, to include Army, Navy and Air Force) and their locations.
8. Ammunition available (to include different fuses).
9. Priority of fires.
10. Control measures.
    a. Checkpoints.
    b. Boundaries.
    c. Phase lines.
    d. Fire support coordination measures.
    e. Priority targets (target list).
    f. RFA (restrictive fire area).
    g. RFL (restrictive fire line).
    h. NFA (no-fire area).
    i. Precoordinated authentication.
11. Communication (include primary and alternate means, emergency signals and code words).

COORDINATION WITH FORWARD UNIT CHECKLIST

A platoon/ squad that requires foot movement through a friendly forward unit must coordinate with that unit's commander for a safe and orderly passage. If no time and place has been designated for coordination with the forward unit, the platoon/squad leader should set a time and place to coordinate with the S-3. He must talk with someone at the forward unit who has the authority to commit the forward unit to assist the platoon/ squad during departure. Coordination is a two-way exchange of information:

1. Identification (yourself and your unit).
2. Size of platoon/squad.
3. Time(s) and place(s) of departure and return, location(s) of departure point(s), ERRP, and de-trucking points.
4. General area of operations.
5. Information on terrain and vegetation.
6. Known or suspected enemy positions or obstacles.
7. Possible enemy ambush sites.
8. Latest enemy activity.
9. Detailed information on friendly positions such as crew-served weapons, FPF.
10. Fire and barrier plan.
    a. Support the unit can furnish. How long and what can they do?
        (1) Fire support.
        (2) Litter teams.
        (3) Navigational signals and aids.
        (4) Guides.
        (5) Communications.
        (6) Reaction units.
        (7) Other.
    b. Call signs and frequencies.
    c. Pyrotechnic plan.

d. Challenge and password, running password, number combination.
e. Emergency signals and code words.
f. If the unit is relieved, pass the information to the relieving unit.
g. Recognition signals.

ADJACENT UNIT COORDINATION CHECKLIST

Immediately after the OPORD or mission briefing, the platoon/squad leader should check with other platoon/squad leaders who will be operating in the same areas. If the leader is unaware of any other units operating is his area, he should check with the S-3 during the operations coordination. The S-3 can help arrange this coordination if necessary. The platoon/squad leaders should exchange the following information with other units operating in the same area:

1. Identification of the unit.
2. Mission and size of unit.
3. Planned times and points of departure and reentry.
4. Route(s).
5. Fire support and control measures.
6. Frequencies and call signs.
7. Challenge and password, running password, number combination.
8. Pyrotechnic plan.
9. Any information that the unit may have about the enemy.
10. Recognition signals.

REHEARSAL AREA COORDINATION CHECKLIST

The assistant patrol leader coordinates the use of the rehearsal area to facilitate the unit's safe, efficient, and effective use of the rehearsal area before its mission:

1. Identification of your unit.
2. Mission.
3. Terrain similar to objective site.
4. Security of the area.
5. Availability of aggressors.
6. Use of blanks, pyrotechnics, and ammunition.
7. Mock-ups available.
8. Time the area is available (preferably when light conditions approximate light conditions of patrol).
9. Transportation.
10. Coordination with other units using the area.

ARMY AVIATION COORDINATION CHECKLIST

The patrol leader coordinates this with the company commander or S-3 Air to facilitate the time and detailed and effective use of aviation assets as they apply to the tactical mission:

1. **SITUATION**.
   a. **Enemy**.
      (1) Air capability.
      (2) ADA capability.
      (3) Include in Weather: Percent Illum, Illum angle, NVG Window, Ceiling, and Visibility.
   b. **Friendly**.
      (1) Unit(s) supporting operation, Axis of movement/corridor/routes.
      (2) ADA status.

2. **MISSION**.

3. **EXECUTION**.
   a. **Concept of the Operation**. Overview of what requesting unit wants to accomplish with the air assault/ air movement.

b. **Tasks to Combat Units**.
(1) Infantry.
(2) Attack aviation.

c . **Tasks to Combat Support Units**.
(1) Artillery.
(2) Aviation (lift).

d. **Coordinating Instructions**.
(1) Pickup Zone.
- Direction of landing.
- Time of landing/ flight direction.
- Locations of PZ and alternate PZ.
- Loading procedures.
- Marking of PZ (panel, smoke, SM, lights).
- Flight route planned (SP, ACP, RP).
- Formations: PZ, en route, LZ.
- Code words:
-- PZ secure (before landing), PZ clear (lead bird and last bird).
-- Alternate PZ (at PZ, en route, LZ), names of PZ/ alt PZ.
- TAC air/ artillery.
- Number of pax per bird and for entire lift.
- Equipment carried by individuals.
- Marking of key leaders.
- Abort criteria (PZ, en route, LZ).

(2) Landing Zone.
- Direction of landing.
- False insertion plans.
- Time of landing (LZ time).
- Locations of LZ and ALT LZ.
- Marking of LZ (panel, smoke, SM, lights).
- Formation of landing.
- Code words, LZ name, alternate LZ name.• TAC air/ artillery preparation, fire support coordination.
- Secure LZ or not?

4. SUSTAINMENT. Only brief specifics not covered in base order to include number of aircraft per lift and number of lifts, whether the aircraft will refuel/rearm during mission, special equipment carried by personnel, aircraft configuration, and bump plan.

a. Logistics.
(1) Sustainment Overlay.
(2) Maintenance.
(3) Transportation.
(4) Supply.
Class I:
Class III:
Class V:
Class VII:
Class VIII:
Class IX:
Distribution Methods:
(5) Field Services.

b. Personnel Services Support.

(1) Method of marking and handling EPWs.
(2) Religious Services.
c. Army Health System Support.
(1) Medical Command and Control.
(2) Medical Treatment.
(3) Medical Evacuation.
(4) Preventive Medicine.

5. **COMMAND AND CONTROL**.
a. Command.
(1) Location of Commander/Patrol Leader. State where the commander intends to be during the operation, by phase if the operation is phased. Also include locations of air missions commander, ground tactical commander and air assault task force commander.
(2) Succession of Command. State the succession of command if not covered in the unit's SOP.
b. Control.
(1) Command Posts. Describe the employment of command posts (CPs), including the location of each CP and its time of opening and closing, as appropriate. Typically at platoon level the only reference to command posts will be the company CP.
(2) Reports. List reports not covered in SOPs.
c. Signal. Describe the concept of signal support, including current SOI edition or refer to higher OPORD.
(1) Identify the ***SOI index*** that is in effect
(2) Identify ***methods of communication by priority***
(3) Describe ***pyrotechnics and signals***, to include arm and hand signals (demonstrate)
(4) Give ***code words*** such as OPSKEDs
(5) Give ***challenge and password*** (use behind friendly lines)
(6) Give ***number combination*** (use forward of friendly lines)
(7) Give ***running password***
(8) Give ***recognition signals*** (near/ far and day/ night)

VEHICULAR MOVEMENT COORDINATION CHECKLIST

The platoon sergeant or first sergeant coordinates this with the supporting unit to facilitate the effective, detailed, and efficient use of vehicular support and assets:
1. Identification of the unit.
2. Supporting unit identification.
3. Number and type of vehicles and tactical preparation.
4. Entrucking point.
5. Departure time.
6. Preparation of vehicles for movement.
a. Driver responsibilities.
b. Platoon/ squad responsibilities.
c. Special supplies/ equipment required.
7. Availability of vehicles for preparation/ rehearsals/ inspection (times and locations).
8. Routes.
a. Primary.
b. Alternate.
c. Checkpoints.
9. Detrucking points.
a. Primary.
b. Alternate.
10. Order of march.
11. Speed.

11. Communications (frequencies, call signs, codes).
12. Emergency procedures and signals.

2-8. TASK. This is a specific, clearly defined, decisive and measurable activity/action (**Figure 2-13**), accomplished by a Ranger or organization, that contributes to the accomplishment of encompassing missions or other requirements.
2-9. PURPOSE. This is the desired or intended result of the tactical operation stated in terms related to the enemy or the desired situation. Purpose is the Why? of the mission statement and often follows the words "in order to." It is the most important component of the mission statement (**Figure 2-14**).

2-10. OPERATION. A military action or the carrying out of a military action to gain the objectives of any battle or campaign. **Figure 2-15** shows the types of operations.

Figure 2-13. TASKS

ACTIONS BY FRIENDLY FORCE		EFFECT ON ENEMY FORCE
Assault	Follow and Assume	Block
Attack by Fire	Follow and Support	Canalize
Breach	Linkup	Contain
Bypass	Occupy	Defeat
Clear	Reconstitute	Destroy
Combat Search and Rescue	Reduce	Disrupt
Consolidation and Reorganization	Retain	Fix, Interdict
Control	Secure	Isolate
Counterreconnaissance	Seize	Neutralize
Disengagement	Support by Fire	Penetrate
Exfiltration	Suppress	Turn

Figure 2-14. PURPOSE

Allow	Divert	Prevent
Cause	Enable	Protect
Create	Envelop	Support
Deceive	Influence	Surprise
Deny	Open	

Figure 2-15. TYPES OF OPERATION

Movement to Contact:
  Search and Attack
Attack:
  Ambush
  Demonstration
  Feint
  Raid
  Spoiling Attack
	Exploitation
	Pursuit
	Forms of Offensive
	  Maneuver:
	  Envelopment
	  Frontal Attack
	  Infiltration
	  Penetration
	  Turning Movement
		Area Defense
		Mobile Defense
		Retrograde Ops:
		Delay
		Withdrawal
		Retirement
		Recon Ops
		Security Ops
		Information Ops
			Information Ops
			Combined Arms
			Breach
			Passage of Lines
			Relief in Place
			River Crossing Ops
			Troop Movement
			Admin
			Approach March
			Road March

2-11. TERRAIN MODEL. During the planning process, the terrain model (**Figure 2-16**) offers an effective way to visually communicate the patrol routes and also detailed actions on the objective. At a minimum, the model is used to display routes to the objective and to highlight prominent terrain features the patrol will encounter during movement. A second terrain model of the objective area is prepared. It should be large enough and detailed enough to brief the patrol's actions on the objective.

a. Checklist. Make sure you include these on your terrain models:

(1) North seeking arrow.
(2) Scale.
(3) Grid lines.
(4) Objective location.
(5) Exaggerated terrain relief and water obstacles.
(6) Friendly patrol locations.
(7) Targets (indirect fires, including grid and type of round).
(8) Routes, primary and alternate.
(9) Planned RPs (ORP, L/ URP, RP).
(10) Danger areas (roads, trails, open areas).
(11) Legend.
(12) Blowup of objective area.

b. Construction. Here are some field expedient techniques to help you construct your terrain models:

(1) Use a 3 x 5 card, MRE box, or piece of paper to label the objective or key sites.
(2) Use string from the guts of 550 cord or use colored tape to make grid lines. Identify the grids with numbers written on small pieces of paper.
(3) Replicate trees and vegetation using moss; green or brown spray paint; pine needles; crushed leaves; or cut grass.
(4) Use blue chalk, blue spray paint, blue yarn, tin foil, or MRE creamer to designate bodies of water.
(5) Make North seeking arrows from sharpened twigs, pencils, or colored yarn.
(6) Use red yarn, M16 rounds, toy Rangers, or poker chips to designate enemy positions.
(7) Construct friendly positions such as security elements, support by fire, and assault elements using M16 rounds, toy Rangers, poker chips, small MRE packets of sugar and coffee, or preprinted acetate cards.
(8) Use small pieces of cardboard or paper to identify target reference points (TRPs) and indirect fire targets. Show the grids for each point.
(9) Construct breach, support by fire, and assault positions using the same methods, again using colored yarn or string for easy identification.
(10) Construct bunkers and buildings using MRE boxes or tongue depressors/sticks.
(11) Construct perimeter wire from a spiral notebook.
(12) Construct key phase lines with colored string or yarn.

(13) Use colored tape or yarn to replicate trench lines, by digging a furrow and coloring it with colored chalk or spray paint.

NOTE: Clearly identify in a legend all symbols used on the terrain model.

Figure 2-16. TERRAIN MODEL

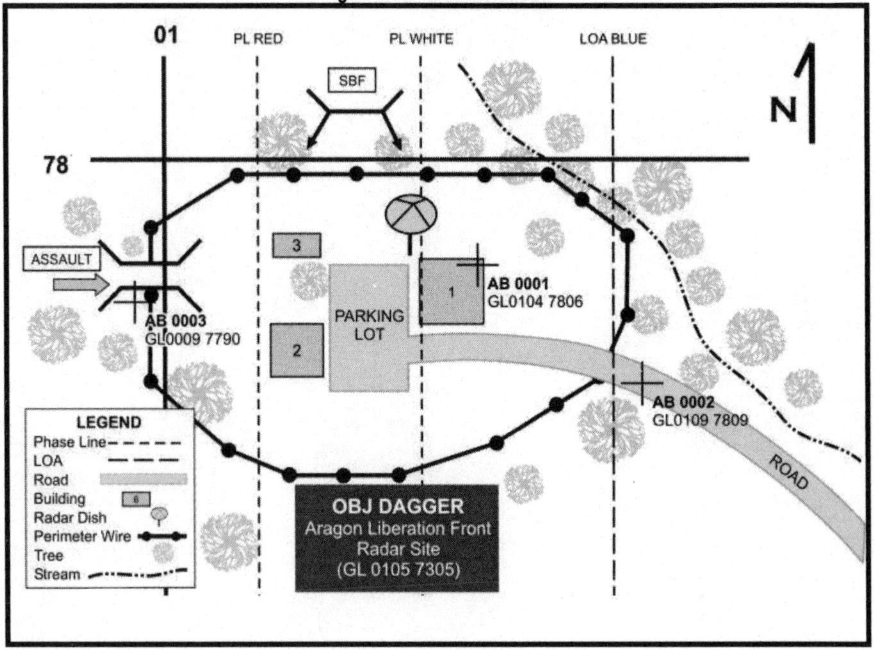

Chapter 3
FIRE SUPPORT

Indirect fire support can greatly increase the combat effectiveness and survivability of any Infantry unit. The ability to plan for and effectively use this asset is a task that every Ranger and small unit leader should master. Fire support assets can help a unit by suppressing, fixing, destroying, or neutralizing the enemy. Leaders should consider employing indirect fire support throughout every offensive and defensive operation. This chapter discusses plans, tasks, capabilities, risk estimate distances, target overlays, close air support, elements and sequence of calls for fire, and example call for fire transmissions.

3-1. BASIC FIRE SUPPORT TASKS. The effectiveness of the fire support system depends on successful performance of its four basic tasks:

- Support forces in contact
- Support the battle plan
- Synchronize the fire support system
- Sustain the fire support system

3-2. TARGETING. Objectives are the overall effects the leader hopes to achieve through the use of fire support assets.

a. ***Decide*** is the first functional step in the targeting process. A decision defines the overall focus and sets priorities for collecting intelligence and planning the attack. The leader must address targeting priorities for each phase or critical event of an operation. At all echelons, he analyzes one or more alternative COAs. Each is based on–

- Mission analysis.
- Current and projected battle situations.
- Anticipated opportunities.

b. ***Detect*** is the second critical function. The G-2 or S-2 directs the effort to detect the HPTs identified in the Decide step. To identify the exact Who, What, When, and How of target acquisition, he works closely with the–

- Analysis and control element.
- FAIO.
- Targeting officer and/ or FSO.

3-3. INTERDICTION. This is an action to divert, disrupt, delay, intercept, board, detain, or destroy the enemy's military surface capabilities, such as vessels, vehicles, aircraft, people, and cargo, before they can be used effectively against friendly forces, or to otherwise achieve friendly objectives.

a. **Limit.** Reduce enemy options. For example, direct air interdiction and fire support to limit enemy avenue(s) of approach and fire support.

b. **Disrupt.** Stop effective interaction between the enemy and his support systems. Reduce enemy efficiency and increase his vulnerability.

c. **Delay.** Disrupt, divert, or destroy enemy capabilities or targets. In other words, change when the enemy reaches a point on the battlefield, or change his ability to project combat power from it.

d. **Divert.** Create a distraction that forces the enemy to tie up critical resources. For example, attack targets that cause the enemy to move capabilities or assets from one area or activity to another.

e. **Destroy.** Ruin the structure or condition of a vital enemy target. You can define destruction as an objective by stating a do able number or percentage of an enemy asset or target that the weapon system(s) can realistically achieve. For example, artillery normally says that destruction comprises a 30 percent reduction in capability or structural integrity; maneuver combat forces normally use 70 percent.

f. **Damage.** This can be a subjective or objective assessment of battle damage, or it can describe the damage to the objective as light, moderate, or severe.

3-4. CAPABILITIES. Table 3-1 and **Table 3-2** show capabilities of field artillery and mortars.

Table 3-1. CAPABILITIES OF FIELD ARTILLERY

WEAPON	MAX RANGE (meters)	MIN RANGE (meters)	MAX RATE (rds per min)	Burst Radius (meters)	SUSTAINED RATE (rds per min)
105-mm Howitzer M119, Towed	14,000m	0m	6 for 2 min	35m	3 rounds for 30 min then 1 round per min
155-mm Howitzer M198, Towed	18m,100m 30,000m (RAP)	0m	4 for 3 min 2 for 30 min	50m	1 round per min temp dependent
155-mm Howitzer M109A6 SP	18m, 100m 30,000m (RAP)	0m	4 for 3 min	50m	1 round for 60 min 0.5

Table 3-2. CAPABILITIES OF MORTARS

WEAPON	MUNITION AVAILABLE	MAX RANGE (meters)	MIN RANGE (meters)	MAX RATE (rds per min)	Burst Radius (meters)	SUSTAINED RATE (rds per min)
60mm	HE,WP,Illum	3,500m (HE)	70 m (HE)	30 for 4 min	30 m	20
81mm	HE,WP,Illum	5,600m (HE)	70 m (HE)	25 for 2 min	38 m	8
120mm	HE,Smoke, Illum	7,200m (HE)	180 m (HE)	15 for 1 min	60 m	5

DANGER

DANGER CLOSE

1. WHEN THE TARGET IS WITHIN 600 METERS OF ANY FRIENDLY TROOPS (FOR MORTARS AND FIELD ARTILLERY), ANNOUNCE DANGER CLOSE IN THE *METHOD OF ENGAGEMENT* PORTION OF THE CALL FOR FIRE.

2. WHEN ADJUSTING 5 INCH OR SMALLER NAVAL GUNS ON TARGETS WITHIN 750 METERS, ANNOUNCE DANGER CLOSE. FOR LARGER NAVAL GUNS, ANNOUNCE DANGER CLOSE FOR TARGETS WITHIN 1,000 METERS. FAILURE TO ADHERE TO THIS GUIDANCE CAN RESULT IN FRATRICIDE.

3. AVOID MAKING CORRECTIONS USING THE BRACKETING METHOD OF ADJUSTMENT, BECAUSE DOING SO CAN CAUSE SERIOUS INJURY OR DEATH. USE ONLY THE CREEPING METHOD OF ADJUSTMENT DURING DANGER CLOSE MISSIONS. MAKE CORRECTIONS OF NO MORE THAN 100 METERS BY CREEPING THE ROUNDS TO THE TARGET.

3-5. RISK ESTIMATE DISTANCES. RED applies to combat only. Minimum safe distances (**Table 3-3**) apply to training IAW AR 350 1. RED takes into account the bursting radius of particular munitions and the characteristics of the delivery system. It associates this combination with a percentage representing the likelihood of becoming a casualty, that is, the percentage of risk. RED is defined as the minimum distance friendly troops can approach the effects of friendly fires without suffering appreciable casualties of 0.1 percent PI or higher.

WARNING

[Commanders] use RED formulas to determine acceptable risk levels in combat only. Specifically, use them to identify the risk to your Rangers at various distances from their targets. Risk estimate distances apply only in combat. In training, use minimum safe distances (MSD).

a. **Casualty Criterion.** The casualty criterion is the 5 minute assault criterion for a prone Ranger in winter clothing and helmet. Physical incapacitation means that a Ranger is physically unable to function in an assault within a 5 minute period after an attack. A PI value of less than 0.1 percent can be interpreted as being less than or equal to one chance in one thousand.

Table 3-3. RISK ESTIMATE DISTANCES FOR MORTARS AND CANNON ARTILLERY

	Risk Estimate Distances (Meters)					
	10 % PI			0.1 % PI		
Description	1/3 Range	2/3 Range	Max Range	1/3 Range	2/3 Range	Max Range
60-mm mortar	60	65	65	100	150	175
81-mm mortar	75	80	80	165	185	230
120-mm mortar	100	100	100	150	300	400
105-mm howitzer	85	85	90	175	200	275
155-mm howitzer	100	100	125	200	280	450
155-mm DPICM	150	180	200	280	300	475

b. **Risk**. Using echelonment of fires within the specified RED for a delivery system requires the unit to assume some risks. The maneuver commander determines by delivery system how close to his forces he will allow fires to fall. Although he makes the decision at this risk level, he relies heavily on the FSO's expertise.

3-6. TARGET OVERLAYS

a. **Fire Support Overlay. Figure 3-1** shows contents of fire support overlay.

(1) **Non Sterile Fire Support Overlay (Figure 3-2)**.

(2) **Sterile Fire Support Overlay (Figure 3-3)**. This includes–

(a) Index marks to position overlay on map

(b) Target symbols

- Point target
- Linear target
- Circular target

Figure 3-1. CONTENTS OF FIRE SUPPORT OVERLAY

Unit and official capacity of person making overlay	Routes – primary and alternate
Date the overlay was prepared	Phase lines and checkpoints used by the patrol
Map sheet number	Spares
Effective period of overlay (DTG)	Index marks to position overlay on map
Priority target	Objective
ORP location	Target symbols
Call signs and frequencies (PRI/ ALT)	Description, location and remarks column, complete

Figure 3-2. NON-STERILE FIRE SUPPORT OVERLAY

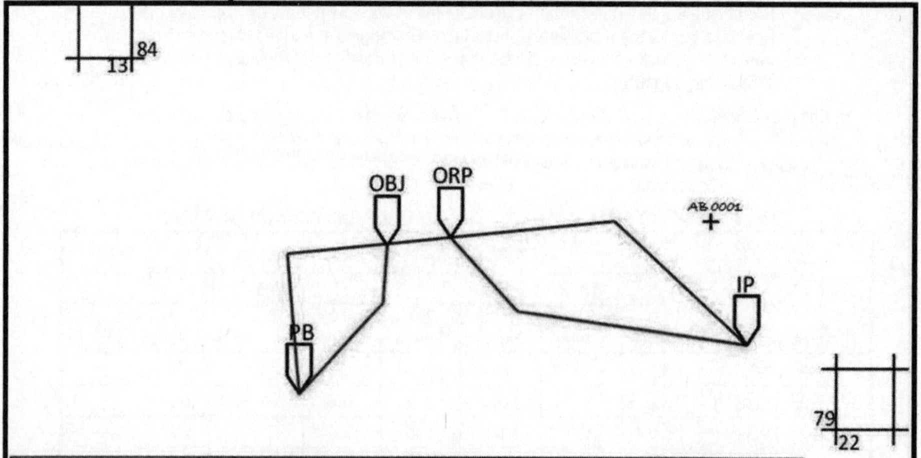

Unit: A -1/1 RTB Prepared By: RGR FO	Map Sheet: CDS 1:50,000 Effective Date: 11-12 Jan 2010	Call Sign: A 1/1 Freq: Pri-44.950 Alt-33.900		DTG Prepared: 110800JAN2010
TGT #	**TGT Location (GL & Terr. Feat)**	**Task & Purpose**	**# and Type of Rnds**	**Type of Target/ Additional Remarks**
AB 0001	GA195821 Hill	Disrupt enemies ability to mass combat power IOT facilitate squad's undetected insertion.	6- 81mm HE/ 6- 81mm WP	Area Target/ will be fired 4 minutes prior to arrival at IP. 1 HE, 1 WP till rounds complete

Figure 3-2. STERILE FIRE SUPPORT OVERLAY

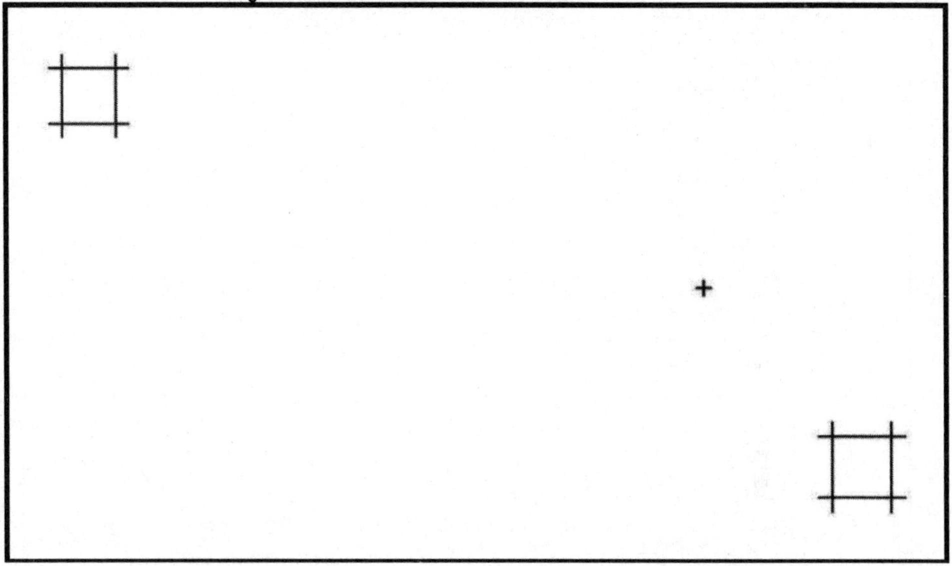

b. **PLOT CR Checklist.** Using one of these (**Table 3-4**) helps ensure the leader's fire support plan is complete. He uses it in identifying all aspects of individual targets before coordination and the OPORD.

Table 3-4. PLOT CR CHECKLIST

Purpose	Planned indirect fires.
Location	Plan targets with an 8-digit grid (minimum)
Observer-Planned Observer	See the impact of the rounds and adjust
Trigger	Method of initiating fires
Communication	Method of communicating between observer and the supporting unit
Resources	Planned allocated resource for each target

3-7. CALL FOR FIRE. Asterisks indicate required elements for a basic call for fire mission. Example call for fire transmissions are shown in **Table 3-5**.

a. **Observer's Identification - Call Signs.****

b. **Warning Order.****

(1) Type of mission.

- Adjust fire
- Fire for effect
- Suppress
- Immediate suppression/immediate smoke

(2) Size of element to fire for effect. When observer does not specify size element to fire, battalion FDC decides.

c. **Method of Target Location.****

- Polar plot
- Shift from a known point
- Grid

d. **Location of Target.****

(1) ***Grid Coordinate.*** Six or, if greater accuracy is required, eight digit.

(2) ***Shift from a Known Point.*** Send OT direction:

- Mils (nearest 10).
- Degrees.
- Cardinal direction.
- Send lateral shift, right/left, nearest 10m
- Send range shift, add/drop, nearest 100m
- Send vertical shift, up/down, nearest 5m; use only if it exceeds 35m)

(3) ***Polar Plot.***

- Send direction to nearest 10 mils
- Send distance to nearest 100m
- Send vertical shift to nearest 5m

e . **Description of Target.****

(1) Type.

(2) Activity.

(3) Number.

(4) Degree of protection.

(5) Size and shape (length/width or radius).

f. **Method of Engagement.**

(1) ***Type of Adjustment.*** When observer does not request a specific type of fire control adjustment, issue area fire.

(a) Precision fire point target.

(b) Area fire moving target.

(2) ***Danger Close.*** This condition exists when friendly troops are within–

(a) 600 meters for mortars.

(b) 600 meters for artillery.

(c) 750 meters for naval guns 5 inches or smaller.

(3) ***Mark.*** Used to orient observer or to indicate targets.

(4) *Trajectory.*

- Low angle (standard).
- High angle (mortar fire or if requested).

(5) ***Ammunition.*** Use HE quick unless specified by the observer.

- Projectile (HE, ILLUM, ICM, SMOKE and so on).
- Fuse (quick, timed, and so on).
- Volume of fire (observer may request the number of rounds to be fired).

(6) ***Distribution.***

- 100 meter sheaf (standard).
- Converged sheaf (used for small hard targets).
- Special sheaf (any length, width and attitude).
- Open sheaf (separate bursts).
- Parallel sheaf (linear target).

g. **Method of Fire and Control.**

(1) ***Method of Fire.*** Specific guns and a specific interval between rounds. Normally adjust fire, one gun is used with a 5 second interval between rounds.

(2) ***Method of Control:***

(a) AT MY COMMAND, FIRE. Remains in effect until observer orders CANCEL AT MY COMMAND.
(b) CANNOT OBSERVE. Observer cannot see the target.
(c) TIME ON TARGET. Observer tells FDC when he wants the rounds to impact.
(d) CONTINUOUS ILLUMINATION. If this was not already calculated by the FDC, the observer indicates interval between rounds in seconds.
(e) COORDINATED ILLUMINATION. Observer tells FDC to set interval between ILLUM and HE shells.
(f) CEASE LOADING.
(g) CHECK FIRING. Halt immediately.
(h) CONTINUOUS FIRE. Load and fire as fast as possible.
(i) REPEAT. Fire another round(s), with or without adjustments.

h. **Correction of Errors.** When FDC has made an error when reading back he fire support data, the observer announces "CORRECTION" and transmits the correct data in its entirety.

i. **Message to Observer.**
(1) Battery(ies) to fire for effect.
(2) Adjustment of battery.
(3) Changes to the initial call for fire.
(4) Number of rounds (per tube) to be fired for effect.
(5) Target numbers.
(6) Additional information:
(a) Time of flight. Moving target mission.
(b) Probable error in range. 38 meters or greater (normal mission).
(c) Angle "T." 500 mils or greater.

j. **Authentication.** Challenge and reply.

Table 3-5. EXAMPLE CALL FOR FIRE TRANSMISSIONS

GRID MISSION	
Observer	**Firing Unit**
F24, this is J42, ADJUST FIRE, OVER.	J42, this is F24, AJUST FIRE, OUT.
	GRID WM180513, DIRECTION 0530, OUT.
Infantry platoon dug in, OVER.	Infantry platoon dug in, OUT.
	SHOT OVER.
SHOT OUT.	
	SPLASH, OVER.
SPLASH OUT.	
End of mission, 15 casualties, platoon dispersed, OVER.	End of mission, 15 casualties, platoon dispersed, OUT.
SHIFT FROM KNOWN POINT	
Observer	**Firing Unit**
J42, this is F24, ADJUST FIRE, SHIFT AB1001, OVER. DIRECTION 2420, RIGHT 400, ADD 400, OVER. Five T-72 Tanks at POL site, OVER. I AUTHENTICATE Tango, OVER. SHOT OUT. SPLASH OUT. End of mission, 2 tanks destroyed, 3 in woodline, OVER	F24, this is J42, ADJUST FIRE, SHIFT AB1001, OUT. DIRECTION 2420, RIGHT 400, ADD 400, OUT. Five T-72 Tanks at POL site, AUTHENTICATE Juliet November, OVER. SHOT, OVER. SPLASH, OVER. End of mission, 2 tanks destroyed, 3 in woodline, OUT
POLAR	
Observer	**Firing Unit**
J42, this is F24, ADJUST FIRE, POLAR OVER. DIRECTION 2300, DISTANCE 4,000, OVER. Infantry platoon dug in, OVER SHOT OUT. SPLASH OUT. End of mission, 15 casualties, platoon dispersed, OVER.	F24, this is J42, ADJUST FIRE, POLAR, OUT DIRECTION 2300, DIST 4,000, OUT. Infantry platoon dug in, OUT SHOT OVER. SPLASH, OVER. End of mission, 15 casualties, platoon dispersed, OUT.

3-8. CLOSE AIR SUPPORT. The two types of close air support requests are planned and immediate. Planned requests are processed by the Army chain to Corps for approval. Immediate requests are initiated at any level and processed by the battalion S-3, FSO, and Air Liaison Officer.

a. **Format for Requesting Immediate CAS**. (**Figure 3-4**)

(1) Observer identification.

(2) Warning order (request close air).

(3) Target description. This must include, as a minimum, type and number of targets, activity or movement, and point or area targets.

(4) Target location (grid) should include elevation.

(5) Desired time on target (TOT).

(6) Desired effects on target.

(7) Final control.

(8) Remarks.

(a) Friendly locations.

(b) Wind direction, hazards.

(c) Threats such as ADA, small arms.

Figure 3-4. EXAMPLE CAS REQUEST

(1. INITIAL POINT (IP);) **NP459854 (or) XRAY**

(2. HEADING (IP TO TARGET)) **069** MAGNETIC
(OFFST: LEFT/RIGHT)

(3. DISTANCE (IP TO TARGET)) **9.8** (NAUTICAL MILES)

(4. TARGET ELEVATION) **1,140** (FEET ABOVE MEAN SEA LEVEL)

(5. TARGET DESCRIPTION) **5 TANKS ATTACKING WEST**

(6. TARGET LOCATION) **NP675920** (UTM, LAT/LONG, VISUAL REFERENCES, ETC.)

(7. TYPE OF MARK) **LASER** (CODE) **372**

(8. LOCATION OF FRIENDLIES) **1,000M SW OF TARGET**

(9. EGRESS **NW TO AVOID ARTILLERY SUPPRESSION**
(REMARKS) ____________________

____

(TIME ON TARGET) TOT ____________________
(TIME TO TARGET (TTT)) STANDBY __________ PLUS ______
(minutes) (seconds)

OMIT DATA NOT REQUIRED. LINE NUMBERS ARE NOT TRANSMITTED. ALL UNITS OF MEASURE ARE STANDARD. SPECIFY IF OTHER UNITS OF MEASURE ARE BEING USED.

b. **Close Air Support Capabilities. (Table 3-6)**

Table 3-6. CLOSE AIR SUPPORT

AIRCRAFT	SERVICE	CHARACTERISTICS
A-10 *	AF	Specialized CAS aircraft, 30-mm gun; subsonic Typical load 6,000 lbs; max load 16,000 lbs
F-16 *	AF	Multirole aircraft; complements the F-4 and F-15 in an air-to—air role; supersonic Most accurate ground delivery system in the inventory Typical load 6,000 lbs; max load 10,600 lbs.
F-18 *	N/ MC	Multirole fighter scheduled to replace the F-4 20-mm gun mounted in the nose fires air-to-air missiles Wide variety of air-to-surface weapons Typical load 7,000 lbs; max load 17,000 lbs
AC-130	AF/ R	Specialized CAS/ RACO aircraft, propeller driven Two models, both of which have advanced sensors and a target-acquisition system, including FLIR and low light TV. Very accurate. Vulnerable to enemy air defense systems, so must operate in a low-threat environment. A model Two 40-mm guns Two 20-mm guns Two 7.62-mm miniguns H model 105-mm howitzer replaces one of the 40-mm guns Lacks the 7.62-mm minigun
*FM capability.		

3-9. CLOSE COMBAT ATTACK AVIATION. Close combat attack (CCA) is defined as a hasty or deliberate attack in support of units engaged in close combat. During CCA, armed helicopters engage enemy units with direct fire that impacts nearby friendly forces. Targets may range from a few hundred meters to a few thousand meters. Close combat attack is coordinated and directed by a team, platoon, or company-level ground unit soldiers using standardized CCA procedures in unit SOPs.

a. **Aircraft Capabilities and Limitations.**

(1) ***AH 64D (Apache).***

(a) *Capabilities.*

- *Air speed (knots)*: 164 max, 120 cruise combat radius: 200 km
- *30 mm chain gun:* 1,200 rds, 3,500 m range
- *2.75 inch rockets:* 19 per pod (76), 3 to 5 km range
- *Hellfire missile:* 8 per side (16), 5 to 8 km range
- *Mobility:* AO can cover corps or division area
- *Speed:* 100 to 120 knots day/80 to 100 knots night
- *Versatility:* scout weapons teams vs. pure attack
- *Lethality:* Attack battalion can engage 288 targets
- *Video reconnaissance:* Provide near real-time intelligence

APACHE

(b) *Limitations.* Threat ID with FLIR, low ceilings (clouds) less than 500 feet AGL degrade hellfire capability, combat service support consumes large amounts of Classes III, V, and IX

(2) ***OH 58D (Kiowa).***

(a) *Speed and Armament.*

- Air speed (knots): 125 max, 100 cruise combat radius: 120 km
- .50 cal MG: 500 rds, 2,000 m range
- 2.75 inch rockets: 7 per pod (14), 3 to 5 km range
- Hellfire: 2 per side (4), 5 to 8 km range
- Stinger: 2 per side (4), 4 km range

KIOWA

(b) *Capabilities.*

- Mast mounted sight
- Ability to designate targets while remaining masked
- Thermal imaging system (day/ night)
- Laser designator/ aim laser
- Video image crosslink (VIXL)
- Moving map display
- Video

(c) *Limitations.* Power limited, infrared crossover, battlefield obscurants, low ceilings (Hellfire), remote designation constraints, instrument meteorological conditions

b. **CCA Call for Fire Format.**

(1) IP/ BP/ ABF or friendly location_________
- Grid/ Lat Long

(2) Known point
- Terrain feature

(3) HDG to TGT _________ (mag)
- Specify from IP/ BP/ ABF or friendly location

(4) DST to TGT __________ (m) (Specify from IP/ BP/ ABF or friendly location.)

(5) TGT elevation_________ (ft msl)

(6) TGT Description _________________

(7) TGT location _____________
- Grid
- Known point
- Terrain Feature

(8) Type TGT Mark __________ (day/ night)

(9) Location of friendly __________
- Omit if previously given
- Grid/ Lat Long
- Known point
- Terrain feature

(10) Egress direction___________
- Cardinal, to avoid overflying threats

Chapter 4
COMMUNICATIONS

The basic requirement of combat communications is to provide rapid, reliable, and secure interchange of information. Communications are vital to mission success. This chapter helps the Ranger squad/ platoon maintain effective communications and correct any radio antenna problems.

Section I. EQUIPMENT

This section discusses military radio communications equipment and automated net control devices (ANCDs).

4-1. MILITARY RADIOS. Each military radio has a receiver and transmitter. Rangers use several different types of radios (**Table 4-1**), with various features. Knowing what each radio has and can be crucial in planning and requesting the most reliable and effective communications equipment for a particular mission. Military operations use five primary frequency ranges (**Table 4-2**).

Table 4-1. MILITARY RADIOS

CHARACTERISTICS	MODELS				
	AN/PRC-117F(c)	**AN/PRC-152**	**AN/PRC-148**	**AN/PRC-119F**	**AN/PRC-150C**
Description	MultiBand Man-Pack Receiver/ Transmitter	Multiband Hand-Held Receiver/ Transmitter	Multiband interteam or intrateam radio	Multiband, Multimission Man-pack	Advanced Man-Pack Transceiver
Frequency(ies) Ranges					
• HF					Yes
• VHF Low	Yes	Yes	Yes	Yes (Up to 5W)	Partial (1.6 to 60 MHz)
• VHF High	Yes	Yes	Yes		
• UHF	Yes	Yes	Yes		
• TACSAT	Yes	Yes	Yes (Up to 5W)		20W
Power Output	Up to 20W	Up to 10W TACSAT Up to 5W all else	Up to 5W	Up to 10W	Up to 20W
Battery Requirements	Two of any of these: • BB-390 • BB-2590 • BB-590 • BA-5590	Rechargeable Lithium-Ion (included with Radio)	Rechargeable Lithium-Ion Battery (included with Radio)	Any one of these: • BB-390 • BB-2590 • BB-590 • BA-5590	Two of any of these: • BB-390 • BB-2590 • BB-590 • BA-5590
Scanning: Can scan up to–	10 user-programmed nets (TACSAT or LOS frequencies)	10 user-programmed nets (TACSAT or LOS frequencies)	10 user-programmed nets (TACSAT or LOS frequencies)	4 channels in FM mode	100 channels in FM mode

CHARACTERISTICS	MODELS				
	AN/PRC-117F(c)	AN/PRC-152	AN/PRC-148	AN/PRC-119F	AN/PRC-150C
Data Transmission					
• LOS AM/FM	Yes [1]	Yes	Yes	Yes	Yes
• GPS:					
– Commercial [2]	Yes [2]	Yes [2]	Yes [2]		Yes [2]
– Optional Internal		Yes		Yes	
• TACSAT [1]	Yes		Yes		
– DAGR	Yes [3]	Yes [3]	Yes		
– PLGR	Yes [3]	Yes [3]	Yes		
– NMEA-183	Yes [3]	Yes [3]	Yes		
– Internal		Yes			
– Optional Internal					
Dimensions and Weight	3.2 x 10.5 x 13.5 inches with two batteries and battery box 15.9 lb with batteries	2.9 x 9.6 x 2.5 inches with battery attached 2.6 lbs with internal GPS and battery	2.7 x 7.8 x 1.5 inches 2.2 lbs with battery	3.4 x 5.3 x 10.2 inches 7.7 lbs without battery	3.5 x 10.5 x 13.2 inches with battery pack 10 lbs without batteries
Disadvantages	Requires additional batteries for extended operations Heavier than AN/ PRC-119F	Lower Power Output than AN/ PRC-117F(c)	Lower power output than AN/ PRC-117F(c) and AN/ PRC-152	Lower power output than AN/ PRC-117F(c) Limited frequency range Cannot communicate with USAF aircraft	Limited frequency range HF communication poor in mountainous terrain

CHARACTERISTICS	MODELS				
	AN/PRC-117F(c)	AN/PRC-152	AN/PRC-148	AN/PRC-119F	AN/PRC-150C
Encryption • Type:					
– ANDVT	Yes	Yes	Yes		Yes
– Vinson	Yes	Yes	Yes	Yes	Yes
– KG–84	Yes	Yes			Yes
– Fascinator	Yes	Yes	Yes		

CHARACTERISTICS	MODELS				
	AN/PRC-117F(c)	AN/PRC-152	AN/PRC-148 MBITR	AN/PRC-119F	AN/PRC-150C
• Method:					
– SINCGARS	Yes	Yes	Yes	Yes	
– Havequick I	Yes	Yes	Yes		
– Havequick I	Yes	Yes	Yes		
– Serial Tone					Yes
– ECCM	Yes	Yes	Yes		
– Freq Hop	Yes	Yes	Yes	Yes	Yes
• Fill Devices:					
– KYK-13	Yes	Yes	Yes	Yes	Yes
– KOI-18	Yes	Yes	Yes	Yes	Yes
– KYX-15	Yes	Yes	Yes	Yes	Yes
– SKL	Yes	Yes	Yes	Yes	Yes
– AN/CYZ-10	Yes	Yes	Yes	Yes	Yes
Immersion Depth	1 meter	20 meters	2 to 20 meters	1 meter	0.9 meter
SA Reporting Capable	Yes	Yes	Yes w/ RGU	If equipped with optional internal GPS	No
Special features	Ship-to-Shore, Ground-to-Ground, and Air-to-Ground Capable	Ship-to-Shore, Ground-to-Ground, and Air-to-Ground Capable Submersible to 20 meters.	Ship-to-Shore, Ground-to-Ground, and Air-to-Ground Capable Submersible to 20 meters	NA	Advanced Automatic Link Establishment (ALE),
Terrain Restrictions	**LOS:** open to slightly rolling terrain **TACSAT:** any terrain	**LOS:** open to slightly rolling terrain **TACSAT:** any terrain	**LOS:** open to slightly rolling terrain **TACSAT:** any terrain	**LOS:** open to slightly rolling terrain	**Ground wave:** wide open, flat terrain **Sky wave:** long-range communication

1 When coupled with Windows Messaging Terminal software-equipped laptop.
2 Commercial GPS capable (used with SA reporting feature).
3 DAGR Is compatible, PLGR requires optional RGU (Remote Control Unit with GPS User Notification).

Table 4-1. FREQUENCY RANGES

High Frequency (HF)	Very High Frequency (VHF) Low	Very High Frequency	Ultra High Frequency (UHF)	TACSAT	UHF SATCOM
1.6 to 29.999 MHz	30.000 to 89.999 MHz	90.000 to 224.999 MHz	225.000 to 512.000 MHz	243.000 to 270.000 MHz	292.000 to 318.00 MHz
long range LOS [1]	LOS [2]	LOS [2]	LOS [2]	Satellite	Satellite

[1] Long-range LOS; capable of round-the–world communication due to longer physical wavelengths, which cause HF transmissions to "bounce" off terrain and be reflected by the Earth's ionosphere instead of absorbed like VHF and UHF transmissions. This keeps the transmission bouncing, essentially trapped, between the ground and the Earth's atmosphere. For this volatile capability to offer effective communications, several factors must be ideal.

[2] Line of sight (LOS) frequencies; meaning that the range of radios is limited to direct line of sight for maximum effectiveness. Curvature of the earth, mountainous terrain, and dense vegetation will degrade LOS radio maximum range capabilities.

[3] Modern military communications rely on UHF Satellite Communication (SATCOM) or Dedicated Tactical Satellite Communication (TACSAT) for round-the–world real time secure voice and data communication.

4-2. MAN-PACK RADIO ASSEMBLY (AN/PRC 119F). To assemble a man-pack radio, you must first check and install a battery.

a. Inspect the battery box for dirt or damage.
b. Stand radio on its side with the battery cover facing up.
c. Check battery life condition (you will be using the rechargeable BB 390 batteries).
d. Place battery in box.
e. Close and latch the battery cover.
f. Return radio to upright position.
g. If you installed a used battery, then enter the battery life condition into the radio:
   (1) Set FCTN to LD.
   (2) Press BAT, then CLR.
   (3) Enter number recorded on side of battery.
   (4) Press STO.
   (5) Set FCTN to SQ ON.
h. Inspect and position the antenna.
   (1) Inspect whip antenna connector on antenna and on radio for damage.
   (2) Screw whip antenna into base.
   (3) Hand tighten.
   (4) Carefully mate antenna base with RT ANT connector.
   (5) Hand tighten.
   (6) Position antenna as needed by bending goose neck.

> **NOTE:** Keep the antenna straight, if possible. If the antenna is bent to a horizontal position, you might have to turn the radio before you can receive and transmit messages.

i. Set up the handset.
   (1) Inspect the handset for damage.
   (2) Push handset on AUD/ DATA and twist clockwise to lock in place.
j. Pack.

(1) Place RT in field pack with antenna on the left shoulder.
(2) Fold top flap of field over RT and secure flap to field pack using straps and buckles.

k. Set Presets. Set--

(1) CHAN: 1
(2) MODE: SC
(3) RF PWR: HI
(4) VOL: Mid Range
(5) DIM: Full clockwise
(6) FCTN: LD
(7) DATA RATE: OFF

l. Single-Channel Loading Frequencies.

(1) Obtain Ranger SOI:
(2) Set FCTN: LD
(3) Set mode: SC
(4) Set CHAN: MAN, Cue, or set channel (1 to 6) where you want to store frequency
(5) Press FREQ: (Display will show "00000" or frequency RT is currently turned on)
(6) Press CLR: (Display will show five lines)
(7) Enter: The number of the new frequency. If you make a mistake with a number press CLR
(8) Press STO: (Display will blink)
(9) Set FCTN: SQ ON

m. Clearing of frequencies.

(1) Set Mode: SC
(2) Set CHAN: MAN, Cue or desired channel where frequency is to be cleared
(3) Press FREQ:
(4) Press CLR:
(5) Press Load, STO
(6) Set FCTN: SQ ON

n. Scanning of multiple frequencies.

(1) Load: All desired frequencies using "Single Channel Loading Frequencies" instructions
(2) Set CHAN: CUE
(3) Set SC: FH
(4) Set FCTN: SQ ON
(5) Press STO: (Display will say SCAN)
(6) Press 8: You can now scan more than one frequency

4-3. AUTOMATED NET CONTROL DEVICE. The RTO retrieves all necessary COMSEC and information from the ANCD. To retrieve the SOI from an ANCD:

a. Press the On button on the ANCD keypad.
b. Press the Letter Lock button to unlock the keys on the ANCD.
c. Press Main Menu key (onscreen)

Appl Date Time Setup Util Bit

d. Using the arrow keys, scroll over to Appl and press the Enter button on the keypad.

e. Once you have entered into the Appl (applications) menu using the arrow keys, scroll over to SOI and press the Enter button on the keypad. For example–

RDS SOI Radio.

(**appl**) < (Indicates that you are in the applications menu).

SOI menu will look like this:

Qref Group Net Sufx Pyro
Tmpd Set C/ S Find Memo

f. Using the arrow keys, scroll over to Set and press the enter button on the keypad.

g. In the Set menu, press Choose once.

Choose Send Receive

h. When the Choose menu opens, press the Enter key on the set, either 1 5 or 6 10.

i. After you choose the applicable set, the ANCD will return you to the SOI menu.

j. Using the keypad, scroll to Tmpd. Enter the number for the set you need that day. After you do this, the ANCD will automatically return you to the SOI menu.

k. Using the arrow keys on the keypad, scroll over to Net and press Enter. The Net menu should look like this:

T05 < (time period) 1 < (platoon) / A < (company) / A3Q < (prefix).C 60.000.

NOTE: Get your frequency and the first part of your call sign from the Net menu.

l. After you get the information you need from Net, press the abort button on the keypad to return to the SOI menu.

m. Using the keypad, scroll to Sufx and press the Enter button.

n. Once you are in the Sufx menu, use the up and down arrow keys on the ANCD to locate your two digit designator. These two digits go at the end of your Prefix and together they are your call sign.

o. Once you finish, press the Off button on the ANCD to end operation.

4-4. BASIC TROUBLESHOOTING. You need basic troubleshooting skills in order to correct the simple communications problems that occur during a mission. Being able to quickly troubleshoot can make the difference in successful accomplishment of the mission and mission failure.

a. **Check Radio Settings.**

(1) Radio frequency: load proper frequency.
(2) Power output: set to HIGH power.
(3) Time if using frequency hop (FH): reset time.
(4) Crypto fill if using cipher text (CT): reload crypto from ANCD.
(5) Control knob: ensure radio is in ON position.

b. **Check Radio Assembly/ Battery.**

(1) Check antenna fitting: attach long whip or field expedient antenna
(2) Check hand mike fitting: ensure contacts are clean and fitting is properly secured to radio
(3) Check battery: install fresh battery

c. With line of sight (LOS) radios, you might have to move to higher ground to make radio contact, especially in densely vegetated or uneven terrain.

Section II. ANTENNAS

This section discusses repair techniques, construction and adjustment, field expedient antennas, antenna length and orientation, and improvement of marginal communications.

4-5. REPAIRS. Antennas are sometimes broken or damaged, causing communications degradation or failure. If you have a spare antenna, replace the bad one. When you have no spare, you [Ranger squad/ platoon] might have to construct an emergency antenna. The following paragraphs suggest some ways to repair antennas and antenna supports, and to construct and adjust emergency antennas.

DANGER

RADIO TRANSMITTER

SERIOUS INJURY OR DEATH CAN RESULT FROM CONTACT WITH THE RADIATING ANTENNA OF A MEDIUM-POWER OR HIGH-POWER TRANSMITTER.

***TURN OFF* THE TRANSMITTER WHILE ADJUSTING THE ANTENNA.**

a. **Whip Antennas.** When a whip antenna breaks in two, connect the broken part to the part attached to the base by joining the sections. To restore the antenna to its original length, add a piece of wire that is nearly the same length as the missing part of the whip. Lash the pole support securely to both sections of the antenna. Before connecting the two antenna sections to the pole support, clean them well to ensure good contact. If possible, solder the connections.

b. **Wire Antennas.** Emergency repair of a wire antenna may involve the repair or replacement of the wire used as the antenna or transmission line; or the repair or replacement of the assembly used to support the antenna.

(1) When one or more wires of an antenna are broken, you can repair the antenna by reconnecting the broken wires. To do this, lower the antenna to the ground, clean the ends of the wires, and twist the wires together. Whenever possible, solder the connection.

(2) If the antenna receives damage beyond repair, construct a new one. Make sure that the length of the substitute antenna wires are the same length as those of the original.

(3) Antenna supports may also require repair or replacement. You can use anything as a substitute for the damaged support provided it is strong enough and insulated. If the radiating element were not properly insulated, then field antennas could short to ground, and will no longer work. Many common items make good field expedient insulators.

(4) The best are plastic or glass, for example, plastic spoons, buttons, bottle necks, and plastic bags. Though wood and rope are less effective insulators than plastic or glass, they are better than nothing. The radiating element (the antenna wire) should touch only this supporting (nonconductive) insulator and the antenna terminal. It should remain physically separated from everything else.

4-6. CONSTRUCTION AND ADJUSTMENT. Ranger squad/ platoons may use the following methods to construct and adjust antennas:

a. **Construction.** The best wire for antennas is copper or aluminum. However, in an emergency, use any wire you can find.

(1) The exact length of most antennas is critical. Make sure that the emergency antenna is the same length as the original antenna.

(2) Antennas can usually survive heavy wind storms if supported by a tree trunk or strong branch. To keep the antenna tight and keep it from breaking or stretching when the trees sway, attach a spring or old inner tube to one end of the antenna. Another technique is to pass a rope through a pulley or eyehook. Attach the rope to the end of the antenna, and heavily weight the rope to keep the antenna tight.

(3) To ensure the rope or wire guidelines do not interfere with the operation of the antenna, cut the wire into several short lengths and connect the pieces with insulators.

b. **Adjustment.** An improvised antenna may change the performance of a radio set. The following methods can be used to determine if the antenna is operating properly:

(1) A distant station may be used to test the antenna. If the signal received from this station is strong, the antenna is operating satisfactorily. If the signal is weak, adjust the height and length of the antenna and the transmission line to receive the strongest signal at a given setting on the volume control of the receiver. This is the best method of tuning an antenna when transmission is dangerous or forbidden.

(2) In some radio sets, use the transmitter to adjust the antenna. First, set the controls of the transmitter to normal; then, tune the system by adjusting the antenna height, the antenna length, and the transmission line length to obtain the best transmission output.

4-7. FIELD EXPEDIENT (FE) OMNI DIRECTIONAL ANTENNAS. Vertical antennas are omni directional. The omni directional antenna transmits and receives equally well in all directions. Most tactical antennas are vertical; for example, the man pack portable radio uses a vertical whip and so do the vehicular radios in tactical vehicles. A vertical antenna can be made by using a metal pipe or rod of the correct length, held erect by means of guidelines. The lower end of the antenna should be insulated from the ground by placing it on a large block of wood or other insulating material. A vertical antenna may also be a wire supported by a tree or a wooden pole. For short vertical antennas, a pole may be used without guidelines (if properly supported at the base). If the length of the vertical mast is not long enough to support the wire upright, it may be necessary to modify the connection at the top of the antenna.

a. **End Fed, Quarter, Half, or Full Wave Antenna.** An emergency, end fed half wave antenna (**Figure 4-1**) can be constructed from available materials such as field wire, rope, and wooden insulators. Compute the length of the (one quarter, one half, or full wave) antenna by using the formula provided previously. Cut the wires as close as possible to the correct length (better the wire is too long than too short). The electrical length of this antenna is measured from the antenna terminal on the radio set to the far end of the antenna. The best performance can be obtained by constructing the antenna longer than necessary and then shortening it, as required, until the best results are obtained. Connect the antenna to the radio using either method (**Figure 4-2A** and **Figure 4-2B**).

Figure 4-1. FIELD EXPEDIENT, END-FED QUARTER, HALF, OR FULL WAVE ANTENNA

Figure 4-2A. COBRA HEAD

Figure 4-2B. ANTENNA BASE

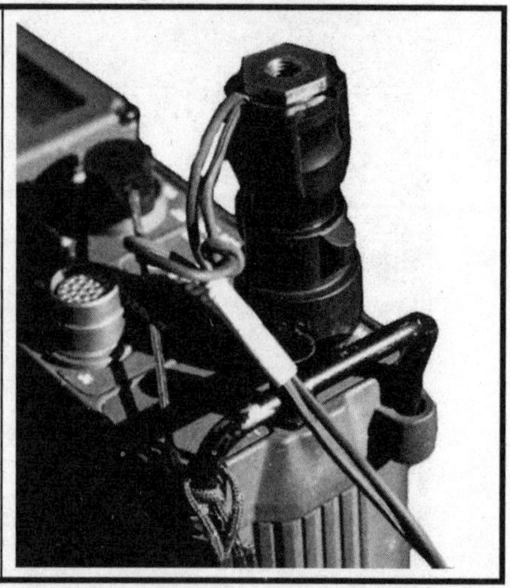

b. **Expedient 292-Type Antenna.** Developed for jungle, these antennas, properly used, can improve communications. Their weight and bulk render them impractical for most squad or platoon operations, but the unit can carry the masthead and antenna sections only, and mount them on wood poles or from trees; or they can construct an expedient version (**Figure 4-3**, **Figure 4-4**, and **Figure 4-5**) using any insulated wire and other available material. For example, most any plastic, glass, or rubber items or, if these are unavailable, dry wood, can serve as insulators:

(1) Use the planning considerations discussed in the next paragraph to determine the length of the elements (one radiating wire and three ground plane wires) for the desired frequency. Cut these elements (**A**) from claymore or similar wire. The heavier the gauge, the better, but insulated copper core wire works best. Cut spacing sticks (**B**) the same length as the ground plane wires. Place the sticks in a triangle and tie their ends together with wire, tape, or rope. Attach an insulator (**C**) to each corner and one end of each ground-plane wire to each insulator. Bring the loose ends of the ground-plane wires together, attach them to an insulator (**C**), and tie securely. Strip about 3 inches of insulation from each wire and twist them together.

(2) Tie one end of the radiating element wire to the other side of insulator and the other end to another insulator (**B**). Strip about 3 inches of insulation from the radiating element (**C**).

(3) Cut enough wire to reach from the proposed location of the antenna to the radio set. Keep this line as short as possible, because excess length reduces the efficiency of the system. Tie a knot at each end to identify it as the "hot" lead. Remove insulation from the "hot" wire and tie it to the radiating element wire at insulator (**C**). Remove insulation from the other wire and attach it to the bare ground plane element wires at insulator (**C**). Tape all connections and do not allow the radiating element wire to touch the ground plane wires.

(4) Attach a rope to the insulator on the free end of the radiating element and toss the rope over the branches of a tree. Pull the antenna as high as possible, keeping the lead in routed down through the triangle. Secure the rope to hold the antenna in place.

(5) At the radio set, remove about 1 inch of insulation from each end of the wire. Connect the ends to the positive side of the cobra head connector. Be sure the connections are tight or secure.

(6) Set up correct frequency, turn on the set, and proceed with communications.

Figure 4-3. COMPLETED EXPEDIENT 292-TYPE ANTENNA

Figure 4-4. COMPLETED EXPEDIENT 292-TYPE ANTENNA

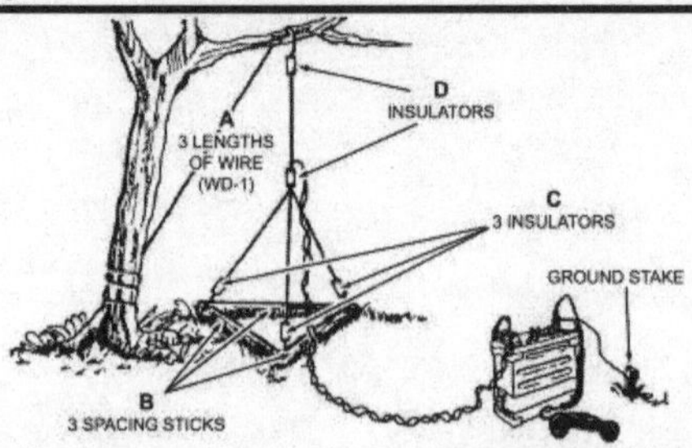

Table 4-2. QUICK-REFERENCE TABLE

OPERATING FREQUENCY IN MHz	ELEMENT LENGTH (RADIATING ELEMENTS AND GROUND-PLANE ELEMENTS)
30	2.38 m (7 ft 10 in)
32	2.23 m (7 ft 4 in)
34	2.1 m (6 ft 11 in)
36	1.98 m (6 ft 6 in)
38	1.87 m (6 ft 2 in)
40	1.78 m (5 ft 10 in)
43	1.66 m (5 ft 5 in)
46	1.55 m (5 ft 1 in)
49	1.46 m (4 ft 9 in)
52	1.37 m (4 ft 6 in)
55	1.3 m (4 ft 3 in)
58	1.23 m (4 ft 0 in)
61	1.17 m (3 ft 10 in)
64	1.12 m (3 ft 8 in)
68	1.05 m (3 ft 5 in)
72	0.99 m (3 ft 3 in)
76	0.94 m (3 ft 1 in)

Figure 4-5. COMPLETED 292-TYPE ANTENNA

4 8. **ANTENNA LENGTH PLANNING CONSIDERATIONS.** The length of an antenna must be considered in the construction of field expedients. At a minimum a quarter of the frequency wavelength should be used as the length of the FE Antenna. Another important factor in LOS communications is the height of the antenna with relation to the receiving station. The higher the antenna the greater the range the radio transmission will have. Terrain and curvature of the Earth affect LOS communication by absorbing

VHF and UHF communications into the Earth's surface. This can be overcome by increasing antenna height, power output, and radio frequency. Since radio frequencies are pre designated and power output is limited by the capabilities of the radio set, antenna length and height are the two variables that can be manipulated to increase radio communication range. Using the following formulas it will be possible to plan for the use of field expedient antennas, determine the best location to gain/ maintain communication, and plan for communication windows as necessary.

a. To calculate the physical length of an antenna in feet, use the following equation. It will give you the antenna length in feet for a 1/ 4 wavelength of the frequency. To determine the antenna length in feet for a full wavelength antenna, multiply the antenna length by 4:

X = 234/ Freq
(X = the length of the antenna in feet; Freq = the radio frequency used)

EXAMPLE

234/ 38.950 = 6.01 feet (Quarter Wavelength Antenna)
6.01 feet x 2 = 12.02 feet (Half Wavelength Antenna)
6.01 feet x 4 = 24.04 feet (Full Wavelength Antenna)

b. Curvature of the earth allows for a person 5 foot 7 inches looking across a flat surface to see a distance of about 4.7 km (**Figure 4-6**), anything beyond this distance is below the horizon and essentially dead space. To overcome this, the person must move to a higher elevation to see beyond 4.7 km. LOS communication is subject to this same principle. Use the following formula to calculate the required antenna height for a given distance (keep in mind that if you are in low ground such as a valley, draw, depression, the height of the antenna will be greater; when on high ground the antenna height may be shorter). Use the following formula to compute height of antenna to compensate for curvature of the earth:

X = 234/ Freq
(X = the length of the antenna in feet; Freq = the radio frequency used

Distance in km (**Dkm**) from receiving station = square root of $\sqrt{(12.7 \times Am)}$
where Am is the antenna height in meters.

EXAMPLE

Known height: Dkm = $\sqrt{(12.7 \times Am)}$, Dkm = $\sqrt{(12.7 \times 1.7m)}$ or
Unknown height: Am= $0.07874 \times (Dkm)^2$, Am= $0.07874 \times (4.7km)^2$

Figure 4-6. CURVATURE OF THE EARTH

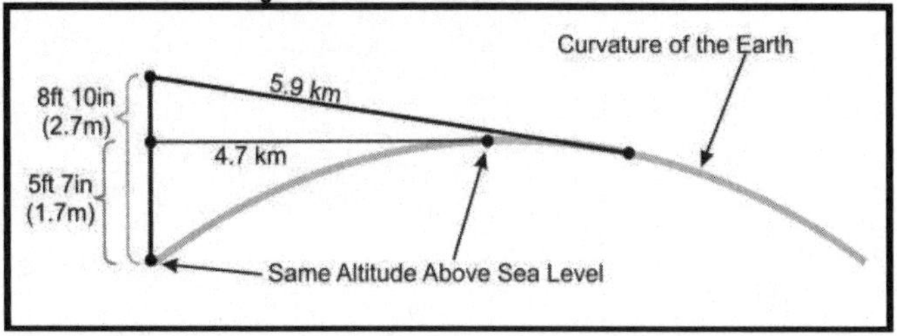

Chapter 5
DEMOLITIONS

This chapter introduces Rangers to the characteristics of explosives (low and high, **Table 5-1**), to initiation systems, modernized demolition initiator (MDI) components, detonation systems, safety considerations, expedient explosives, breaching charges, and timber cutting charges (FM 5-250).

- *Low explosives* have a detonating velocity up to 1,300 feet per second, which produces a pushing or shoving effect.
- *High explosives* have a detonating velocity of 3,280 to 27,888 feet per second, which produces a shattering effect.

Table 5-1. CHARACTERISTICS OF US DEMOLITIONS EXPLOSIVES

NAME	APPLICATIONS	DETONATION VELOCITY		RE FACTOR*	FUME TOXICITY	WATER RESISTANCE
		MIN/ SEC	FT/ SEC			
Ammonium nitrate	Cratering charge	2,700	8.800	0.42	Dangerous	Poor
PETN	Det cord blasting caps demolition charges	8,300	27,200	1.66	Slightly dangerous	Excellent
RDX	Blasting caps composition explosive	8,350	27,400	1.60	Dangerous	Excellent
Trinitrotoluene (TNT)	Demolition charge composition explosive	6,900	22,600	1.00	Dangerous	Excellent
Tetryl	Booster charge composition explosive	7,100	23,300	1.25	Dangerous	Excellent
Nitroglycerin	Commercial dynamite	7,700	25,200	1.50	Dangerous	Good
Black powder	Time fuse	400	1,300	0.55	Dangerous	Poor
Amatol 80 / 20	Bursting charge	4,900	16,000	1.17	Dangerous	Poor
Composition A3	Booster charge bursting charge	8,100	26,500	---	Dangerous	Good
Composition B	Bursting charge	7,800	25,600	1.35	Dangerous	Excellent
Composition C4 (M112)	Cutting and breaching charges	8.040	26,400	1.34	Slightly dangerous	Excellent
Composition H6	Cratering charge	7,190	23,600	1.33	Dangerous	Excellent
Tetrytol 75 / 25	Demolition charge	7,000	23,000	1.20	Dangerous	Excellent
Pentolite 50 / 50	Booster and bursting charges	7,450	24,400	---	Dangerous	Excellent
M1 Dynamite	Demolition charge	6,100	20,000	0.92	Dangerous	Fair
Det cord	Priming demolition charge	6,100 to 7,300	20,000 to 24,000	---	Slightly dangerous	Excellent
Sheet explosive M118 and M186	Cutting charge	7,300	24,000	1.14	Dangerous	Excellent
Bangalore torpedo M1A2	Demolition charge	7,800	25,600	1.17	Dangerous	Excellent
Shaped charges M2A3, M2A4, and M3A1	Cutting charge	7,800	25,600	1.17	Dangerous	Excellent

** TNT = 1.00 relative effectiveness*

5-1. INITIATING (PRIMING) SYSTEMS. The best way to prime demolition systems is with MDIs. These are blasting caps attached to various lengths of time fuse or shock tube. They can be used with a fuse igniter and detonating cord to create many firing systems. In the absence of MDI, field expedient methods may be used.

a. **Shock Tube.**

(1) Thin, plastic tube of extruded polymer with a layer of special explosive material on the interior surface.

(2) Explosive material propagates a detonation wave that moves along the shock tube to a factory crimped and sealed blasting cap.

(3) Detonation is normally contained within the plastic tubing. However, burns may occur if the shock tube is held.

WARNING
Burns may occur if the shock tube is held.

(4) Advantages of shock tube:

(a) It is extremely reliable.

(b) It offers instant electric initiation, and it also prevents radio transmitters, static electricity, and such from accidentally causing an initiation.

(c) It may be extended using leftover sections from previous operations.

b. **Blasting Caps.** Five types of MDI blasting caps are available to replace the M6 electric and M7 non-electric blasting cap. Three are high-strength, and two are low-strength. High-strength blasting caps can prime all standard military explosives (including detonating cord) or and can initiate the shock tube for other MDI blasting caps.

(1) ***M11.***

- Factory crimped to 30 feet of shock tube.
- A movable "J" hook is attached for quick and easy attachment to det cord.
- A red flag is attached 1 meter from the blasting cap and a yellow flag 2 meters from the blasting cap.

(2) ***M14.***

- Factory crimped to 7.5 feet of time fuse.
- May be initiated using a fuse igniter or match.
- Burn-time for total length is about five minutes.
- Yellow bands indicate calibrated one-minute time intervals.

(3) ***M15.***

- Two blasting caps factory crimped to 70 feet of shock tube.
- Each blasting cap has delay elements to allow for staged detonations.
- Low-strength blasting caps. Used as a relay device to transmit a shock tube detonation impulse from an initiator to a high strength-blasting cap.

(4) ***M12.*** This is factory crimped to 500 feet of shock tube on a cardboard spool.

(5) ***M13.*** This is factory crimped to 1,000 feet of shock tube.

c. **Matches.** If fuse igniter is unavailable, light the time (blasting) fuse with a match. Split the fuse at the end (**Figure 5-1**), and place the head of an unlit match in the powder train. Light the inserted match head with a flaming match, or rub the abrasive on the match box against it. You may have to retry this in windy conditions.

NOTE: High altitudes and colder temperatures increase burn time.

d. **M81 Fuse Igniter**. This is used to ignite time blasting fuse or to initiate the shock tube of MDI blasting caps.

Figure 5-1. TECHNIQUE FOR LIGHTING TIME FUSE WITH A MATCH

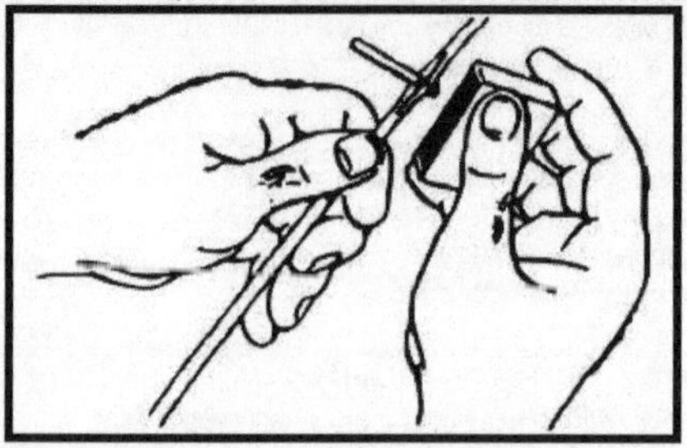

5-2. DETONATION (FIRING) SYSTEMS. The two types of firing systems are MDI alone, or MDI plus detonating cord.

a. MDI Alone. An MDI firing system is one in which the initiation set, transmission and branch lines are constructed using MDI components and the explosive charges are primed with MDI blasting caps. Construct the charge in the following manner.

(1) Emplace and secure explosive charge, such as C4, TNT, cratering charge, on target.
(2) Place a sandbag or other easily identifiable marker over the M11, M14, or M15 blasting cap to be used.
(3) Connect to an M12 or M13 transmission line if desired.
(4) Connect blasting cap with shock tube to an M14 cap with time fuse. Cut time blasting fuse to desired delay time.
(5) Prime the explosive charge by inserting the blasting cap into the charge.
(6) Visually inspect firing system for possible misfire indicators such as cracks, bulges, or corrosion.
(7) Return to the firing point and secure a fuse igniter to the cut end of the time fuse.
(8) Remove the safety cotter pin from the igniter's body.
(9) Actuate the charge by grasping the igniter body with one hand while sharply pulling the pull ring.

b. **MDI and Detonating Cord.** Construct the charge using the above steps for MDI stand-alone system. Incorporate detonating cord branch lines into the system using the "J" hooks of the M11 shock tube. Taping the ends of the detonation cord reduces the effect of moisture on the system.

5-3. SAFETY. MDI is not recommended for below ground use, except in quarry operations with water-gel or slurry explosives. Use detonating cord when it is necessary to bury primed charges.

a. Do not handle misfires downrange until the required 30 minute waiting period for both primary and secondary initiation systems has elapsed and other safety precautions have been accomplished.

b. Never yank or pull hard on the shock tube. This may actuate the blasting cap.

c. Do not dispose of used shock tubes by burning because of potentially toxic fumes given off from the burning plastic.

d. Do not use M1 dynamite with the M15 blasting cap. The M15 delay blasting cap should be used only with water-gel or slurry explosives.

e. Always use protective equipment when handling demolitions. Minimum protection consists of leather gloves, ballistic eye protection, and helmet.

5-4. EXPEDIENT EXPLOSIVES- IMPROVISED SHAPED CHARGE. An improvised shaped charge (**Figure 5-2**) concentrates the energy of the explosion released on a small area, making a tubular or linear fracture in the target.

a. The versatility and simplicity of these charges make them effective against targets, especially those made of concrete or those with armor plating.

(1) Bowls, funnels, cone shaped glasses, (champagne glasses with stem removed) used as cones. Champagne or cognac bottles are excellent.

(2) Charge characteristics.

(a) *Cavity Liners*. These are made of copper, tin, or zinc. If none is available, cut a cavity out of the plastic explosive.

(b) *Cavity Angle.* This will work with 30 to 60 degree angles. The cavity angle in most high-explosive antitank (HEAT) ammunition is 42 to 45 degrees.

(c) *Explosive Height (In Container).* This is 2 times the height of the cone measured from the base of the cone to the top of the explosive.

(d) *Standoff.* Normal standoff is one and one half times the cone's diameter.

(e) *Detonation Point.* The exact top center of the charge is the detonation point. Cover the blasting cap with a small amount of C4 if any part of the blasting cap is exposed.

b. Remove the narrow neck of a bottle or the stem of a glass by wrapping it with a piece of soft, absorbent twine or by soaking the string in gasoline and lighting it. Place two bands of adhesive tape, one on each side of the twine, to hold the twine firmly in place. The bottle or stem must be turned continuously with the neck up, to heat the glass uniformly.

c. A narrow band of plastic explosive placed around the neck and burned gives the same result. After the twine or plastic has burned, submerge the neck of the bottle in water and tap it against some object to break it off. Tape the sharp edge of the bottle to prevent cutting hands while tamping the explosive in place.

d. Do not immerse the bottle in water before the plastic has been completely burned, or it could detonate.

Figure 5 2. IMPROVISED SHAPED CHARGE

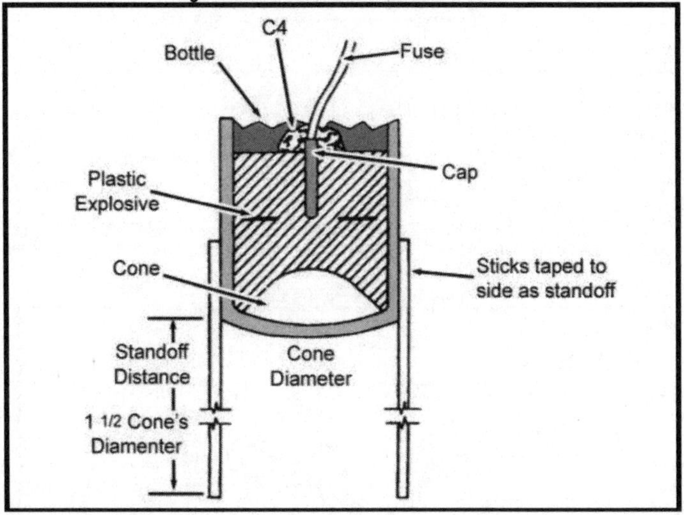

5-5. EXPEDIENT EXPLOSIVES-PLATTER CHARGE. This device (**Figure 5-3**) turns a metal plate into a powerful blunt-nosed projectile. The plate should be steel–preferably round, but square will work–and it should weigh from 2 to 6 pounds.

a. The weight of the explosive should equal the weight of the platter.

b. Uniformly pack the explosive behind the platter. You will only need a container if the explosives fail to remain firmly against the platter. You can use tape to anchor the explosives, if needed.

c. Prime the charge at the exact, rear center of the charge. If any part of the blasting cap is exposed, cover it with a small quantity of C4.

d. Aim charge at the direct center of the target, and ensure that the charge is on the opposite side of the platter from the target. Effective range is 35 yards for a small target. With practice, you might hit a 55-gallon drum at 25 yards 90 percent of the time. A gutted fuse igniter can serve as an expedient aiming device.

Figure 5-3. PLATTER CHARGE

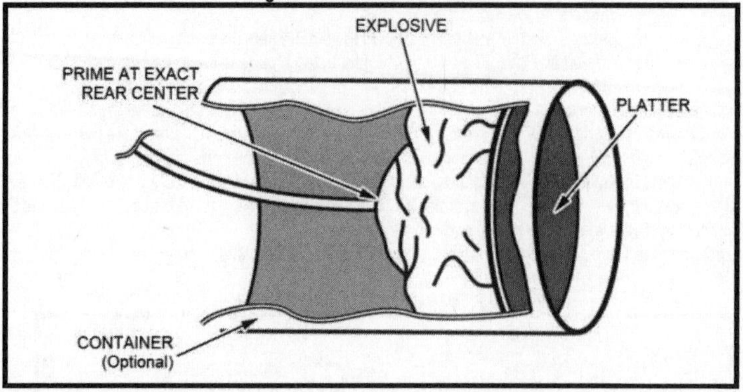

5-6. EXPEDIENT EXPLOSIVES-GRAPESHOT CHARGE. To use this antipersonnel fragmentation mine (**Figure 5-4**)--

a. **Hole.** Create a hole in the center, bottom of the container, for the blasting cap.

b. **Explosives.** Place explosives evenly on the bottom of the container. Remove all voids and air pockets by pressing the C4 into place using a non-sparking instrument.

c. **Buffer.** Place buffer material directly over the top of the explosives.

d. **Projectiles.** Place projectiles over top of the buffer materials, then cover to prevent spilling from movement.

e. **Aim.** Aim at target from about 100 feet. Use a small amount of C4 on any exposed portion of the blasting cap.

Figure 5-4. GRAPESHOT CHARGE

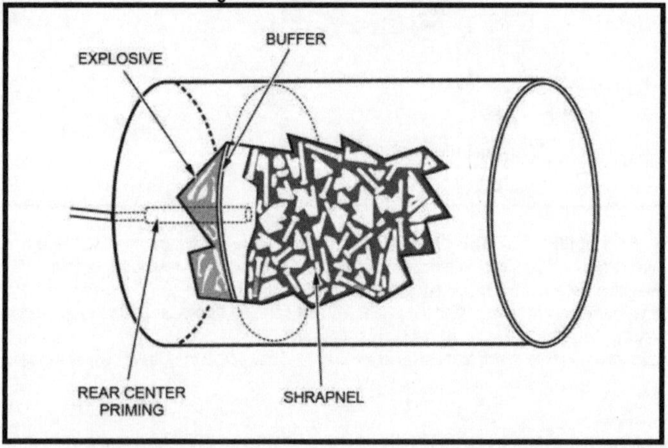

5-7. DEMOLITION KNOTS. Several knots are used in demolitions. **Figure 5-5** and **Figure 5-6** show a few simple knots that can join demolitions to detonation cord.

Figure 5-5. VARIOUS JOINING KNOTS USED IN DEMOLITIONS

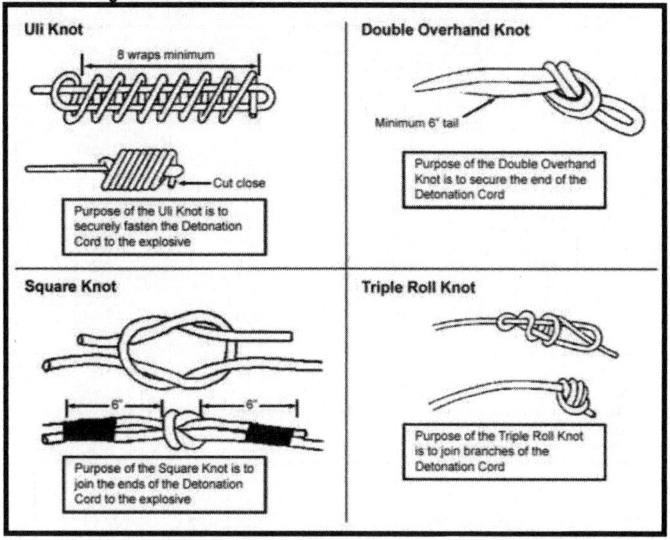

Figure 5-6. BRITISH JUNCTION

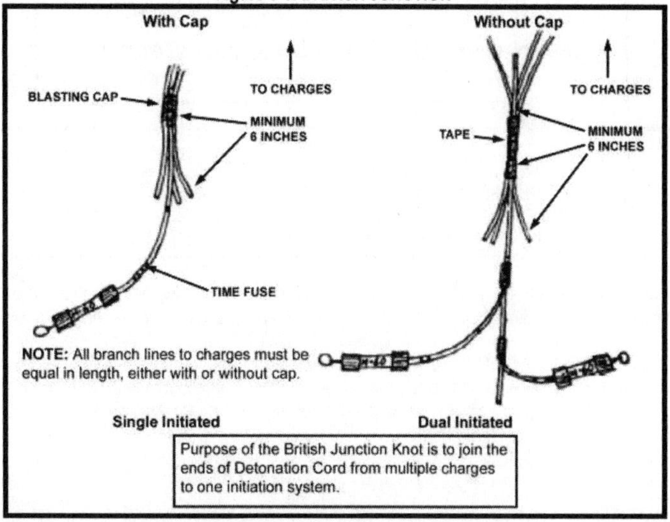

5-8. **MINIMUM SAFE DISTANCES.** Rangers must remain especially aware of their situations when using demolitions. **Table 5-2** shows minimum safe distances for employing up to 500 pounds. For charges over 500 pounds, see **Figure 5-7.**

Table 5-2. MINIMUM SAFE DISTANCE FOR PERSONNEL IN OPEN (BARE CHARGE)

EXPLOSIVE WEIGHT (LB)	SAFE DISTANCE FEET	SAFE DISTANCE METERS	EXPLOSIVE WEIGHT (LB)	SAFE DISTANCE FEET	SAFE DISTANCE METERS
27 OR LESS	985	300	175	1,838	560
30	1,021	311	200	1,920	585
35	1,073	327	225	1,999	609
40	1,123	342	250	2,067	630
45	1,168	356	275	2,136	651
50	1,211	369	300	2,199	670
60	1,287	392	325	2,258	688
70	1,355	413	350	2,313	705
80	1,415	431	375	2,369	722
90	1,474	449	400	2,418	737
100	1,526	465	425	2,461	750
125	1,641	500	500	2,625	800
150	1,752	534	---	---	---

Figure 5-7. MINIMUM SAFE DISTANCE FOR CHARGES OVER 500 POUNDS

$$\text{Safe distance (meters)} = 100 \sqrt[3]{\text{pounds of explosive}}$$

$$\text{Safe distance (feet)} = 300 \sqrt[3]{\text{pounds of explosive}}$$

DANGER
CHARGES ON TARGETS
FOR CHARGES ON TARGETS, THE MINIMUM RADIUS OF DANGER IS 1,000 METERS. MINIMUM SAFE DISTANCE WHEN IN A MISSILE-PROOF SHELTER FROM THE POINT OF DETONATION IS 100 METERS.

5-9. **BREACHING CHARGES.** For **Table 5-3**, the left column represents the thickness of reinforced concrete wall. The remaining 7 columns show the number of packages of C4 required to breach the wall using the charge placements shown in the drawings above the columns.

a. Use **Table 5-3**, **Table 5-4**, and **Table 5-5** for breaching charges.

b. Use the formula in **Figure 5-8** to calculate the charges (**Table 3-5** and **Figure 3-15** for more information).

c. Multiply number of packages of C4 from **Table 5-3** by conversion factor from **Table 5-4** for materials other than reinforced concrete.

Table 5-3. BREACHING CHARGES FOR REINFORCED CONCRETE

PLACEMENT METHODS

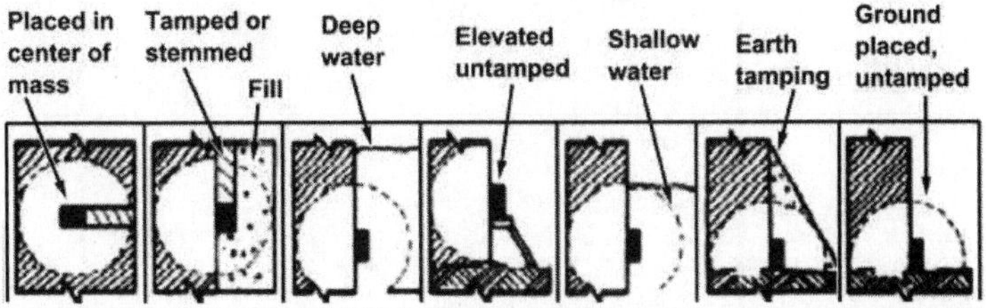

Reinforced Concrete Thickness (ft)	C = 1.0	C = 1.0	C = 1.0	C = 1.8	C = 2.0	C = 2.0	C = 3.6
	Packages of M112 (C4)						
2.0	1	5	5	9	10	10	17
2.5	2	9	9	17	18	18	33
3.0	2	13	13	24	26	26	47
3.5	4	21	21	37	41	41	74
4.0	5	31	31	56	62	62	111
4.5	7	44	44	79	88	88	157
5.0	9	48	48	85	95	95	170
5.5	12	63	63	113	126	126	226
6.0	13	82	82	147	163	163	293
6.5	17	104	104	186	207	207	372
7.0	21	111	111	200	222	222	399
7.5	26	137	137	245	273	273	490
8.0	31	166	166	298	331	331	595

Table 5-4. CONVERSION FACTORS FOR MATERIALS OTHER THAN REINFORCED CONCRETE

MATERIAL	*CONVERSION FACTOR*
• EARTH • ORDINARY MASONRY • HARD PAN • SHALE	0.1
• ORDINARY CONCRETE • ROCK • GOOD TIMBER • EARTH CONSTRUCTION	0.5
• DENSE CONCRETE • FIRST-CLASS MASONRY	0.7

Table 5-5. MATERIAL FACTOR (K) FOR BREACHING CHARGES

Material	R	K
Earth	All values	**0.07**
Poor masonry, shale, hardpan, good timber, and earthen construction	Less than 1.5 m (5 ft) 1.5 m (5 ft) or more	**0.32** **0.29**
Good masonry, concrete block, and rock	0.3 m (1 ft) or less Over 0.3 m (1 ft) to less than 0.9 m (3 ft) 0.9 m (3 ft) to less than 1.5 m (5 ft) 1.5 M (5 ft) to less than 2.1 m (7 ft) 2.1 m (7 ft) or more	**0.88** **0.48** **0.40** **0.32** **0.27**
Dense concrete and first-class masonry	0.3 m (1 ft) or less Over 0.3 m (1 ft) to less than 0.9 m (3 ft) 0.9 m (3 ft) to less than 1.5 m (5 ft) 1.5 M (5 ft) to less than 2.1 m (7 ft) 2.1 m (7 ft) or more	**1.14** **0.62** **0.52** **0.41** **0.35**
Reinforced concrete (factor does not consider cutting steel)	0.3 m (1 ft) or less Over 0.3 m (1 ft) to less than 0.9 m (3 ft) 0.9 m (3 ft) to less than 1.5 m (5 ft) 1.5 M (5 ft) to less than 2.1 m (7 ft) 2.1 m (7 ft) or more	**1.76** **0.96** **0.80** **0.63** **0.54**

Figure 5-8. FORMULA FOR COMPUTING SIZE OF CHARGE TO BREACH CONCRETE, MASONRY, AND ROCK

$$P = R^3KC$$

Where–

P = TNT required (in pounds)

R = Breaching radius (in feet)

K = Material factor, which reflects the strength, hardness, and mass

of the material to be demolished

C = Tamping factor, which depends on the location and tamping

of the charge

5-10. TIMBER CUTTING CHARGES. Table 5-6 shows timber-cutting charge sizes. **Figure 5-9** through **Figure 5-15** show the types of charges and the formulas to use with each.

Table 5-6. TIMBER-CUTTING CHARGE SIZE

	Packages of C4 Required (1.25-lb Packages) by Timber Diameter (in)											
Charge Type	6	8	10	12	15	18	21	24	27	30	33	36
Internal	1	1	1	1	1	1	2	2	2	3	3	4
External	1	1	2	3	4	5	7	9	11	14	17	20
Abatis	---	---	---	---	---	---	---	7	9	11	14	16
NOTE: Packages required are rounded UP to the next whole package.												

Figure 5-9. ABATIS

Figure 5-10. FORMULA FOR FALLEN TREE OBSTACLES OR TEST SHOT

$$P = D^2/50 = P = 0.02D^2$$

Where–

P = TNT required per tree (in pounds)

D = Diameter or least dimension of dimensioned timber; in inches

Figure 5-11. TIMBER-CUTTING RING CHARGE

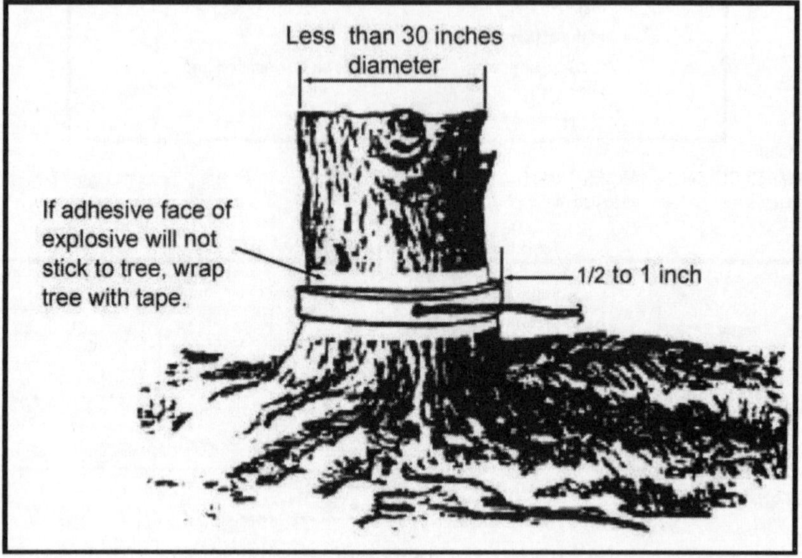

Figure 5-12. TIMBER-CUTTING CHARGE (EXTERNAL)

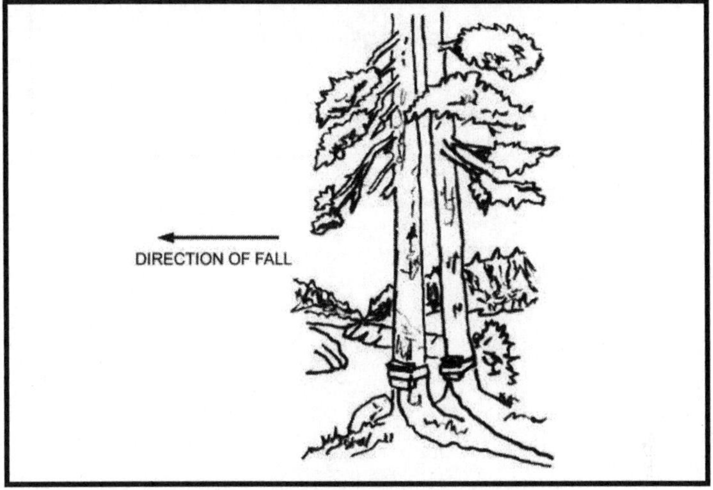

Figure 5-13. FORMULA FOR EXTERNAL TIMBER-CUTTING CHARGE

$$P = D^2/40 \text{ or } P = 0.025D^2$$

Where–

P = TNT required per target (in pounds)

D = Diameter or least dimension of dimensioned timber; in inches

Figure 5-14. TIMBER-CUTTING CHARGE (INTERNAL)

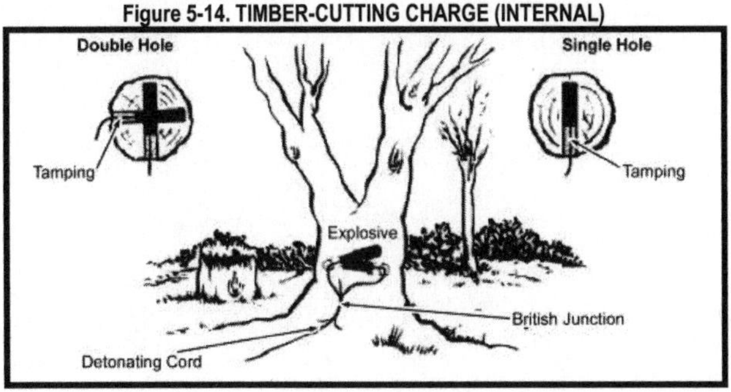

Figure 5-15. FORMULA FOR INTERNAL TIMBER-CUTTING CHARGE

$$P = D^2/250 \text{ or } P = 0.004D^2$$

Where–

P = TNT required per target (in pounds)

D = Diameter or least dimension of dimensioned timber; in inches

Chapter 6
MOVEMENT

To survive on the battlefield, stealth, dispersion, and security must be enforced in all tactical movements. The leader must be skilled in all movement techniques (**FM 3-21.8**).

6-1. FORMATIONS. Movement formations include elements and Rangers arranged in relation to each other. Fire teams, squads, and platoons use several formations. Formations give the leader control based on a METT-TC analysis. Leaders position themselves where they can best command and control the formations, which are shown in **Figure 6-1**. Formations–

a. **Allow** the fire team leader to lead by example, "Follow me and do as I do." All Rangers in the team must be able to see their leader.

b. **Reflect** fire team formations. Squad formations are very similar with more Rangers. Squads can operate in lines and files similar to fire teams. When squads operate in wedges or in echelon, the fire teams use those formations, and simply arrange themselves in column or with one team behind the other. Squads may also use the "vee," where one team forms the lines of the vee with the squad leader at front (at the point of the vee) for command and control. Platoons use the same formations as squads. When the unit operates as a platoon, the platoon leader must carefully select the location for his machine guns in the movement formation.

Figure 6-1. FORMATIONS

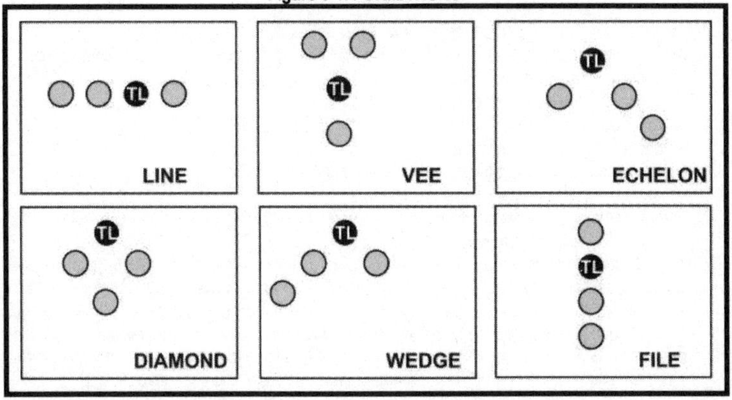

6-2. MOVEMENT TECHNIQUES. Selecting a movement technique is based on the likelihood of enemy contact and the relative need for speed. Specifically, the factors to consider include ***control, dispersion, speed,*** and ***security.*** Movement techniques are neither fixed ***nor are they formations.*** Instead, movement techniques are distinguished by a set of criteria such as distance between individual Rangers and between teams or squads. Movement techniques vary depending on METT-TC. However, *Rangers always must be able to see their fire team leaders, and the platoon leader should be able to see his lead squad leader.* Leaders control movement with hand and arm signals and use radios only when needed. Leaders match the movement technique to the situation as follows:

a. **Traveling**. Use when enemy contact is not likely, but speed is necessary. Leave 10 meters between Rangers, and 20 meters between squads.

- More control than traveling overwatch but less than bounding overwatch.
- Minimum dispersion.
- Maximum speed.

• Minimum security.

b. **Traveling Overwatch.** Use when enemy contact is possible. This is the most used movement technique. Leave 20 meters between Rangers, and 50 meters between teams.

(1) Only the lead squad should use traveling overwatch; however, in cases where greater dispersion is desired, all squads may use it.

(2) In other formations, all squads use traveling overwatch unless the platoon leader specifies not to. Traveling overwatch offers good control, dispersion, speed, and security forward.

(3) The lead squad must be far enough ahead of the rest of the platoon to detect or engage any enemy before the enemy observes or fires on the main body. However, the lead squad must stay between 50 and 100 meters in front of the platoon so the platoon can support them with small arms fires. This is normally between 50 to 100 meters, depending on terrain, vegetation, and light and weather conditions.

c. **Bounding Overwatch.** Use when enemy contact is likely, or when crossing a danger area. Both squad and platoon have bounding and overwatch elements. The bounding element moves while the other one occupies a position where it can overwatch by fire the bounding element's route. *The bounding element must remain within firing range of the overwatching element at all times.*

(1) ***Characteristics.*** Bounding overwatch offers maximum control, dispersion, and security with minimum speed.

(2) ***Types of Bounds.***

(a) <u>*Successive bounds*</u>. One element moves to a position, then the overwatching element moves to a position generally online with the first element.

(b) <u>*Alternating Bounds.*</u> One element moves <u>*into*</u> position, then the overwatching element moves to a position <u>*in front of*</u> the first element.

(3) ***Length.*** The length of a bound depends on the terrain, visibility, and control.

(4) ***Instructions.*** Before a bound, the leader gives the following instructions to his subordinates:

- Direction of the enemy if known
- Position of overwatch elements
- Next overwatch position
- Route of the bounding element
- What to do after the bounding element reaches the next position
- How the elements receive follow on orders

(5) ***Squad Bounding Overwatch.*** Rangers leave about 20 meters between them. The distance between teams and squads varies (**Figure 6-2**).

(6) ***Platoon Bounding Overwatch.*** When platoons use bounding overwatch (**Figure 6-3**), one squad bounds, a second squad overwatches, and a third awaits orders. Rangers leave about 20 meters between them. The distance between teams and squads varies. Forward observers stay with the overwatching squad to call for fire. Platoon leaders normally stay with the overwatching squad, which uses machine guns and attached weapons to support the bounding squad. Another way is to have one squad use bounding overwatch while the other two use traveling or traveling overwatch. When deciding where to move the bounding element, consider–

- The enemy's likely location
- The mission
- The routes to the next overwatch position
- The weapons ranges of the overwatching unit
- The responsiveness of the rest of the unit
- The fields of fire at the next overwatch position

Figure 6-2. SQUAD BOUNDING OVERWATCH

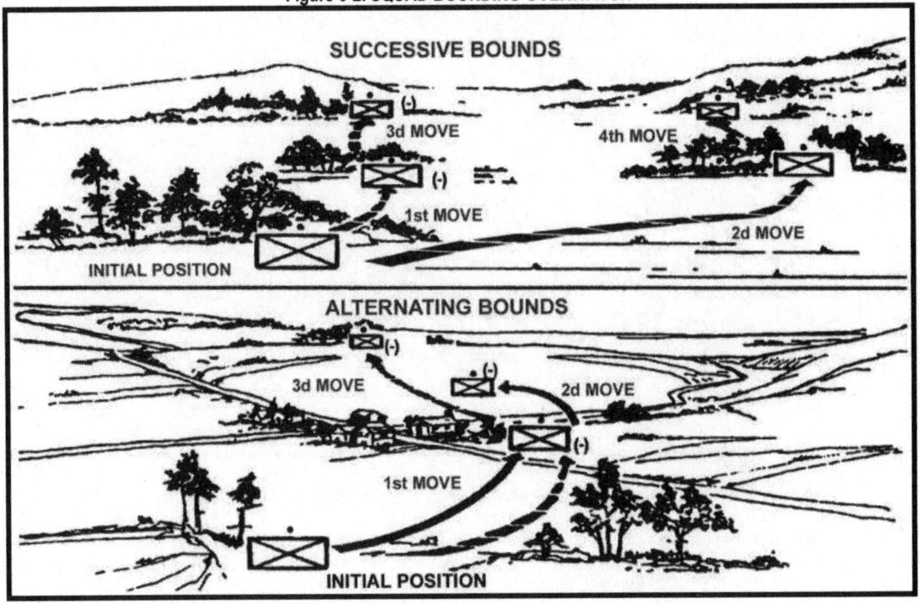

Figure 6-3. PLATOON BOUNDING OVERWATCH

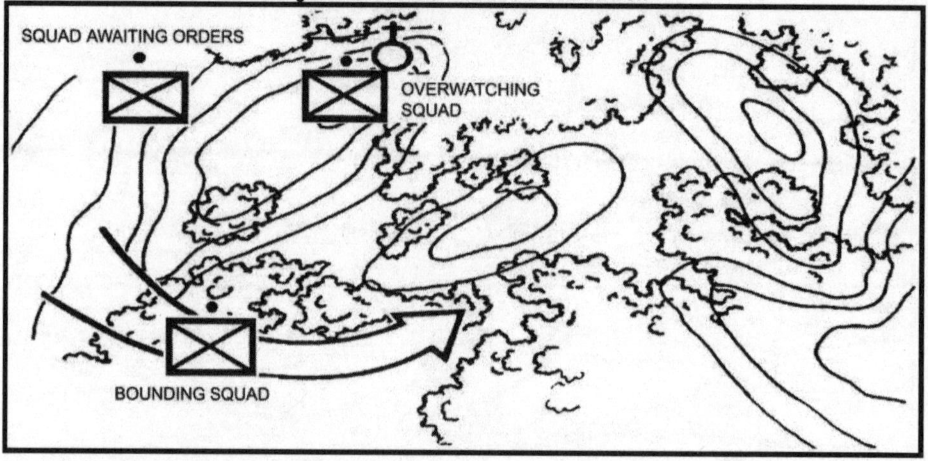

6-3. STANDARDS. Unit moves on designated route or arrives at specified location IAW OPORD, maintaining accountability of all assigned/ attached personnel.

a. Unit uses the movement formation and technique ordered by the leader (based on METT-TC).

b. Leaders remain oriented (within 200 meters) and follow a planned route, unless METT-TC dictates otherwise.

c. (During movement) unit maintains 360-degree security and remains 100 percent alert.

d. (During halts) unit maintains 360-degree security and at least 75 percent security.

e. If the unit makes contact with the enemy, they do so with the smallest element possible.

f. The unit uses control measures during movement such as head counts, rally points, or phase lines.

6-4. FUNDAMENTALS.

a. Mission Accomplishment. Mission accomplishment depends on successful land navigation. The patrol should use stealth and vigilance to avoid chance contact. Designate a primary and alternate compass and pace man per patrol. All leaders except fire team leaders move inside their formations to best control the platoon.

> **NOTE:** The point man is never tasked to perform compass or pace duties. His sole responsibility is forward security for the element.

(1) ***Stealth.*** Patrols must use stealth, and use the cover and concealment of the terrain to its maximum advantage. Whenever possible, the patrol should move during limited visibility to maximize the technological advantages of night vision devices and hinder the enemy's ability to detect the patrol. They exploit the enemy's weaknesses, and try to time movements to coincide with other operations that distract the enemy.

(2) ***Security.*** The patrol must continue to use both active and passive security measures. The leader assigns subunit responsibilities for security at danger areas, patrol bases and, most of all, in the objective area.

(3) ***Fire Support.*** The leader plans fire support (mortars, artillery, tactical air, attack helicopter, naval gunfire).

(4) ***Choice of Technique.*** The enemy threat and terrain determines which of the three movement techniques to use:

(a) Fire teams maintain visual contact, but the distance between them is such that the entire patrol does not become engaged if it makes contact. Fire teams can spread their formations as necessary to gain better observation to the flanks. Although widely spaced, men retain their relative positions in their wedge, and follow their team leader. Only in extreme situations should the file be used.

(b) The lead squad must secure the front, and is responsible for navigation. For a long movement, the PL may rotate lead squad responsibilities. The fire team/squad in the rear is charged with rear security.

(c) Vary movement techniques to meet the changing situation.

(5) ***Three Dimensional Battlefield.*** The patrol achieves 360-degree security high and low. Within a fire team, squad and so on, the leader assigns appropriate sectors of fire to subordinates. This ensures the battlefield is covered. This includes trees, multiple storied structures, tunnels, sewers, and ditches.

6-5. TACTICAL MARCHES. Platoons conduct two types of marches with the company: foot marches and motor (road) marches.

a. **Purpose/ General.** A foot march is successful when troops arrive at the destination at the prescribed time, physically able to execute their tactical mission.

b. **Standard.**

- The unit crosses the start point and release point at the times specified in the order.
- The unit follows the prescribed route, rate of march, and interval without deviation unless required otherwise by enemy action or higher headquarters action.

c. **Fundamentals.**

- Effective control
- Detailed planning
- Rehearsals

d. **Considerations.**

(1) ***METT-TC.***

- Mission--Task and purpose
- Enemy--Intentions, capabilities, and course of action
- Terrain and Weather--Road condition/ trafficability, and visibility
- Troops/Equipment--Condition of Rangers and their loads; number and types of weapons and radios
- Time--Start time, release time, rate of march, and time available
- Civilians--Movement through populated areas, refugees, OPSEC

(2) ***Task Organization***

- Security – Advance and trail teams
- Main Body – Two remaining line squads and weapons squad
- Headquarters Command and Control
- Control measures

(3) ***Start Point and Release*** Point (given by higher)

- Check Points – at check points report to higher and use to remain oriented
- Rally or rendezvous points – used when elements become separated
- Locations of Leaders – where they can best control their elements
- Communications Plan – location of radios, frequencies, call signs, and OPSKEDs
- Dispersion between Rangers
  - 3 to 5 meters/day
  - 1 to 3 meters/night

(4) ***March Order.*** May be issued as an OPORD, FRAGO, or annex to either (must use operational overlay or strip map). The march order includes–

- Formations and order of movement
- Route of march, assembly area, start point, release point, rally points, check points, break/ halt points
- Start point time, release point time, and rate of march
- March interval for squads, teams and individuals
- Actions on enemy contact—air and ground
- Actions at halts

- Fires—detailed plan of fire support for the march
- Water supply plan
- MEDEVAC Plan

e. **Duties and Responsibilities.**

(1) ***Platoon Leader:***

- Before – Issues WARNO and FRAGO; inspects and supervises
- During – Ensures unit makes movement time, maintains interval, and remains oriented; maintains security; checks condition of Rangers; enforces water discipline and field sanitation
- After – Ensures Rangers are prepared to accomplish their mission; supervises SLs; ensures Rangers receive medical coverage as needed

(2) ***Platoon Sergeant:***

- Before – Helps PL; makes recommendations; enforces uniform and packing lists
- During –Controls stragglers, assist platoon leader maintain proper interval and security
- At Halts – Enforces security, ensures welfare of men, enforces field sanitation, litter discipline, and preventive medicine.
- After – Coordinates for water, rations, and medical supplies. Recovers casualties.

(3) ***Squad Leaders:***

- Before – Provides detailed instruction to TLs, inspects boots and socks for serviceability and proper fit, adjustment of equipment, full canteens, and equal distribution of loads
- During – Controls squad, maintains proper interval between men and equipment, enforces security, remains oriented
- At halts – Ensures security is maintained, provides Rangers for water resupply as detailed. Physically checks the Rangers in his squad; ensures they drink water and change socks as necessary. Rotates heavy equipment. Units should plan the latter in detail to avoid confusion before, during, and after halts
- After Occupies squad sector of assembly area, conducts foot inspection and reports condition of men to PL; prepares men to accomplish the mission.

(4) ***Security Squad:***

- Lead Team
  - Serves as point element for platoon, reconnoiters route to SP
  - Calls in check points and provides early warning
  - Maintains rate of march, moves 10 to 20 meters in front of main body
- Trail Team
  - Provides rear security, moves 10 to 20 meters behind main body

(5) ***Medic:***

- Assesses and treats march casualties
- Advises chain of command on evacuation and transportation requirements of casualties

(6) ***Individual:***

- Maintains interval and follows TL's examples
- Relays hand and arm signals; remains alert during movement and at halts
- Remains alert during movement and at halts

6-6. MOVEMENT DURING LIMITED VISIBILITY CONDITIONS. During hours of limited visibility, the platoon uses surveillance, target acquisition, and night observation (STANO) devices to enhance effectiveness. Leaders must be able to control, navigate, maintain security, and move during limited visibility.

a. **Control.** When visibility is poor, the following methods aid in control:

- Leaders move closer to front
- Platoon reduces speed
- Platoon uses luminescent tape on equipment
- Leader reduces intervals between men and elements
- Leader counts heads often

b. **Navigation.** While navigating during limited visibility, the unit uses the same techniques are they do in daylight, but leaders exercise more care to keep the patrol oriented.

c. **Security.**

- Enforce strict noise and light discipline.
- Use radio listening silence.
- Use camouflage.
- Use terrain to avoid detection by enemy surveillance or night vision devices.
- Make frequent listening halts; conduct SLLS (Stop, Look, Listen, Smell).
- Mask the sounds of movement when possible. (Rain, wind, and flowing water will mask the sounds of movement.)

d. **Rally Points.** Leaders plan actions to be taken at rally points in detail. All elements must maintain communications at all time. The two techniques for actions at rally points follow:

- ***Minimum Force:*** Patrol members assemble at the rally point, and the senior leader assumes command. When the minimum force (designated in the OPORD) is assembled and organized, the patrol will continue the mission.
- ***Time Available:*** The senior leader determines if the patrol has enough time remaining to accomplish the mission.

e. **Actions at Halts.** During halts, the unit posts security and covers all approaches into the sector with key weapons.

- ***Short Halt.*** This typically takes 1 to 2 minutes long. Rangers seek immediate cover and concealment and take a knee. Leaders assign sectors of fire.
- ***Long Halt.*** This typically takes more than 2 minutes. Rangers assume the prone position behind cover and concealment. Leaders ensure Rangers have clear fields of fire, and assign sectors of fire.

6-7. DANGER AREAS. A danger area is any place on a unit's route where the leader determines his unit may be exposed to enemy observation or fire. Some examples of danger areas are open areas, roads and trails, urban terrain, enemy positions, and natural and manmade obstacles. ***Bypass danger areas whenever possible.***

a. **Standards.**

- The unit prevents the enemy from surprising the main body.
- The unit moves all personnel and equipment across the danger area.
- The unit prevents decisive engagement by the enemy.

b. **Fundamentals.**

- Designate near and far side rally points.
- Secure near side, left and right flank, and rear security.
- Recon and secure the far side.
- Cross the danger area.
- Plan for fires on all known danger areas.

c. **Technique for Crossing Danger Areas.**

(1) ***Linear Danger Are***a (LDA; **Figure 6-4**) ***Actions for a Squad.***

STEP 1. The alpha team leader (ATL) observes the linear danger area and sends the hand and arm signal to the SL, who determines to bound across.

STEP 2. SL directs the ATL to move his team across the LDA far enough to fit the remainder of the squad on the far side of the LDA. Bravo team moves to the LDA to the right or left to provide an overwatch position prior to A team crossing.

STEP 3. SL receives the hand and arm signal that it is safe to move the rest of the squad across (B team is still providing overwatch).

STEP 4. SL moves himself, RTO and B team across the LDA. (A team provides overwatch for squad missions.)

STEP 5. A team assumes original azimuth at SLs command or hand and arm signal.

(2) ***LDA Crossing for a Platoon.***

(a) The lead squad halts the platoon and signals danger area.

(b) The platoon leader moves forward to the lead squad to confirm the danger area, and then decides if current location is suitable for crossing.
(c) The platoon leader confirms danger area/ crossing site and establishes near and far side rally points.
(d) On the platoon leader's signal, the trail squad moves forward to establish left and right near side security.
(e) Once near side security is established, the A team of the lead squad with the squad leader, moves across to confirm there is enough room to fit the rest of the platoon on the far side of the LDA.
(f) Once he conducts Stop, Look, Listen, and Smell (SLLS), squad leader signals platoon leader "All Clear."

- Day time—hand and arm signal such as a "thumbs up"
- Night time—clandestine signal such as infrared, red lens

(g) The platoon leader then directs the B team of the lead squad to bound across by team and link up with the A team of the lead squad and pick up a half step while the rest of the platoon crosses.
(h) Platoon leader then crosses with RTO, FO, WSL, and two gun teams.
(i) Once across, PL signals the 2nd squad in movement to cross.
(j) PSG with medic and one gun team crosses after second squad is across (sterilizing central crossing site).
(k) PSG signals security squad to cross at their location.
(l) PSG calls PL via FM to confirm all elements are across.
(m) PL directs lead squad to pick up normal rate of movement.

Figure 6-4. LINEAR DANGER AREA

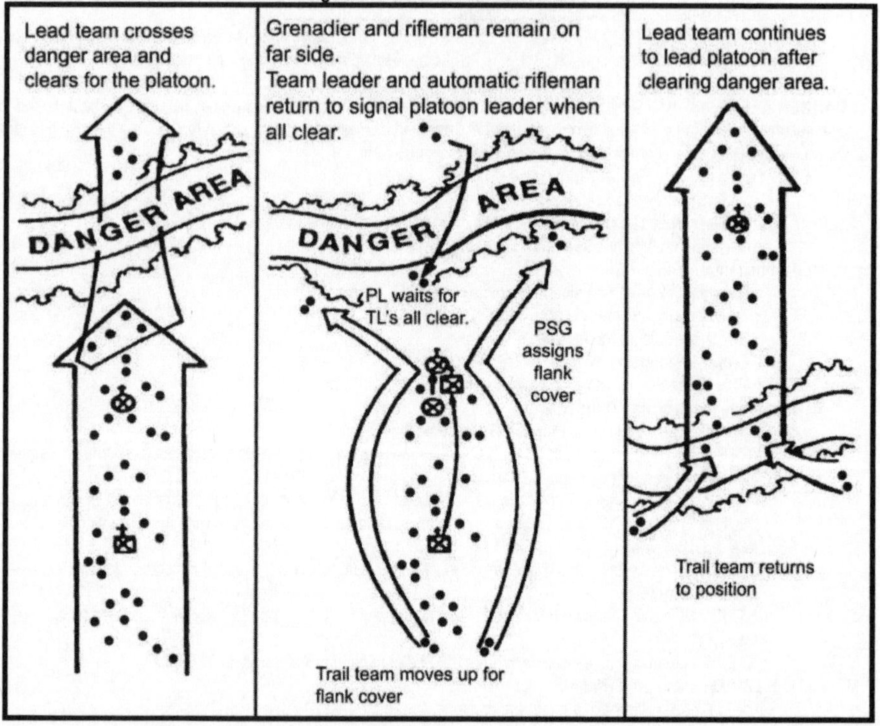

> **NOTE:** Platoon leader plans fires on all known LDA crossing sites. Near side security in overwatch sterilizes signs of the patrol.

(3) ***Danger Area (Small/ Open)***

(a) The lead squad halts the platoon and signals "Danger area."

(b) The PL moves forward to the lead squad to confirm the danger area.

(c) The platoon leader confirms danger area and establishes near and far side rally points.

(d) The PL designates lead squad to bypass danger area using the detour bypass method.

(e) The paceman suspends current pace count and initiates an interim pace count. Alternate pace/compass man moves forward and offsets compass 90 degrees left or right as designated and moves in that direction until clear of danger area.

(f) After moving set distance (N meters as instructed by PL). Lead squad assumes original azimuth, and primary pace man resumes original pace.

(g) After the open area, the alternate pace/ compass man offsets his compass 90 degrees left or right, and leads the platoon/ squad the same distance (N meters) back to the original azimuth.

(4) ***Danger Areas (Series).*** A series of danger areas is two or more danger areas within an area that can be either observed or covered by fire.

- Double linear danger area (use linear danger area technique and cross as one LDA)
- Linear/ small open danger area (use by pass/ contour technique. **Figure 6-5**)
- Linear/ large open danger area (use platoon wedge when crossing).

> **NOTE: A series of danger areas is crossed using the technique which provides the most security.**

Figure 6-5. SMALL OPEN DANGER AREA

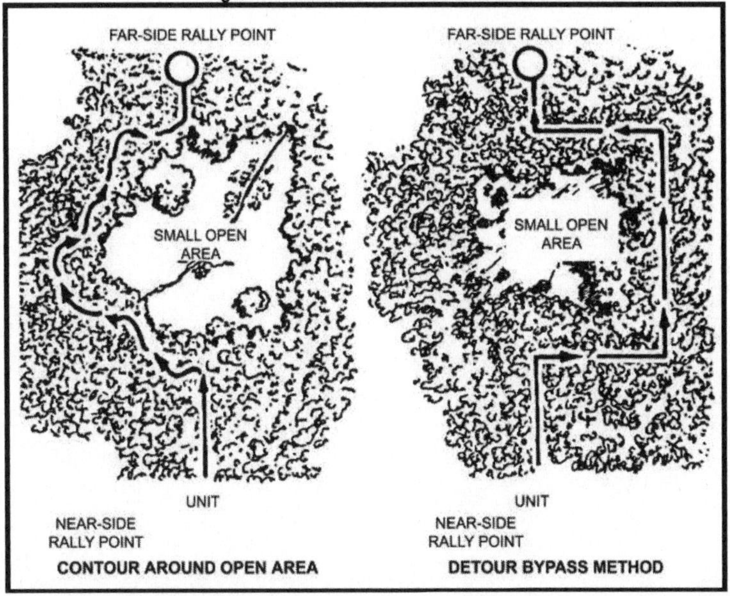

(5) ***Danger Area (Large).***

(a) Lead squad halts the platoon, and signals danger area.

(b) The platoon leader moves forward with RTO and FO and confirms danger area.

(c) The platoon leader confirms danger area and establishes near and far side rally points.

(d) PL designates direction of movement.

(e) PL designates change of formation as necessary to ensure security.

> **NOTE:** Platoon leader will plan for fires at all known LDA crossing sites. Nearside security in overwatch will sterilize signs of the patrol.

> **NOTE:** Before point man steps into danger area, PL and FO adjust targets to cover movement.

> **NOTE:** If far side of danger area is within 250 meters, PL establishes overwatch, and designates lead squad to clear woodline on far side.

Chapter 7
PATROLS

This chapter describes the principles and types of (*reconnaissance and combat*), and planning considerations and supporting tasks for patrols by Infantry platoons and squads. It also discusses *patrol bases and movements to contact* (**FM 3-21.8, FM 3-0,** and **FM 1-02**). Here, the terms *"element"* and *"team"* refer to the squad's fire teams, or buddy teams that perform the tasks as described.

Section I. PRINCIPLES

All patrols are governed by five principles.

7-1. PLANNING. Quickly make a simple plan and effectively communicate it to the lowest level. A great plan that takes forever to complete and is poorly disseminated isn't a great plan. Plan and prepare to a realistic standard and rehearse everything.

7-2. RECONNAISSANCE. Your responsibility as a Ranger leader is to confirm what you think you know, and to learn that which you do not already know.

7-3. SECURITY. Preserve your force as a whole. Every Ranger and every rifle counts; anyone could be the difference between victory and defeat.

7-4. CONTROL. Clarify the concept of the operation and commander's intent, coupled with disciplined communications, to bring every man and weapon available to overwhelm the enemy at the decisive point.

7-5. COMMON SENSE. Use all available information and good judgment to make sound, timely decisions.

Section II. PLANNING

This section provides the planning considerations common to most patrols. It discusses task organization, initial planning and coordination, completion of the plan, and contingency planning.

7-6. TASK ORGANIZATION. A patrol is a detachment sent out by a larger unit to conduct a specific mission. Patrols operate semi-independently and return to the main body upon completion of their mission. Patrolling fulfills the Infantry's primary function of finding the enemy to either engage him or report his disposition, location, and actions. Patrols act as both the eyes and ears of the larger unit and as a fist to deliver a sharp devastating jab and then withdraw before the enemy can recover.

a. **Definitions.**

(1) ***Patrol.*** Sent out by a larger unit to conduct a specific combat, reconnaissance, or security mission. A patrol's organization is temporary and specifically matched to the immediate task. Because a patrol is an organization, not a mission, it is not correct to speak of giving a unit a mission to "Patrol."

(2) ***Patrolling or Conducting a Patrol.*** The semi-independent operation conducted to accomplish the patrol's mission. A patrol requires a specific task and purpose.

(3) ***Employment.*** A commander sends a patrol out from the main body to conduct a specific tactical task with an associated purpose. Upon completion of that task, the patrol leader returns to the main body, reports to the commander and describes the events that took place, the status of the patrol's members and equipment, and any observations.

(4) ***Leadership.*** If a patrol is made up of an organic unit, such as a rifle squad, the squad leader is responsible. If a patrol is made up of mixed elements from several units, an officer or NCO is designated as the patrol leader. This temporary title defines his role and responsibilities for that mission. The patrol leader may designate an assistant, normally the next senior man in the patrol, and any subordinate element leaders he requires.

(5) ***Size.*** A patrol can be a unit as small as a fire team. Squad- and platoon-sized patrols are normal. Sometimes, for combat tasks such as a raid, the patrol can consist of most of the combat elements of a

rifle company. Unlike operations in which the Infantry platoon or squad is integrated into a larger organization, the patrol is semi-independent and relies on itself for security. Elements and teams for platoons conducting patrols include—

b. **Common Elements of Patrols.**

(1) ***Headquarters Element.*** The headquarters consists of the platoon leader, RTO, platoon sergeant (PSG), FO, RTO, and medic. It may include any attachments that the PL decides that he or the PSG must control directly.

(2) ***Aid and Litter Team.*** Aid and litter teams are responsible for buddy aid and evacuation of casualties.

(3) ***Enemy Prisoner of War Team.*** EPW teams control enemy prisoners using the five S's and the leader's guidance.

(4) ***Surveillance Team.*** The surveillance team keeps watch on the objective from the time that the leader's reconnaissance ends until the unit deploys for actions on the objective. They then rejoin their parent element.

(5) ***En Route Recorder.*** Part of the HQ element, maintains communications with higher and acts as the recorder for all CCIR collected during the mission.

(6) ***Compass Man.*** The compass man assists in navigation by ensuring the patrol remains on course at all times. Instructions to the compass man must include initial and subsequent azimuths. As a technique, the compass man should preset his compass on the initial azimuth before the unit moves out, especially if the move will be during limited visibility conditions. The platoon or squad leader should also designate an alternate compass man.

(7) ***Point/ Pace Man.*** As required, the PL designates a primary and alternate point man and a pace man for the patrol. The pace man aids in navigation by keeping an accurate count of distance traveled. The point man selects the actual route through the terrain, guided by the compass man or team leader. In addition, the point man also provides frontal security.

c. **Common Elements of Combat Patrols.**

(1) ***Assault Element.*** The assault element seizes and secures the objective and protects special teams as they complete their assigned actions on the objective.

(2) ***Security Element.*** The security element provides security at danger areas, secures the ORP, isolates the objective, and supports the withdrawal of the rest of the patrol once actions on the objective are complete. The security element may have separate security teams, each with an assigned task or sequence of tasks.

(3) ***Support Element.*** The support element provides direct and indirect fire support for the unit. Direct fires include machine guns, medium and light antiarmor weapons, small recoilless rifles. Indirect fires available may include mortars, artillery, CAS, and organic M203 weapon systems.

(4) ***Demolition Team.*** Demolition teams are responsible for preparing and detonating the charges to destroy designated equipment, vehicles, or facilities on the objective.

(5) ***EPW and Search Teams.*** The assault element may provide two-Ranger (buddy teams) or four-Ranger (fire team) search teams to search bunkers, buildings, or tunnels on the objective. These teams will search the objective or kill zone for any PIR that may give the PL an idea of the enemy concept for future operations. Primary and alternate teams may be assigned to ensure enough prepared personnel are available on the objective.

(6) ***Breach Element.*** The breach team conducts initial penetration of enemy obstacles to seize a foothold and allow the patrol to enter an objective. This is typically done IAW METT-TC and the steps outlined in the "Conduct an Initial Breach of a Mined Wire Obstacle" battle drill in Chapter 6 of this Handbook.

d. **Common Elements, Recon Patrols.**

(1) ***Reconnaissance Team***. Reconnaissance teams reconnoiter the objective area once the security teams are in position. Normally these are two-Ranger teams (buddy teams) to reduce the possibility of detection.

(2) ***Reconnaissance and Security Teams.*** R&S teams are normally used in a zone reconnaissance, but may be useful in any situation when it is impractical to separate the responsibilities for reconnaissance and security.

(3) ***Security Element.*** When the responsibilities of reconnaissance and security are separate, the security element provides security at danger areas, secures the ORP, isolates the objective, and supports the withdrawal of the rest of the platoon once the recon is complete. The security element may have separate security teams, each with an assigned task or sequence of tasks.

7-7 INITIAL PLANNING AND COORDINATION. Leaders plan and prepare for patrols using the troop leading procedures and the estimate of the situation, as described in Chapter 2. Through an estimate of the situation, leaders identify required actions on the objective (mission analysis) and plan backward to departure from friendly lines and forward to reentry of friendly lines. Because patrolling units act independently, move beyond the direct fire support of the parent unit, and operate forward of friendly units, coordination must be thorough and detailed. Coordination is continuous throughout planning and preparation. PLs use checklists to preclude omitting any items vital to the accomplishment of the mission.

a. **Coordination with Higher Headquarters**. This coordination includes intelligence, operations, and fire support IAW Chapter 2, Coordination Checklists (**page 2-34**). This initial coordination is an integral part of Step 3 of Troop Leading Procedures, Make a Tentative Plan.

b. **Coordination with Adjacent Units.** The leader also coordinates his unit's patrol activities with the leaders of other units that will be patrolling in adjacent areas at the same time, IAW Coordination Checklists (Chapter 7).

7-8. COMPLETION OF PLAN. As the PL completes his plan, he considers

a. **Specified and Implied Tasks.** The PL ensures that he has assigned all specified tasks to be performed on the objective, at rally points, at danger areas, at security or surveillance locations, along the route(s), and at passage lanes. These make up the maneuver and tasks to maneuver units subparagraphs of the Execution paragraph.

b. **Key Travel and Execution Times.** The leader estimates time requirements for movement to the objective, leader's reconnaissance of the objective, establishment of security and surveillance, completion of all assigned tasks on the objective, and passage through friendly lines. Some planning factors are

- Movement: Average of 1 kmph during daylight hours in woodland terrain; average limited visibility 1/2 kmph. Add additional time for restrictive, or severely restrictive terrain such as mountains, swamps, or thick vegetation.
- Leader's recon: NLT 1.5 hour.
- Establishment of security and surveillance: 0.5 hour.

c. **Primary and Alternate Routes.** The leader selects primary and alternate routes to and from the objective. The return routes should differ from the routes to the objective. The PL may delegate route selection to a subordinate, but is ultimately responsible for the routes selected.

d. **Signals.** The leader should consider the use of special signals. These include hand and arm signals, flares, voice, whistles, radios, and infrared equipment. Primary and alternate signals must be identified and rehearsed so that all Rangers know their meaning.

e. **Challenge and Password Forward of Friendly Lines.** The challenge and password from the unit's ANCD must not be used beyond the FLOT.

(1) ***Odd Number System.*** The leader specifies an odd number. The challenge can be any number less than the specified number. The password will be the number that must be added to it to equal the specified number, for example, the number is 7, the challenge is 3, and the password is 4.

(2) ***Running Password.*** ANCDs may also designate a running password. This code word alerts a unit that friendly Rangers are approaching in a less than organized manner and possibly under pressure. The number of Rangers approaching follows the running password. For example, if the running password is *"Ranger,"* and five friendly Rangers are approaching, they would say *"Ranger five."*

f. **Location of Leaders.** The PL considers where he and the PSG and other key leaders are located during each phase of the mission. The PL positions himself where he can best control the actions of the patrol. The PSG is normally located with the assault element during a raid or attack to help the PL control the use of additional assaulting squads, and will assist with securing the OBJ. The PSG will locate himself at the CCP to facilitate casualty treatment and evacuation. During a reconnaissance mission, the PSG will stay behind in the ORP to facilitate the transfer of Intel to the higher headquarters, and control the recon elements movement into and out of the ORP.

g. **Actions on Enemy Contact.** Unless required by the mission, the unit avoids enemy contact. The leader's plan must address actions on chance contact at each phase of the patrol mission. The unit's ability to continue will depend on how early contact is made, whether the platoon is able to break contact successfully (so that its subsequent direction of movement is undetected), and whether the unit receives any casualties because of the contact. The plan must address the handling of seriously wounded Rangers and KIAs. The plan must also address the handling of prisoners who are captured because of chance contact and are not part of the planned mission.

h. **Contingency Plans.** The leader leaves his unit for many reasons throughout the planning, coordination, preparation, and execution of his patrol mission. Each time the leader departs the patrol main body, he must issue a five point contingency plan to the leader left in charge of the unit. The patrol leader will additionally issue specific guidance stating what tasks are to be accomplished in the ORP in his absence. The contingency plan is remembered using the memory aid GOTWA shown in **Figure 7-1.**

Figure 7-1. GOTWA.

G *Where the leader is* **GOING.**
O **OTHERS** *he is taking with him.*
T **TIME** *he plans to be gone.*
W **WHAT** *to do if the leader does not return in time.*
A **ACTIONS** *by unit in the event contact is made while the leader is gone.*
(The leader will designate a stay-behind leader until he returns)

i. **Rally Points**. The leader considers the use and location of rally points. A rally point is a place designated by the leader where the unit moves to reassemble and reorganize if it becomes dispersed. Rangers must know which rally point to move to at each phase of the patrol mission should they become separated from the unit. They must also know what actions are required there and how long they are to wait at each rally point before moving to another.

(1) ***Criteria.*** Rally points must be

- Easily identifiable in daylight and limited visibility.
- Show no signs of recent enemy activity.
- Covered and concealed.
- Away from natural lines of drift and high speed avenues of approach.
- Defendable for short periods of time.

(2) ***Types.*** The most common types of rally points include initial, en route, objective, and near and far side rally points.

j. **Objective Rally Point.** The ORP typically lies 200 to 400m from the objective, or at a minimum, one major terrain feature away. Actions at the ORP include

- Conduct SLLS and pinpoint location.
- Conduct a leaders recon of the objective.
- Issue a FRAGO, if needed.
- Make final preparations before continuing operations, for example, recamouflage, prepare demolitions, line up rucksacks for quick recovery; prepare EPW bindings, first aid kits, and litters; and inspect weapons.
- Account for Rangers and equipment after completing actions at the objective.
- Reestablish the chain of command after actions at the objective are complete.
- Disseminate information from reconnaissance, if no contact was made.

k. **Leader's Reconnaissance of the Objective.** The plan must include a leader's reconnaissance of the objective once the platoon or squad establishes the ORP. Before departing, the leader must issue a 5 point contingency plan. During his reconnaissance, the leader pinpoints the objective, selects reconnaissance, security, support, and assault positions for his elements, and adjusts his plan based on his observation of the objective. Each type of patrol requires different tasks during the leader's reconnaissance. The platoon leader will bring different elements with him. (These are discussed separately under each type of patrol). The leader must plan time to return to the ORP, complete his plan, disseminate information, issue orders and instructions, and allow his squads to make any additional preparations. During the Leader's Reconnaissance for a Raid or Ambush, the PL will leave surveillance on the OBJ.

l. **Actions on the Objective.** Each type of patrol requires different actions on the objective. Actions on the objective are discussed under each type of patrol.

Section III. RECONNAISSANCE PATROLS

This section discusses the fundamentals of reconnaissance, task standards for the two most common types of recon (area and zone), and actions on the objective for those types of recon. Both types of recon patrol provide timely and accurate information on the enemy and terrain and confirm the leader's plan before it is executed. Units on reconnaissance operations collect specific information (*priority intelligence requirements* [PIR]) or general information *(information requirements* [IR]) based on the instructions from their higher commander.

7-9. FUNDAMENTALS OF RECONNAISSANCE. In order to have a successful area reconnaissance, the platoon leader applies the fundamentals of the reconnaissance to his plan during the conduct of the operation.

a. **Obtain Required Information.** The parent unit tells the patrol leader what information is required. This is in the form of the information requirement and priority intelligence requirements. The platoon's mission is then tailored to what information is required. During the entire patrol, members must continuously gain and exchange all information gathered, but cannot consider the mission accomplished unless all PIR has been gathered.

b. **Avoid Detection by the Enemy.** A patrol avoids letting the enemy know that it is in the objective area. If the enemy knows he is being observed, he may move, change his plans, or increase his security measures. Methods of avoiding detection are

- Minimize movement in the objective area (area reconnaissance).
- Move no closer to the enemy than necessary.
- If possible, use long range surveillance or night vision devices.
- Use camouflage, stealth, and noise and light discipline.
- Minimize radio traffic.

c. **Employ Security Measures.** A patrol must be able to break contact and return to the friendly unit with what information is gathered. If necessary, they break contact and continue the mission. Leaders emplace security elements where they can overwatch the reconnaissance elements. They suppress the enemy so the reconnaissance element can break contact.

d. **Task Organize.** When the platoon leader receives the order, he analyzes his mission to ensure he understands what must be done. Then he task organizes his platoon to best accomplish the mission IAW METT-TC. Reconnaissances are typically squad sized missions.

7-10. ASK STANDARDS

a. ***Area Reconnaissance.*** The area recon patrol collects all available information on PIR and other intelligence not specified in the order for the area. The patrol completes the recon and reports all information by the time specified in the order. The patrol is not compromised.

b. ***Zone Reconnaissance.*** The zone recon patrol determines all PIR and other intelligence not specified in the order for its assigned zone. The patrol reconnoiters without detection by the enemy. The patrol completes the recon and reports all information by the time specified in the order.

7-11. ACTIONS ON THE OBJECTIVE, AREA RECONNAISSANCE. The element occupies the ORP as discussed in the section on occupation of the ORP (**A, Figure 7-2**). The RTO reports to higher that the unit has occupied the ORP. The leader confirms his location on map while subordinate leaders make necessary perimeter adjustments. The PL organizes the platoon in one of two ways: separate recon and security elements, or combined recon and security elements.

a. The PL takes subordinates leaders and key personnel on a leader's recon to confirm the objective and plan.

(1) Issues a 5 point contingency plan before departure.

(2) Establishes a suitable release point that is beyond sight and sound of the objective if possible, but that is definitely out of sight. The RP should also have good rally point characteristics.

(3) Allows all personnel to become familiar with the release point and surrounding area.

(4) Identifies the objective and emplaces surveillance. Designates a surveillance team to keep the objective under surveillance. Issues a contingency plan to the senior man remaining with the surveillance team. The surveillance team is positioned with one man facing the objective, and one facing back in the direction of the release point.

(5) Takes subordinate leaders forward to pinpoint the objective, emplace surveillance, establish a limit of advance, and choose vantage points.

(6) Maintains communications with the platoon throughout the leader's recon.

b. The PSG maintains security and supervises priorities of work in the ORP.

(1) Reestablishes security at the ORP.

(2) Disseminates the PLs contingency plan.

(3) Oversees preparation of recon personnel (personnel recamouflaged, NVDs and binoculars prepared, weapons on safe with a round in the chamber).

c. The PL and his recon party return to the ORP.

(1) Confirms the plan or issues a FRAGO.

(2) Allows subordinate leaders time to disseminate the plan.

d. The patrol conducts the recon by long range observation and surveillance if possible. R&S element (**B, Figure 7-2**).

(1) Moves to observation points that offer cover and concealment and that are outside of small arms range.

(2) Establishes a series of observation posts (OP) if information cannot be gathered from one location.

(3) Gathers all PIR using the SALUTE format.

e. If necessary, the patrol conducts its recon by short range observation and surveillance (**C, Figure 7-2**).

(1) Moves to an OP near the objective.

(2) Passes close enough to the objective to gain information.

(3) Gathers all PIR using the SALUTE format.

f. R&S teams move using a technique such as the cloverleaf method to move to successive OPs (**D, Figure 7-2**). In this method, R&S teams avoid paralleling the objective site, maintain extreme stealth, do not cross the limit of advance, and maximize the use of available cover and concealment.

g. During the conduct of the recon, each R&S team returns to the release point when any of the following occurs (**E, Figure 7-2**):

- They have gathered all their PIR.
- They have reached the limit of advance.
- The allocated time to conduct the recon has elapsed.
- Contact has been made.

h. At the release point, the leader analyzes what information has been gathered and determines if he has met the PIR requirements.

i. If the leader determines that he has not gathered sufficient information to meet the PIR requirements, or if the information he and the subordinate leader gathered differs drastically, he may have to send R&S teams back to the objective site. In this case, R&S teams alternate areas of responsibilities. For example, if one team reconnoitered from the 6 – 3 – 12, then that team will now recon from the 6 – 9 – 12.

j. The R&S element returns undetected to the ORP by the specified time.

(1) Disseminates information to all patrol members through key leaders at the ORP, or moves to a position at least one terrain feature or one kilometer away to disseminate. To disseminate, the leader has the RTO prepare three sketches of the objective site based on the leader's sketch and provides the copies to the subordinate leaders to assist in dissemination.

(2) Reports any information requirements and/ or any information requiring immediate attention to higher headquarters, and departs for the designated area.

k. If contact is made, the R&S element moves to the release point. The recon element tries to break contact and return to the ORP, secure rucksacks, and quickly move out of the area. Once they have moved a safe distance away, the leader informs higher HQ of the situation and takes further instructions from them.

(1) While emplacing surveillance, the recon element withdraws through the release point to the ORP, and follows the same procedures as above.

(2) While conducting the reconnaissance, the compromised element returns a sufficient volume of fire to allow them to break contact. Surveillance can fire an AT 4 at the largest weapon on the objective. All elements pull off the objective and move to the release point. The senior man quickly accounts for all personnel and return to the ORP. Once in the ORP, leadership follows the procedures previously described. **Figure 7-3** shows the critical tasks for a patrol.

Figure 7-2. ACTIONS ON THE OBJECTIVE, AREA RECONNAISSANCE

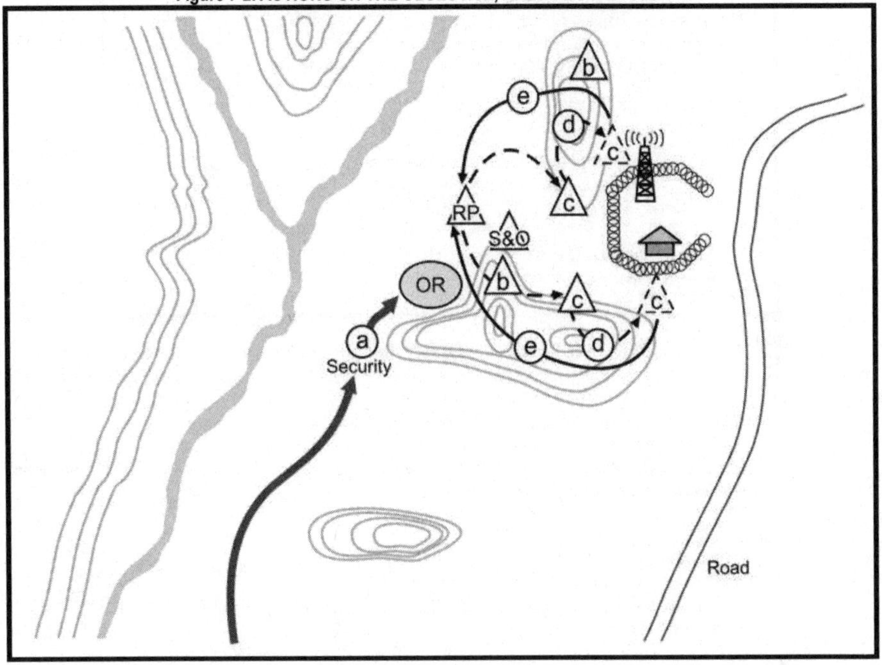

Figure 7-3. ACTIONS ON THE OBJECTIVE, AREA RECONNAISSANCE

Secure and occupy ORP. Conduct a leader's reconnaissance of the objective: – Estimate release point. – Pinpoint objective. – Estimate surveillance (S&O team). Position security element if used. Conduct reconnaissance by long-range surveillance if possible. Conduct reconnaissance by short-range surveillance if necessary.
Teams Move as necessary to successive observation posts. On order, return to release point. Once PIR is gathered, return to ORP. Patrol – Link up as directed in ORP. – Disseminate info before moving.

7-12. ACTIONS ON THE OBJECTIVE, ZONE RECONNAISSANCE. The element occupies the initial ORP as discussed in the section occupation of the ORP. The radio operator calls in spare for occupation of ORP. The leader confirms his location on map while subordinate leaders make necessary perimeter adjustments.

a. **Organization.** The recon team leaders organize their recon elements.
   (1) Designate security and recon elements.
   (2) Assign responsibilities (point man, pace man, en route recorder, and rear security), if not already assigned.
   (3) Designates easily recognizable rally points.
   (4) Ensure local security at all halts.

b. **Actions.** The patrol reconnoiters the zone.
   (1) Moves tactically to the ORPs.
   (2) Occupies designated ORPs.
   (3) Follows the method designated by the PL (fan, converging routes, or box method, **Table 7-1**).
   (4) The recon teams reconnoiter.
   - During movement, the squad will gather all PIR specified by the order.
   - Recon team leaders will ensure sketches are drawn or digital photos are taken of all enemy hard sites, roads, and trails.
   - Return to the ORP, or link up at the rendezvous point on time.
   - When the squad arrives at new rendezvous point or ORP, the recon team leaders report to the PL with all information gathered.

   (5) The PL continues to control the recon elements.
   - PL moves with the recon element that establishes the rendezvous point.
   - PL changes recon methods as required.
   - PL designates times for the elements to return to the ORP or to linkup.
   - PL collects all information and disseminates it to the entire patrol. PL will brief all key subordinate leaders on information gathered by other squads, establishing one consolidated sketch if possible, and allow team leaders time to brief their teams.
   - PL and PSG account for all personnel.

(6) The patrol continues the reconnaissance until all designated areas have been reconnoitered, and returns undetected to friendly lines.

Table 7 1. COMPARISON OF ZONE RECONNAISSANCE METHODS

FAN METHOD	*CONVERGING ROUTES METHOD*	*BOX METHOD*
• Uses a series (fan) of ORPs. • Patrol establishes security at first ORP. • Each recon element moves from ORP along a different fan-shaped route. Route overlaps with that of other recon elements. This ensures recon of entire area. • Leader maintains reserve at ORP. • When all recon elements return to ORP, PL collects and disseminates all info before moving to next ORP.	• PL selects routes from ORP thru zone to a rendezvous point at the far side of the zone from the ORP. • Each recon element moves and reconnoiters along a specified route. They converge (link up) at one time and place.	• PL sends recon elements from the first ORP along routes that form a box. He sends other elements along routes throughout the box. All teams link up at the far side of the box from the ORP.

Section IV. COMBAT PATROLS

Combat patrols are the second type of patrol. Combat patrols are further divided into raids, ambushes, and security patrols. Units conduct combat patrols to destroy or capture enemy soldiers or equipment; destroy installations, facilities, or key points; or harass enemy forces. Combat patrols also provide security for larger units. This section describes overall combat patrol planning considerations, task considerations for each type of combat patrol, and finally actions on the objective for each type.

7-13. PLANNING CONSIDERATIONS. In planning a combat patrol, the PL considers the following:

a. **Tasks to Maneuver Units.** Normally the platoon headquarters element controls the patrol on a combat patrol mission. The PL makes every try to maintain squad and fire team integrity as he assigns tasks to subordinates units.

(1) The PL must consider the requirements for assaulting the objective, supporting the assault by fire, and security of the entire unit throughout the mission.

- For the assault on the objective, the PL considers the required actions on the objective, the size of the objective, and the known or presumed strength and disposition of the enemy on and near the objective.
- The PL considers the weapons available, and the type and volume of fires required to provide fire support for the assault on the objective.
- The PL considers the requirement to secure the platoon at points along the route, at danger areas, at the ORP, along enemy avenues of approach into the objective, and elsewhere during the mission.
- The PL will also designate engagement/ disengagement criteria.

(2) The PL assigns additional tasks to his squads for demolition, search of EPWs, guarding of EPWs, treatment and evacuation (litter teams) of friendly casualties, and other tasks required for successful completion of patrol mission (if not already in the SOP).

(3) The PL determines who will control any attachments of skilled personnel or special equipment.

b. **Leader's Reconnaissance of the Objective.** In a combat patrol, the PL has additional considerations for the conduct of his reconnaissance of the objective from the ORP.

(1) ***Composition of the Leader's Reconnaissance Party.*** The platoon leader will normally bring the following personnel.

- Squad leaders to include the weapons squad leader.
- Surveillance team.
- Forward observer.

- Security element (dependent on time available).

(2) ***Conduct of the Leader's Reconnaissance.*** In a combat patrol, the PL considers the following additional actions in the conduct of the leader's reconnaissance of the objective.

- The PL designates a release point about half way between the ORP and this objective. The PL posts the surveillance team. Squads and fire teams separate at the release point, and then they move to their assigned positions.
- The PL confirms the location of the objective or kill zone. He NOTES the terrain and identifies where he can emplace claymores to cover dead space. Any change to his plan is issued to the squad leaders (while overlooking the objective if possible).
- If the objective is the kill zone for an ambush, the leader's reconnaissance party should not cross the objective; to do so will leave tracks that may compromise the mission.
- The PL confirms the suitability of the assault and support positions and routes from them back to the ORP.
- The PL issues a five point contingency plan before returning to the ORP.

7-14. AMBUSH. An ambush is a surprise attack from a concealed position on a moving or temporarily halted target. Ambushes are categorized as either hasty or deliberate and divided into two types, point or area; and formation linear or L shaped. The leader considers various key factors in determining the ambush category, type, and formation, and from these decisions, develops his ambush plan.

a. **Key Factors.**

- Coverage (ideally whole kill zone) by fire.
- METT-TC.
- Existing or reinforcing obstacles, including claymores, to keep the enemy in the kill zone.
- Security teams, who typically have hand held antitank weapons such as AT 4s or LAWs, claymores, and various means of communication.
- Security elements or teams to isolate the kill zone.
- Protection of the assault and support elements with claymores or explosives.
- Assault through the kill zone to the limit of advance (LOA).

> **NOTE:** The assault element must be able to move quickly through its own protective obstacles.

- Time the actions of all elements of the platoon to preclude loss of surprise. In the event any member of the ambush is compromised, he may immediately initiate the ambush.
- When the ambush must be manned for a long time, use only one squad to conduct the entire ambush and determining movement time of rotating squads from the ORP to the ambush site.

b. **Categories.**

- ***Hasty***. A unit conducts a hasty ambush when it makes visual contact with an enemy force and has time to establish an ambush without being detected. The actions for a hasty ambush must be well rehearsed so that Rangers know what to do on the leader's signal. They must also know what action to take if the unit is detected before it is ready to initiate the ambush.
- ***Deliberate.*** A deliberate ambush is conducted at a predetermined location against any enemy element that meets the commander's engagement criteria. The leader requires the following detailed information in planning a deliberate ambush: size and composition of the targeted enemy, and weapons and equipment available to the enemy.

c. **Types.**

- ***Point.*** In a point ambush, Rangers deploy to attack an enemy in a single kill zone.
- ***Area.*** In an area, Rangers deploy in two or more related point ambushes.

d. **Formations (Figure 7-4).**

- ***Linear.*** In an ambush using a linear formation, the assault and support elements deploy parallel to the enemy's route. This positions both elements on the long axis of the kill zone and subjects the enemy to flanking fire. This formation can be used in close terrain that restricts the enemy's ability to maneuver against the platoon, or in open terrain provided a means of keeping the enemy in the kill zone can be effected.

• ***L Shaped***. In an L shaped ambush, the assault element forms the long leg parallel to the enemy's direction of movement along the kill zone. The support element forms the short leg at one end of and at right angles to the assault element. This provides both flanking (long leg) and enfilading fires (short leg) against the enemy. The L shaped ambush can be used at a sharp bend in a trail, road, or stream. It should not be used where the short leg would have to cross a straight road or trail.

Figure 7-4. AMBUSH FORMATIONS

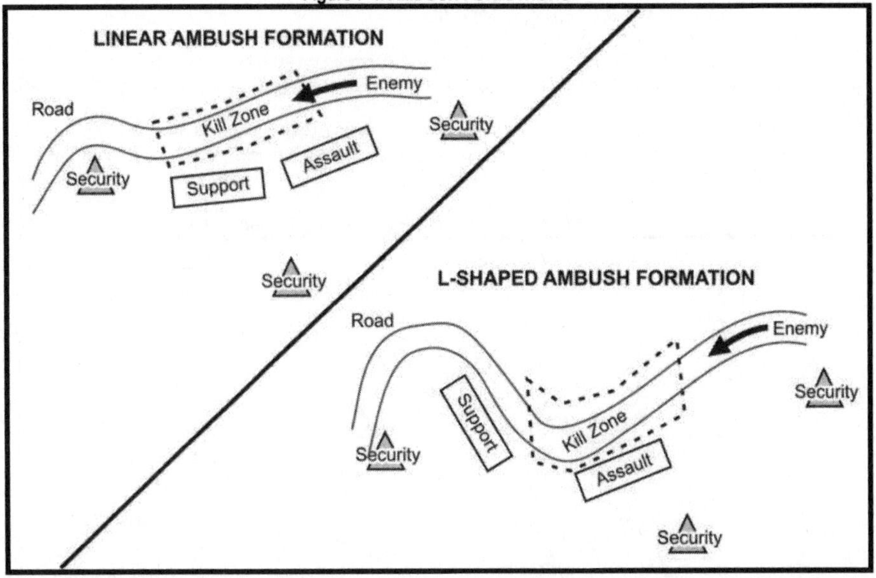

7-15 HASTY AMBUSH. The platoon moves quickly to concealed positions. The ambush is not initiated until the majority of the enemy is in the kill zone. The unit does not become decisively engaged. The platoon surprises the enemy. The patrol captures, kills, or forces the withdrawal of the entire enemy within the kill zone. On order, the patrol withdraws all personnel and equipment in the kill zone from observation and direct fire. The unit does not become decisively engaged by follow on elements. The platoon continues follow on operations. Actions on the objective follow (**Figure 7-5**).

a. Using visual signals, any Ranger alerts the unit that an enemy force is in sight. The Ranger continues to monitor the location and activities of the enemy force until his team or squad leader relieves him, and gives the enemy location and direction of movement.

b. The platoon or squad halts and remains motionless.

- The PL gives the signal to conduct a hasty ambush, taking care not to alert the enemy of the patrol's presence.
- The leader determines the best nearby location for a hasty ambush. He uses arm and hand signals to direct the unit members to covered and concealed positions.

c. The leader designates the location and extent of the kill zone.

d. Teams and squads move silently to covered and concealed positions, ensuring positions are undetected and have good observation and fields of fire into the kill zone.

e. Security elements move out to cover each flank and the rear of the unit. The leader directs the security elements to move a given distance, set up, and then rejoin the unit on order or, after the ambush (the sound of firing ceases). At squad level, the two outside buddy teams normally provide flank security as well as fires into the kill zone. At platoon level, fire teams make up the security elements.

f. The PL assigns sectors of fire and issues any other commands necessary such as control measures.

g. The PL initiates the ambush, using the greatest casualty producing weapon available, when the largest percentage of enemy is in the kill zone. The PL

- Controls the rate and distribution of fire.
- Employs indirect fire to support the ambush.
- Orders cease fire.
- (If the situation dictates) orders the patrol to assault through the kill zone.

h. The PL designates personnel to conduct a hasty search of enemy personnel and process enemy prisoners and equipment.

i. The PL orders the platoon to withdraw from the ambush site along a covered and concealed route.

j. The PL gains accountability, reorganizes as necessary, disseminates information, reports the situation, and continues the mission as directed.

Figure 7-5. ACTIONS ON THE OBJECTIVE—HASTY AMBUSH

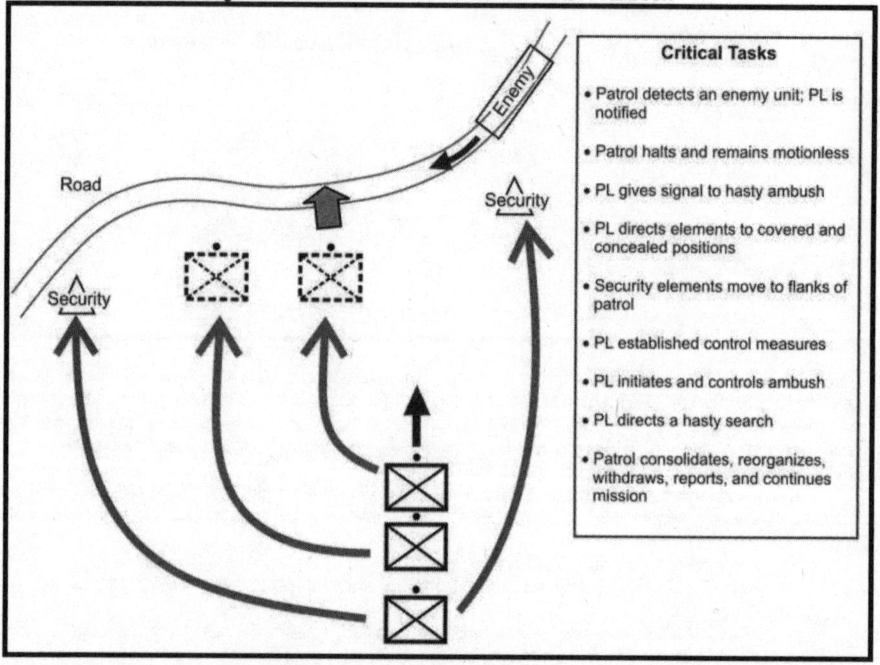

7-16. DELIBERATE (POINT/ AREA) AMBUSH. The ambush is emplaced NLT the time specified in the order. The patrol surprises the enemy and engages the enemy main body. The patrol kills or captures all enemy in the kill zone and destroys equipment based on the commander's intent. The patrol withdraws all personnel and equipment from the objective, on order, within the time specified in the order. The patrol obtains all available PIR from the ambush and continues follow on operations. Actions on the objective follow (**Figure 7-6**).

a. The PL prepares the patrol for the ambush in the ORP.

b. The PL prepares to conduct a leader's reconnaissance. He

- Designates the members of the leader's recon party (typically includes squad leaders, surveillance team, FO, and possibly the security element.
- Issues a contingency plan to the PSG.

c. The PL conducts his leader's reconnaissance. He

- Ensures the leader's recon party moves undetected.
- Confirms the objective location and suitability for the ambush.
- Selects a kill zone.
- Posts the surveillance team at the site and issues a contingency plan.
- Confirms suitability of assault and support positions, and routes from them to the ORP.
- Selects position of each weapon system in support by fire position, then designates sectors of fire.
- Identifies all offensive control measures to be used. Identifies the probably line of deployment (PLD), the assault position, LOA, any boundaries or other control measures. If available, the PL can use infrared aiming devices to identify these positions on the ground.

d. The PL adjusts his plan based on info from the reconnaissance. He

- Assigns positions.
- Designates withdrawal routes.

e. The PL confirms the ambush formation.

f. The security team(s) occupy first, securing the flanks of the ambush site, and providing early warning. The security element must be in position before the support and assault elements move forward of the release point. A security team remains in the ORP if the patrol plans to return to the ORP after actions on the objective. If the ORP is abandoned, a rear security team should be emplaced.

g. Support element leader assigns sectors of fire. He

- Emplaces claymores and obstacles as designated.
- Identifies sectors of fire and emplaces limiting stakes to prevent friendly fires from hitting other elements.
- Overwatches the movement of the assault element into position.

h. Once the support element is in position, or on the PLs order, the assault element

- Departs the ORP and moves into position.
- Upon reaching the PLD, the assault element transitions from the movement formation to the battle formation.
- Identifies individual sectors of fire as assigned by the PL. Emplaces aiming stakes or uses metal to metal contact with the machine gun tripods to prevent fratricide on the objective.
- Emplaces claymores to help destroy the enemy in the kill zone.
- Camouflages positions.

i. The security element spots the enemy and notifies the PL, and reports the direction of movement, size of the target, and any special weapons or equipment carried. The security element also keeps the platoon leader informed if any enemy forces are following the lead force.

j. The PL alerts other elements, and determines if the enemy force is too large, or if the ambush can engage the enemy successfully.

k. The PL initiates the ambush using the highest casualty producing device. He may use a command detonated claymore. He must also plan a backup method for initiating the ambush, in case his primary means fails. This should also be a casualty producing device such as his individual weapon. He passes this information to all Rangers, and practices it during rehearsals.

l. The PL ensures that the assault and support elements deliver fire with the heaviest, most accurate volume possible on the enemy in the kill zone. In limited visibility, the PL may use infrared lasers to further define specific targets in the kill zone.

m. Before assaulting the target, the PL gives the signal to lift or shift fires.

n. The assault element—

- Assaults before the remaining enemy can react.
- Kills or captures enemy in the kill zone.
- Uses individual movement techniques or bounds by fire teams to move.
- Upon reaching the limit of advance, halts and establishes security. If needed, it reestablishes the chain of command and remains key weapon systems. All Rangers will load a fresh magazine or drum of ammunition using the buddy system. ACE reports will be submitted through the chain of command. The PL will submit an initial contact report to higher.

o. The PL directs special teams (EPW search, aid and litter, demo) to accomplish their assigned tasks once the assault element has established its LOA.

- Once the kill zone is clear, collect and secure all EPWs and move them out of the kill zone before searching their bodies. Coordinate for an EPW exchange point to link up with higher to extract all EPWs and treat them IAW the five S's.
- Search from one side to the other and mark bodies that have been searched to ensure the area is thoroughly covered. Units should use the "clear out, search in" technique, clear from the center of the objective out ensuring the area is clear of all enemy combatants; then search all enemy personnel towards the center of the objective. Search all dead enemy personnel using the two-Ranger search technique:
  - As the search team approaches a dead enemy soldier, one Ranger guards while the other Ranger searches. First, the Ranger kicks the enemy's weapon away.
  - Second, he rolls the body over (if on the stomach) by lying on top and when given the go ahead by the guard (who is positioned at the enemy's head), the searcher rolls the body over on him. This is done for protection in case the enemy soldier has a grenade with the pin pulled underneath him.
  - The searchers then conduct a systematic search of the dead soldier from head to toe removing all papers and anything new (different type rank, shoulder boards, different unit patch, pistol, weapon, or NVD). They note if the enemy has a fresh or shabby haircut and the condition of his uniform and boots. They note the radio frequency, and then they secure the SOI, maps, documents, and overlays.
  - Once the body has been thoroughly searched, the search team will continue in this manner until all enemy personnel in and near the kill zone have been searched.
- Identify, collect, and prepare all equipment to be carried back or destroyed.
- Evacuate and treat friendly wounded first, then enemy wounded, time permitting.
- The demolition team prepares dual primed explosives or incendiary grenades and awaits the signal to initiate. This is normally the last action performed before the unit departs the objective and may signal the security elements to return to the ORP.
- Actions on the objective with stationary assault line; all actions are the same with the exception of the search teams. To provide security within the teams to the far side of the kill zone during the search, they work in three Ranger teams. Before the search begins, the Rangers move all KIAs to the near side of the kill zone.

p. If enemy reinforcements try to penetrate the kill zone, the flank security will engage to prevent the assault element from being compromised.

q. The platoon leader directs the unit's withdrawal from the ambush site:

- Elements normally withdraw in the reverse order that they established their positions.
- The elements may return to the RP or directly to the ORP, depending on the distance between elements.
- The security element of the ORP must be alert to assist the platoon's return to the ORP. It maintains security for the ORP while the rest of the platoon prepares to leave.
- If possible, all elements should return to the location at which they separated from the main body. This location should usually be the RP.

r. The PL and PSG direct actions at the ORP, to include accountability of personnel and equipment and recovery of rucksacks and other equipment left at the ORP during the ambush.

s. The platoon leader disseminates information, or moves the platoon to a safe location (no less than one kilometer or one terrain feature away from the objective) and disseminates information.

t. As required, the PL and FO execute indirect fires to cover the platoon's withdrawal.

Figure 7-6. ACTIONS ON THE OBJECTIVE–DELIBERATE AMBUSH

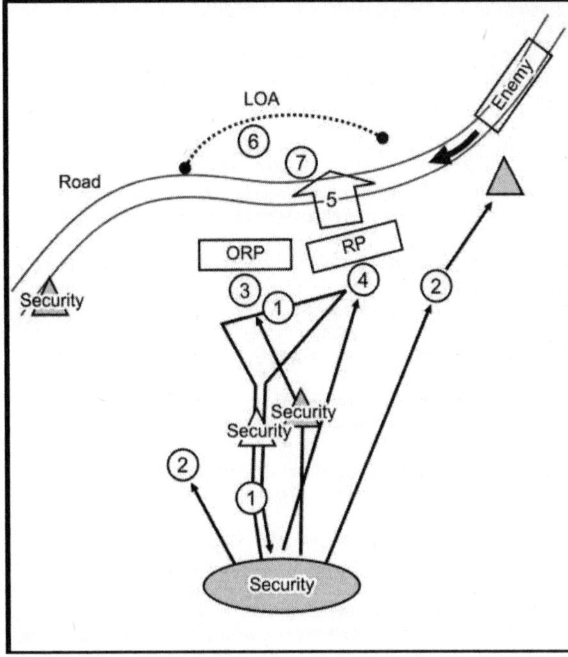

Critical Tasks

- Secure and occupy ORP
- Recon OBJ (Kill Zone) (1)
- Emplace security elements (2)
- Emplace support elements (3)
- Emplace assault elements (4)
- Security notifies PL of enemy
- PL initiates ambush
- Support lifts/shifts fire
- Assault element assaults (5)
- Establish LOA/Security (6)
- Consolidate/Reorganize (7)
  - Reposition as required
  - Search kill zone
  - Treat wounded
- Assault withdraws
- Support withdraws
- Security withdraws
- Patrol consolidates in ORP

7-17. PERFORM RAID. The patrol initiates the raid NLT the time specified in the order, surprises the enemy, assaults the objective, and accomplishes its assigned mission within the commander's intent. The patrol does not become decisively engaged en route to the objective. The patrol obtains all available PIR from the raid objective and continues follow on operations.

a. **Planning Considerations.** A raid is a form of attack, usually small scale, involving a swift entry into hostile territory to secure information, confuse the enemy, or destroy installations followed by a planned withdrawal. Squads do not conduct raids. The sequence of platoon actions for a raid is similar to those for an ambush. Additionally, the assault element of the platoon may have to conduct a breach of an obstacle. It may have additional tasks to perform on the objective such as demolition of fixed facilities. Fundamentals of the raid include

- Surprise and speed. Infiltrate and surprise the enemy without being detected.
- Coordinated fires. Seal off the objective with well synchronized direct and indirect fires.
- Violence of action. Overwhelm the enemy with fire and maneuver.
- Planned withdrawal. Withdraw from the objective in an organized manner, maintaining security.

b. **Actions on the Objective (Raid) (Figure 7-7).**

(1) The patrol moves to and occupies the ORP IAW the patrol SOP. The patrol prepares for the leader's recon.

(2) The PL, squad leaders, and selected personnel conduct a leader's recon.

- PL leaves a five-point contingency plan with the PSG.
- PL establishes the RP, pinpoints the objective, contacts the PSG to prep men, weapons, and equipment, emplaces the surveillance team to observe the objective, and verifies and updates intelligence information. Upon emplacing the surveillance team, the PL will provide a five-point contingency plan.

- Leader's recon verifies location of and routes to security, support, and assault positions.
- Security teams are brought forward on the leader's reconnaissance and emplaced before the leader's recon leaves the RP.
- Leaders conduct the recon without compromising the patrol.
- Leaders normally recon support-by-fire position first, then the assault position.

(3) The PL confirms, denies, or modifies his plan and issues instructions to his squad leaders.

- Assigns positions and withdrawal routes to all elements.
- Designates control measures on the objective (element objectives, lanes, limits of advance, target reference points, and assault line).
- Allows SLs time to disseminate information, and confirm that their elements are ready.

(4) Security elements occupy designated positions, moving undetected into positions that provide early warning and can seal off the objective from outside support or reinforcement.

(5) The support element leader moves the support element to designated positions. The support element leader ensures his element can place well aimed fire on the objective.

(6) The PL moves with the assault element into the assault position. The assault position is normally the last covered and concealed position before reaching the objective. As it passes through the assault position the platoon deploys into its assault formation; that is, its squads and fire teams deploy to place the bulk of their firepower to the front as they assault the objective.

- Makes contact with the surveillance team to confirm any enemy activity on the objective.
- Ensures that the assault position is close enough for immediate assault if the assault element is detected early.
- Moves into position undetected, and establish local security and fire-control measures.

(7) Element leaders inform the PL when their elements are in position and ready.

(8) The PL directs the support element to fire.

(9) Upon gaining fire superiority, the PL directs the assault element to move towards the objective.

- Assault element holds fire until engaged, or until ready to penetrate the objective.
- PL signals the support element to lift or shift fires. The support element lifts or shifts fires as directed, shifting fire to the flanks of targets or areas as directed in the FRAGO.

(10) The assault element attacks and secures the objective. The assault element may be required to breech a wire obstacle. As the platoon, or its assault element, moves onto the objective, it must increase the volume and accuracy of fires. Squad leaders assign specific targets or objectives for their fire teams. Only when these direct fires keep the enemy suppressed can the rest of the unit maneuver. As the assault element gets closer to the enemy, there is more emphasis on suppression and less on maneuver. Ultimately, all but one fire team may be suppressing to allow that one fire team to break into the enemy position. Throughout the assault, Rangers use proper individual movement techniques, and fire teams retain their basic shallow wedge formation. The platoon does not get "on line" to sweep across the objective.

- Assault element assaults through the objective to the designated LOA.
- Assault element leaders establish local security along the LOA, and consolidate and reorganize as necessary. They provide ACE reports to the PL and PSG. The platoon establishes security, operates key weapons, provides first aid, and prepares wounded Rangers for MEDEVAC. They redistribute ammunition and supplies, and they relocate selected weapons to alternate positions if leaders believe that the enemy may have pinpointed them during the attack. They adjust other positions for mutual support. The squad and team leader provide ammunition, casualty, and equipment (ACE) reports to the platoon leader. The PL/PSG reorganizes the patrol based on the contact.
  - On order, special teams accomplish all assigned tasks under the supervision of the PL, who positions himself where he can control the patrol.
  - Special team leaders report to PL when assigned tasks are complete.

(11) On order or signal of the PL, the assault element withdraws from the objective. Using prearranged signals, the assault line begins an organized withdrawal from the objective site, maintaining control and security throughout the withdrawal. The assault element bounds back near the original assault line, and begin a single file withdrawal through the APL's choke point. All Rangers must move through the choke point for an accurate count. Once the assault element is a safe distance from the objective and the headcount is confirmed, the platoon can withdraw the support element. If the support elements were a part of the assault line, they withdraw together, and security is signaled to withdraw. Once the support is a safe distance off the objective, they notify the platoon leader, who contacts the security element and signals them to withdraw. All security teams link up at the release point and notify the platoon leader before moving to the ORP. Personnel returning to the ORP immediately secure their equipment and establish all round security. Once the security element returns, the platoon moves out of the objective area as soon as possible, normally in two to three minutes.

- Before withdrawing, the demo team activates demo devices and charges.
- Support element or designated personnel in the assault element maintain local security during the withdrawal.
- Leaders report updated accountability and status (ACE report) to the PL and PSG.

(12) Squads withdraw from the objective in the order designated in the FRAGO to the ORP.

- Account for personnel and equipment.
- Disseminate information.
- Redistribute ammunition and equipment as required.

(13) The PL reports mission accomplishment to higher and continues the mission.

- Reports raid assessment to higher.
- Informs higher of any IR/ PIR gathered.

Figure 7-7. ACTIONS ON THE OBJECTIVE-RAID

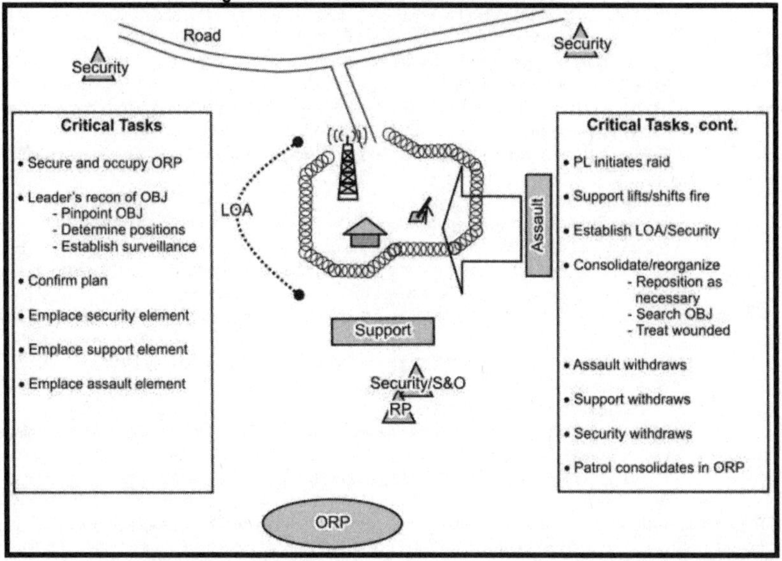

Section V. SUPPORTING TASKS

This section covers linkup, patrol debriefing, and occupation of an ORP.

7-18. LINKUP. A linkup is a meeting of friendly ground forces. Linkups depend on control, detailed planning, communications, and stealth.

a. **Task Standard.** The units link up at the time and place specified in the order. The enemy does not surprise the main bodies. The linkup units establish a consolidated chain of command.

b. **Site Selection.** The leader identifies a tentative linkup site by map reconnaissance, other imagery, or higher headquarters designates a linkup site. The linkup site should have the following characteristics:

- Ease of recognition.
- Cover and concealment.
- No tactical value to the enemy.
- Location away from natural lines of drift.
- Defendable for a short period of time.
- Multiple access and escape routes.

c. **Execution.** Linkup procedure begins as the unit moves to the linkup point. The steps of this procedure are–

(1) The stationary unit performs linkup actions.

- Occupies the linkup rally point (LRP) NLT the time specified in the order.
- Establishes all-round security, establishes communications, and prepares to accept the moving unit.
- The security team clears the immediate area around the linkup point. It then marks the linkup point with the coordinated recognition signal. The security team moves to a covered and concealed position and observes the linkup point and immediate area around it.

(2) The moving unit—

- Performs linkup actions.
- The unit reports its location using phase lines, checkpoints, or other control measures.
- Halts at a safe distance from the linkup point in a covered and concealed position (the linkup rally point).

(3) The PL and a contact team—

- Prepare to make physical contact with the stationary unit.
- Issue a contingency plan to the PSG.
- Maintain communications with the platoon; verify near and far recognition signals for linkup (good visibility and limited visibility).
- Exchange far and near recognition signals with the linkup unit; conduct final coordination with the linkup unit.

(4) The stationary unit

- Guides the patrol from its linkup rally point to the stationary unit linkup rally point.
- Linkup is complete by the time specified in the order.
- The main body of the stationary unit is alerted before the moving unit is brought forward.

(5) The patrol continues its mission IAW the order.

d. **Coordination Checklist.** The PL coordinates or obtains the following information from the unit that his patrol will link up with—

- Exchange frequencies, call signs, codes, and other communication information.
- Verify near and far recognition signals.
- Exchange fire coordination measures.
- Determine command relationship with the linkup unit; plan for consolidation of chain of command.
- Plan actions following linkup.
- Exchange control measures such as contact points, phase lines, contact points, as appropriate.

7-19. DEBRIEF. Immediately after the platoon or squad returns, personnel from higher headquarters conduct a thorough debrief. This may include all members of the platoon or the leaders, RTOs, and any attached personnel. Normally the debriefing is oral. Sometimes a written report is required. Information on the written report should include

- Size and composition of the unit conducting the patrol.
- Mission of the platoon such as type of patrol, location, and purpose.
- Departure and return times.
- Routes. Use checkpoints, grid coordinates for each leg or include an overlay.
- Detailed description of terrain and enemy positions that were identified.
- Results of any contact with the enemy.
- Unit status at the conclusion of the patrol mission, including the disposition of dead or wounded Rangers.
- Conclusions or recommendations.

7-20. OBJECTIVE RALLY POINT. The ORP is a point out of sight, sound, and small arms range of the objective area. It is normally located in the direction that the platoon plans to move after completion of actions on the objective. The ORP is tentative until the objective is pinpointed.

a. **Occupation of the ORP (Figure 7 8).**

(1) The patrol halts beyond sight and sound of the tentative ORP (200 to 400 meters in good visibility, 100 to 200 meters in limited visibility).

(2) The patrol establishes a security halt IAW the unit SOP.

(3) After issuing a five point contingency plan to the PSG, the PL moves forward with a recon element to conduct a leader's recon of the ORP.

(4) For a squad sized patrol, the PL moves forward with a compass man and one member of each fire team to confirm the ORP.

- After physically clearing the ORP location, the PL leaves two Rangers at the 6 o'clock position facing in opposite directions.
- The PL issues a contingency plan and returns with the compass man to guide the patrol forward.
- The PL guides the patrol forward into the ORP, with one team occupying from 3 o'clock through 12 o'clock to 9 o'clock, and the other occupying from 9 o'clock through 6 o'clock to 3 o'clock.

(5) For a platoon-sized patrol, the PL, RTO, WSL, three ammo bearers, a team leader, a SAW gunner, and riflemen go on the leaders recon for the ORP and position themselves at 10, 2, and 6 o'clock.

- The first squad in the order of march is the base squad, occupying from 10 to 2 o'clock.
- The trail squads occupy from 2 to 6 o'clock and 6 to 10 o'clock, respectively.
- The patrol headquarters element occupies the center of the triangle.

b. **Actions in the ORP.** The unit prepares for the mission in the ORP. Once the leader's recon pinpoints the objective, the PSG generally lines up rucksacks IAW unit SOP in the center of the ORP.

Figure 7-8. OCCUPATION OF THE ORP

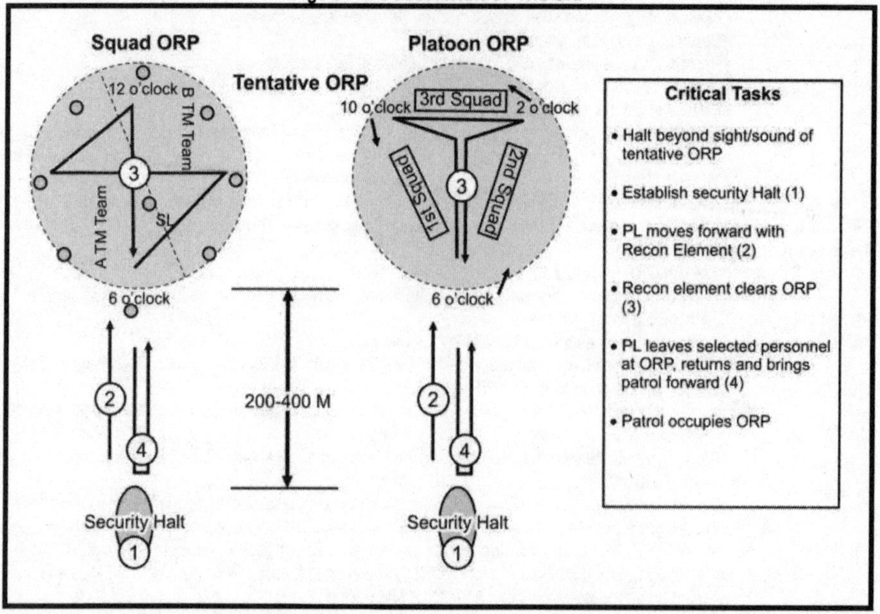

7-21. PATROL BASE. A patrol base is a security perimeter that is set up when a squad or platoon conducting a patrol halts for an extended period. Patrol bases should not be occupied for more than a 24 hour period (except in emergency). A patrol never uses the same patrol base twice.

a. **Use.** Patrol bases are typically used

- To avoid detection by eliminating movement.
- To hide a unit during a long detailed reconnaissance.
- To perform maintenance on weapons, equipment, eat and rest.
- To plan and issue orders.
- To reorganize after infiltrating on an enemy area.
- To establish a base from which to execute several consecutive or concurrent operations.

b. **Site Selection.** The leader selects the tentative site from a map or by aerial reconnaissance. The site's suitability must be confirmed and secured before the unit moves into it. Plans to establish a patrol base must include selecting an alternate patrol base site. The alternate site is used if the first site is unsuitable or if the patrol must unexpectedly evacuate the first patrol base.

c. **Planning Considerations.** Leaders planning for a patrol base must consider the mission and passive and active security measures. A patrol base (PB) must be located so it allows the unit to accomplish its mission.

- Observation posts and communication with observation posts.
- Patrol or platoon fire plan.
- Alert plan.
- Withdrawal plan from the patrol base to include withdrawal routes and a rally point, rendezvous point, or alternate patrol base.
- A security system to make sure that specific Rangers are awake at all times.

- Enforcement of camouflage, noise, and light discipline.
- The conduct of required activities with minimum movement and noise.
- Priorities of Work.

d. **Security Measures.**

- Select terrain the enemy would probably consider of little tactical value.
- Select terrain that is off main lines of drift.
- Select difficult terrain that would impede foot movement, such as an area of dense vegetation, preferably bushes and trees that spread close to the ground.
- Select terrain near a source of water.
- Select terrain that can be defended for a short period and that offers good cover and concealment.
- Avoid known or suspected enemy positions.
- Avoid built up areas.
- Avoid ridges and hilltops, except as needed for maintaining communications.
- Avoid small valleys.
- Avoid roads and trails.

e. **Occupation (Figure 7-9).**

(1) A PB is reconnoitered and occupied in the same manner as an ORP, with the exception that the platoon will typically plan to enter at a 90 degree turn. The PL leaves a two-Ranger OP at the turn, and the patrol covers any tracks from the turn to the PB.

(2) The platoon moves into the PB. Squad sized patrols will generally occupy a cigar shaped perimeter; platoon sized patrols will generally occupy a triangle shaped perimeter.

(3) The PL and another designated leader inspect and adjust the entire perimeter as necessary.

(4) After the PL has checked each squad sector, each SL sends a two-Ranger R&S team to the PL at the CP. The PL issues the three R&S teams a contingency plan, reconnaissance method, and detailed guidance on what to look for (enemy, water, built up areas or human habitat, roads, trails, or possible rally points).

(5) Where each R&S team departs is based on the PLs guidance. The R&S team moves a prescribed distance and direction, and reenters where the PL dictates.

- Squad sized patrols do not normally send out an R&S team at night.
- R&S teams will prepare a sketch of the area to the squad front if possible.
- The patrol remains at 100 % alert during this recon.
- If the PL feels the patrol was tracked or followed, he may elect to wait in silence at 100 % alert before sending out R&S teams.
- The R&S teams may use methods such as the "I," the "Box," or the "T." Regardless of the method chosen, the R&S team must be able to provide the PL with the same information.
- Upon completion of R&S, the PL confirms or denies the patrol base location, and either moves the patrol or begins priorities of work.

f. **Passive (Clandestine) Patrol Base (Squad).**

- The purpose of a passive patrol base is for the rest of a squad or smaller size element.
- Unit moves as a whole and occupies in force.
- Squad leader ensures that the unit moves in at a 90 degree angle to the order of movement.
- A claymore mine is emplaced on route entering patrol base.
- Alpha and Bravo teams sit back to back facing outward, ensuring that at least one individual per team is alert and providing security.

g. **Priorities of Work (Platoon and Squad).** Once the PL is briefed by the R&S teams and determines the area is suitable for a patrol base, the leader establishes or modifies defensive work priorities in order to establish the defense for the patrol base. Priorities of work are not a laundry list of tasks to be completed; to be effective, priorities of work must consist of a task, a given time, and a measurable performance standard. For each priority of work, a clear standard must be issued to guide the element in the successful accomplishment of each task. It must also be designated whether the work will be controlled in a centralized or decentralized manner. Priorities of work are determined IAW METT-TC. Priorities of Work may include, but are not limited to the following tasks:

(1) ***Security (Continuous).***

- Prepare to use all passive and active measures to cover the entire perimeter all of the time, regardless of the percentage of weapons used to cover that all of the terrain.
- Readjust after R&S teams return, or based on current priority of work (such as weapons maintenance).
- Employ all elements, weapons, and personnel to meet conditions of the terrain, enemy, or situation.
- Assign sectors of fire to all personnel and weapons. Develop squad sector sketches and platoon fire plan.
- Confirm location of fighting positions for cover, concealment, and observation and fields of fire. SLs supervise placement of aiming stakes and claymores.
- Only use one point of entry and exit, and count personnel in and out. Everyone is challenged IAW the unit SOP.
- Hasty fighting positions are prepared at least 18 inches deep (at the front), and sloping gently from front to rear, with a grenade sump if possible.

(2) ***Withdrawal Plan.*** The PL designates the signal for withdrawal, order of withdrawal, and the platoon rendezvous point and/ or alternate patrol base.

(3) ***Communication (Continuous).*** Communications must be maintained with higher headquarters, OPs, and within the unit. May be rotated between the patrol's RTOs to allow accomplishment of continuous radio monitoring, radio maintenance, act as runners for PL, or conduct other priorities of work.

(4) ***Mission Preparation and Planning.*** The PL uses the patrol base to plan, issue orders, rehearse, inspect, and prepare for future missions.

(5) ***Weapons and Equipment Maintenance.*** The PL ensures that machine guns, weapon systems, communications equipment, and night vision devices (as well as other equipment) are maintained. These items are not disassembled at the same time for maintenance (no more than 33 percent at a time), and weapons are not disassembled at night. If one machine gun is down, then security for all remaining systems is raised.

(6) ***Water Resupply.*** The PSG organizes watering parties as necessary. The watering party carries canteens in an empty rucksack or duffel bag, and must have communications and a contingency plan prior to departure.

(7) ***Mess Plan.*** At a minimum, security and weapons maintenance are performed prior to mess. Normally no more than half the platoon eats at one time. Rangers typically eat 1 to 3 meters behind their fighting positions.

- *Rest/Sleep Plan Management.* The patrol conducts rest as necessary to prepare for future operations.
- *Alert Plan and Stand to.* The PL states the alert posture and the stand to time. He develops the plan to ensure all positions are checked periodically, OPs are relieved periodically, and at least one leader is always alert. The patrol typically conducts stand to at a time specified by unit SOP such as 30 minutes before and after BMNT or EENT.
- *Resupply.* Distribute or cross load ammunition, meals, equipment, and so on.
- *Sanitation and Personal Hygiene.* The PSG and medic ensure a slit trench is prepared and marked. All Rangers will brush teeth, wash face, shave, wash hands, armpits, groin, feet, and darken (brush shine) boots daily. The patrol will not leave trash behind.

Figure 7-9. PATROL BASE

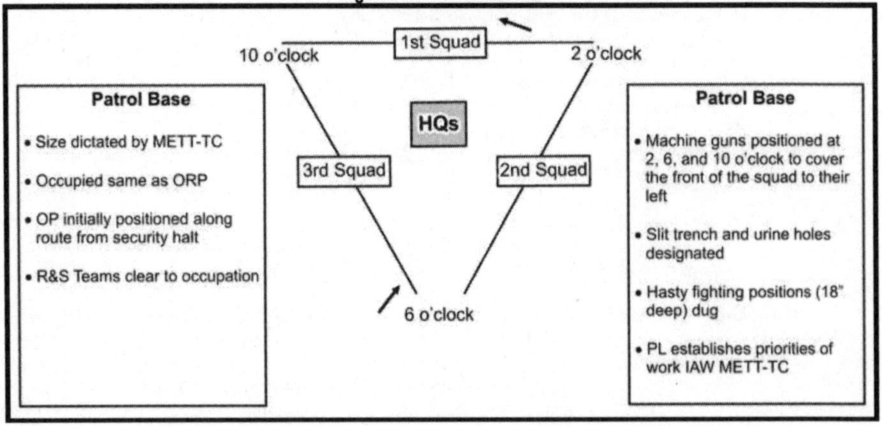

Section VI. MOVEMENT TO CONTACT

The MTC is one of the five types of offensive operations. An MTC gains or regains contact with the enemy. Once contact is made, the unit develops the situation. Normally a platoon conducts an MTC as part of a larger force.

7-22. TECHNIQUES. The two techniques of conducting a movement to contact are search and attack and approach march.

a. **Search and Attack**. The S&A technique is used when the enemy is dispersed, is expected to avoid contact, disengage or withdraw, or you have to deny his movement in an area. The search and attack technique involves the use of multiple platoons, squads, and fire teams coordinating their actions to make contact with the enemy. Platoons typically try to find the enemy and then fix and finish him. They combine patrolling techniques with the requirement to conduct hasty or deliberate attacks once the enemy has been found.

(1) ***Planning Considerations***.

- The factors of METT-TC.
- The requirement for decentralized execution.
- The requirement for mutual support.
- The length of operations.
- Minimize "Soldier's Load" to improve stealth and speed.
- Resupply and MEDEVAC.
- Positioning key leaders and equipment.
- Employment of key weapons.
- Requirement for patrol bases.
- Concept for entering the zone of action.
- The concept for linkups while in contact.
- (2) Critical Performance Measures.
- The platoon locates the enemy without being detected.
- Once engaged, fixes the enemy in position and maneuvers against the enemy.
- Maintains security throughout actions to avoid being flanked.

b. **Approach March.** The concept of the approach march is to make contact with the smallest element, allowing the commander the flexibility of destroying or bypassing the enemy. A platoon uses the approach march method as part of a larger unit. It can be tasked as the advance guard, move as part of the main body, or provide flank or rear security for the company or battalion. They may also receive on order missions as part of the main body.

(1) ***Fundamentals.*** These basics are common to all movements to contact.

- Make enemy contact with smallest element possible.
- Rapidly develop combat power upon enemy contact.
- Provide all round security for the unit.
- Support higher unit's concept.
- Reports all information rapidly and accurately and strives to gain and maintain contact with the enemy.
- Requires decentralized execution.

(2) ***Planning Considerations.*** The following issues should be considered heavily for MTC operations:

- Factors of METT-TC.
- Reduced "Soldier's Load."

(3) ***Critical Performance Measures***.

- PL selects the appropriate movement formation based on likelihood of enemy contact.
- Maintains contact, once contact is made, until ordered to do otherwise.

7-23. TASK STANDARDS. The platoon moves NLT the time specified in the order, the platoon makes contact with the smallest element possible, and the main body is not surprised by the enemy. Once the platoon makes contact, it maintains contact. The platoon destroys squad and smaller sized elements, and fixes elements larger than a squad. The platoon maintains sufficient fighting force capable of conducting further combat operations. Reports of enemy locations and contact are forwarded. If not detected by the enemy, the PL initiates a hasty attack. The platoon sustains no casualties from friendly fire. The platoon is prepared to initiate further movement within 25 minutes of contact, and all personnel and equipment are accounted for.

Chapter 8
BATTLE DRILLS

REACT TO CONTACT (Visual, IED, Direct Fire [includes RPG]) (07-3-D9501)

CONDITIONS (CUES)

This drill begins when any of these three types of cues occur:

Visual Contact

(Dismounted)—The unit is stationary or moving, conducting operations, and makes visual contact with the enemy.

(Mounted)—Unit is stationary or moving, conducting operations, and makes visual contact with the enemy.

IED (Dismounted/Mounted)—The unit is stationary or moving, conducting operations, when it identifies and confirms–or detonates–an IED.

Direct Fire (Dismounted/Mounted)—The unit is stationary or moving, conducting operations, when the enemy initiates contact with a direct-fire weapon.

STANDARDS

Visual Contact

(Dismounted)— The unit destroys the enemy with a hasty ambush or an immediate assault through the enemy position.

(Mounted)— Based on the composition of the mounted patrol, the unit either suppresses and reports the enemy position, and then continues its mission, or the patrol suppresses to fix the enemy position for a follow-on assault to destroy the enemy.

IED (Dismounted/Mounted)— The unit takes immediate action by using the 5 C's procedure (Confirm, Clear, Call, Cordon, and Control).

Direct Fire (Dismounted/Mounted)— The unit immediately returns well aimed fire and seeks cover. The unit leader reports the contact to higher headquarters.

PERFORMANCE MEASURES

1. **Visual (Dismounted)**
   a. ***Hasty Ambush***
      (1) Unit leader determines that the enemy has not seen the unit.
      (2) Unit leader signals Rangers to occupy best available firing positions.
      (3) The leader initiates the ambush with the most casualty producing weapon available, immediately followed by a sustained, well aimed volume of effective fire.
      (4) If prematurely detected, the Ranger(s) aware of the detection initiates the ambush.
      (5) The unit destroys the enemy or forces him to withdraw.
      (6) The unit leader reports the contact to higher headquarters.
   b. ***Immediate Assault***
      (1) The unit and the enemy simultaneously detect each other at close range.
      (2) All Rangers who see the enemy should engage and announce "Contact" with a clock direction and distance to enemy, for example, "Contact three o'clock, 100 meters."
      (3) Elements in contact immediately assault the enemy using fire and movement.
      (4) The unit destroys the enemy or forces him to withdraw.
      (5) The unit leader reports the contact to higher headquarters.
2. **Visual (Mounted)**
   a. The Ranger who spots the enemy announces the contact.
   b. The element in contact immediately suppresses the enemy.
   c. The vehicle commander of the vehicle in contact sends contact report over the radio.
   d. The unit maneuvers on the enemy or continues to move along route.
   e. Vehicle gunners fix and suppress the enemy positions.
   f. The unit leader reports the contact to higher headquarters.

3. **IED (Dismounted/ Mounted)**

a. The unit reacts to a suspected or known IED prior to detonation by using the 5 Cs.

b. While maintaining as safe a distance as possible and 360-degree security, the unit confirms the presence of an IED by using all available optics to identify any wires, antennas, det cord or parts of exposed ordnance.

(1) Conducts surveillance from a safe distance.

(2) Observes the immediate surroundings for suspicious activities.

c. The unit clears all personnel from the area at a safe distance to protect them from a potential second IED.

d. The unit calls higher headquarters to report the IED in accordance with the unit SOP.

e. The unit cordons off the area, directs personnel out of the danger area, prevents all military or civilian traffic from passing and allows entry only to authorized personnel.

(1) Rangers direct people out of the 300 meter minimum danger area.

(2) Identifies and clears an area for an incident control point (ICP).

(3) Rangers occupy positions and continuously secure the area.

f. The unit "controls" the area inside the cordon to ensure only authorized access.

g. The unit continuously scans the area for suspicious activity.

(1) Identifies potential enemy observation, vantage, or ambush points.

(2) Maintains visual observation on the IED to ensure the device is not tampered with.

4. **Direct Fire (Dismounted). (Figure 8-1)**

a. Rangers under direct fire immediately return fire and seek the nearest covered positions. Then, they call out distances and the orientation of direct fire **(Figure 8-2).**

b. Element leaders locate and engage known or suspected enemy positions with well aimed fire and pass information to the unit leader.

c. Element leaders control their Rangers' fire by **(Figure 8-3)**

(1) Marking targets with lasers.

(2) Marking the intended target with tracers or M203 rounds.

d. Rangers maintain contact (visual or oral) with the Rangers on their left or right.

e. Rangers maintain contact with their team leader and relay the location of enemy positions.

Figure 8-1

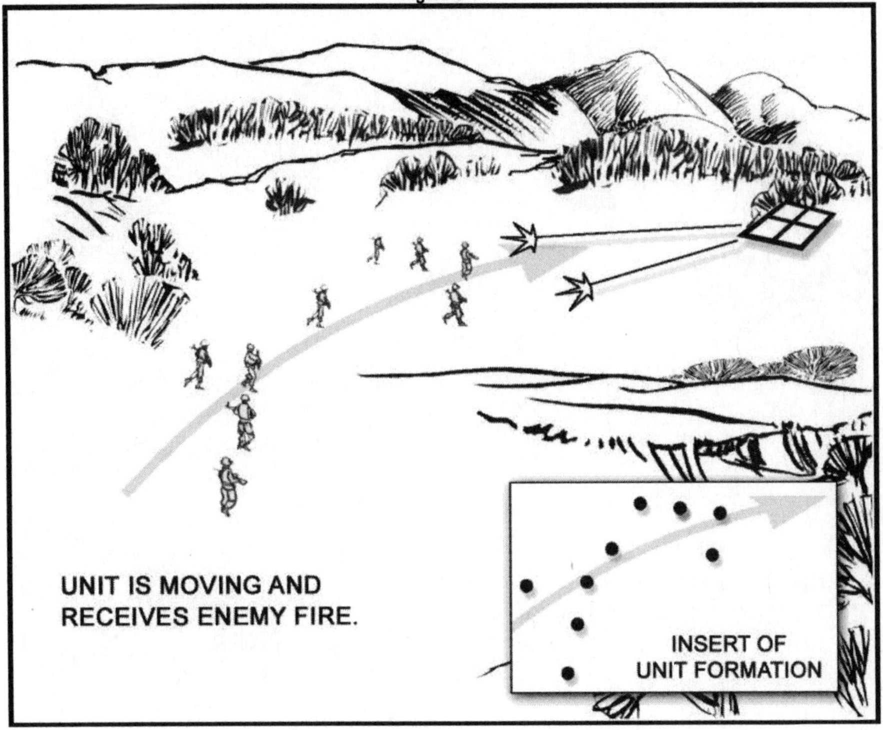
UNIT IS MOVING AND
RECEIVES ENEMY FIRE.
INSERT OF
UNIT FORMATION

Figure 8-2

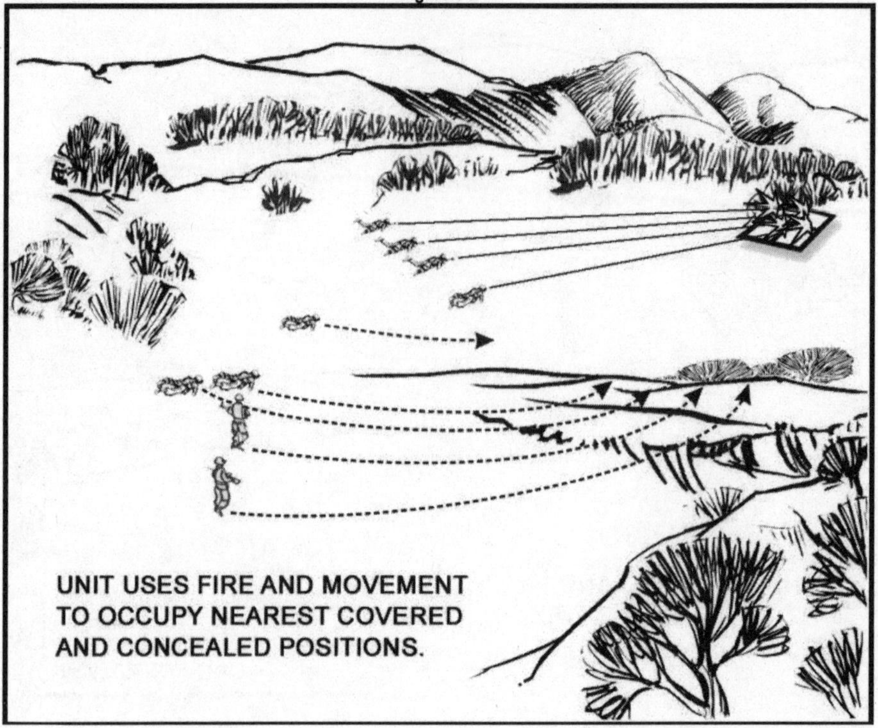
UNIT USES FIRE AND MOVEMENT
TO OCCUPY NEAREST COVERED
AND CONCEALED POSITIONS.

Figure 8-3

f. The element leaders (visually or orally) check the status of their Rangers.
g. The element leaders maintain contact with the unit leader.
h. The unit leader reports the contact to higher headquarters.

5. **Direct Fire (Mounted)**
   a. If moving as part of a logistics patrol, the vehicle gunners immediately suppress enemy positions and continue to move.
   b. The vehicle commanders direct their drivers to accelerate safely through the engagement area.
   c. If moving as part of a combat patrol, vehicle gunners suppress and fix the enemy allowing others to maneuver against and destroy the enemy.
   d. The leaders (visually or orally) check the status of their Rangers and vehicles.
   e. The unit leader reports the contact to higher headquarters.

Supporting Products

The Infantry Rifle Platoon and Squad **(FM 3-21.8)**
The SBCT Infantry Rifle Platoon and Squad **(FM 3-21.9)**
Warrior Ethos and Soldier Combat Skills **(FM 3-21.75)**

BREAK CONTACT (07-3-D9505)

CONDITIONS (CUE) (Dismounted/Mounted)—The unit is stationary or moving, conducting operations. All or part of the unit is receiving enemy direct fire. The unit leader initiates drill by giving the order BREAK CONTACT.

STANDARDS (Dismounted/Mounted)—The unit returns fire. A leader identifies the enemy as a superior force, and makes the decision to break contact. The unit breaks contact using fire and movement. The unit continues to move until the enemy cannot observe or place effective fire on them. The unit leader reports the contact to higher headquarters.

PERFORMANCE MEASURES

1. **Dismounted (Figure 8-4)**
   a. The unit leader designates an element to suppress the enemy with direct fire as the base of fire element.
   b. The unit leader orders distance, direction, a terrain feature, or last rally point for the movement of the first element.
   c. The unit leader calls for and adjusts indirect fire to suppress the enemy positions.
   d. The base of fire element continues to suppress the enemy.

Figure 8-4

   e. The bounding squad/team uses the terrain and/or smoke to conceal its movement, and bounds to an overwatch position.
   f. The bounding element occupies their overwatch position and suppresses the enemy with "well aimed fire" **(Figure 8-5).**

Figure 8-5

g. The base of fire element moves to its next covered and concealed position. (Based on the terrain and volume and accuracy of the enemy's fire, the moving element may need to use fire and movement techniques **(Figure 8-6)**.

Figure 8-6

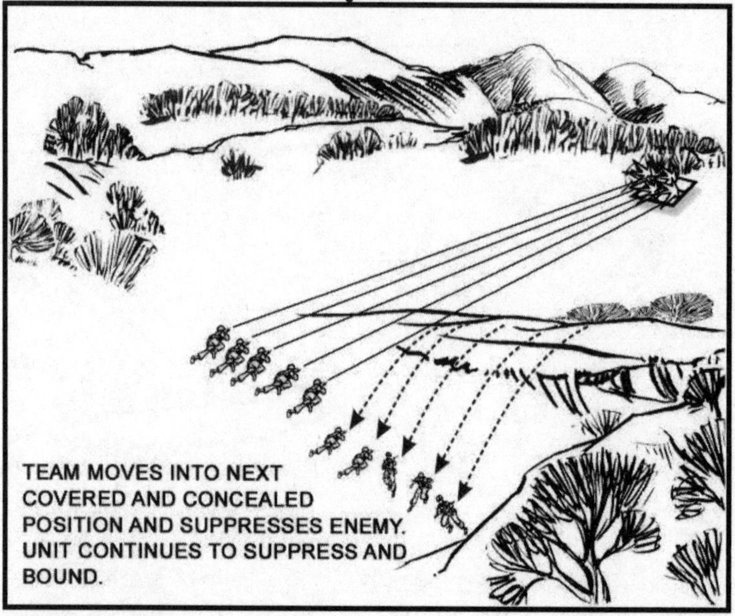

h. The unit continues to suppress the enemy and to bound, until it is no longer in contact with enemy.
i. The unit leader reports the contact to higher headquarters.

2. **Mounted**

a. The unit leader directs the vehicles in contact to place well aimed suppressive fire on the enemy positions.
b. The unit leader orders distance, direction, a terrain feature, or last ORP over the radio for the movement of the first section.
c. The unit leader calls for and adjusts indirect fire to suppress the enemy positions.
d. Gunners in the base of fire vehicles continue to engage the enemy. They attempt to gain fire superiority to support the bound of the moving section.
e. The bounding section moves to assume the overwatch position.
   (1) The section uses the terrain and/ or smoke to mask movement.
   (2) Vehicle gunners and mounted Rangers continue to suppress the enemy.
f. The unit continues to suppress the enemy and bounds until it is no longer receiving enemy fire.
g. The unit leader reports the contact to higher headquarters.

Supporting Products

The Infantry Rifle Platoon and Squad **(FM 3-21.8)**
The SBCT Infantry Rifle Platoon and Squad **(FM 3-21.9)**
Warrior Ethos and Soldier Combat Skills **(FM 3-21.75)**

REACT TO AMBUSH (FAR) (07-3-D9503)

CONDITIONS (CUE) (Dismounted/Mounted) — The platoon/ squad/ section moves tactically, conducting operations. The enemy initiates contact with direct and indirect fire.

NOTE: This drill begins when the enemy initiates ambush with direct and indirect fire.

STANDARDS

(Dismounted)—The unit immediately returns fire and occupies covered and/or concealed positions. The unit moves out of the kill zone, locates the enemy position, and conducts fire and maneuver to destroy the enemy.

(Mounted)—Vehicle gunners immediately return fire on known or suspected enemy positions as the unit continues to move out of the kill zone. The unit leader reports the contact to higher headquarters.

PERFORMANCE MEASURES

1. **Dismounted (Figure 8-7** and **Figure 8-8)**.
   a. Rangers receiving fire immediately return fire, seek cover, establish a support-by-fire position, and suppress the enemy position(s).
   b. Rangers not receiving fire move along a covered and concealed route to the enemy's flank in order to assault his position.

Figure 8-7

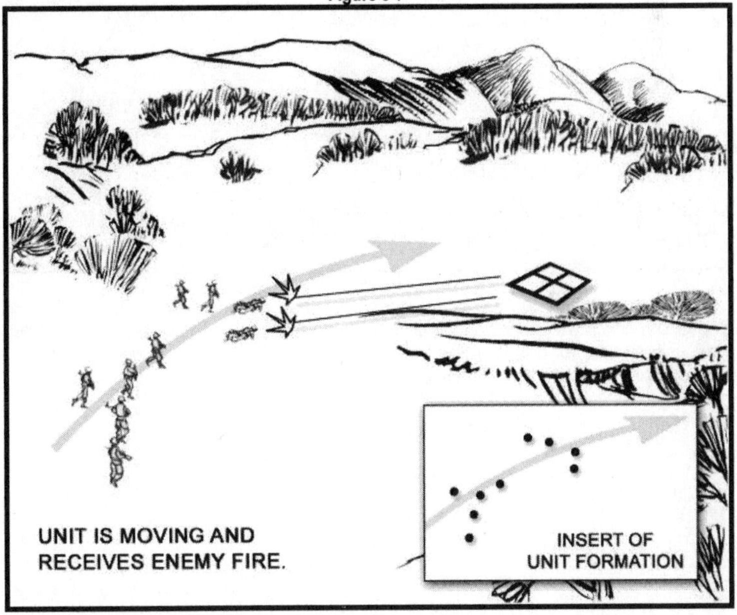

Figure 8-8

c. The unit leader or forward observer calls for and adjusts indirect fires and close air support, if available. On order, lifts or shifts fires to isolate the enemy position or to attack with indirect fires as the enemy retreats (**Figure 8-9**).

d. Rangers in the kill zone shift suppressive fires as the assaulting Rangers fight through and destroy the enemy.

e. The unit leader reports the contact to higher headquarters.

Figure 8-9

2. **Mounted**
   a. Gunners and personnel on vehicles immediately return fire.
   b. If the roadway is clear, all vehicles proceed through the kill zone.
   c. The lead vehicle deploys vehicle smoke to obscure the enemy's view of the kill zone.
   d. Vehicle commanders in disabled vehicles order Rangers to dismount IAW METT-TC, and to set up security while awaiting recovery.
   e. The remainder of the unit follows the lead vehicle out of the kill zone while continuing to suppress the enemy.
   f. The unit leader reports the contact to higher headquarters.

Supporting Products

The Infantry Rifle Platoon and Squad **(FM 3-21.8)**
The SBCT Infantry Rifle Platoon and Squad **(FM 3-21.9)**
Warrior Ethos and Soldier Combat Skills **(FM 3-21.75)**

REACT TO AMBUSH (NEAR) (07-3-D9502)

CONDITIONS (CUE) (Dismounted/Mounted)—The unit moves tactically, conducting operations. The enemy initiates contact with direct fire within hand grenade range. All or part of the unit is receiving accurate enemy direct fire. This drill begins when the enemy initiates ambush within hand grenade range.

STANDARDS

(Dismounted)— Rangers in the kill zone immediately return fire on known or suspected enemy positions, and then assault through the kill zone. Rangers outside the kill zone locate and place well aimed suppressive fire on the enemy. The unit assaults through the kill zone and destroys the enemy.

(Mounted)—Vehicle gunners immediately return fire on known or suspected enemy positions as the unit continues to move out of the kill zone. Rangers on disabled vehicles in the kill zone dismount, occupy covered positions, and engage the enemy with accurate fire. Vehicle gunners and Rangers outside the kill zone suppress the enemy. The unit assaults through the kill zone and destroys the enemy. The unit leader reports the contact to higher headquarters.

PERFORMANCE MEASURES

1. **Dismounted (Figure 8-10).**

Figure 8-10

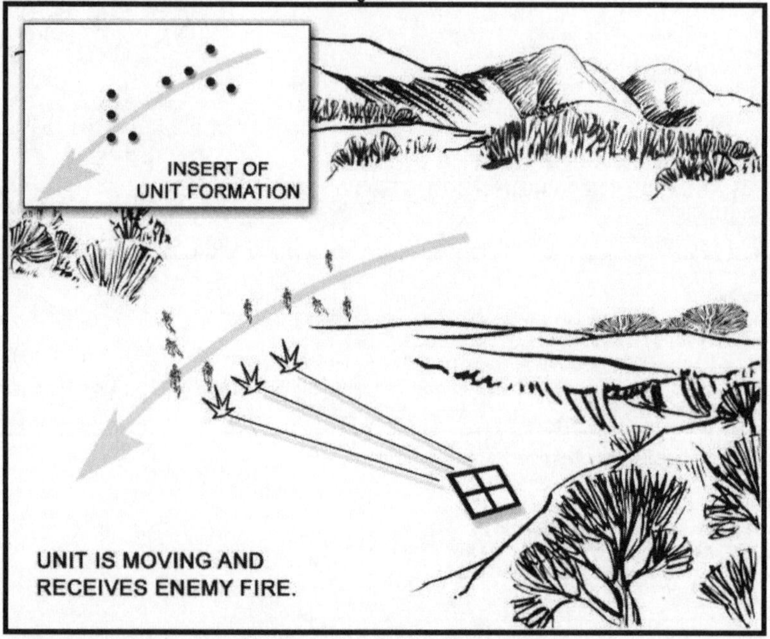

a. Rangers in the kill zone return fire immediately, as follows **(Figure 8-11)**:

(1) No Cover: Immediately, without order or signal, assault through the kill zone.

(2) Cover: Without order or signal, occupy the nearest covered position and throw smoke grenades.

Figure 8-11

b. Rangers in the kill zone assault through the ambush using fire and movement.

c. Rangers not in the kill zone identify the enemy location, place well aimed suppressive fire on the enemy's position and shift fire as Rangers assault the objective (**Figure 8-12**).

d. Rangers assault through and destroy the enemy position.

e. The unit leader reports the contact to higher headquarters.

Figure 8-12

2. **Mounted**
   a. Vehicle gunners in the kill zone immediately return fire and deploy vehicle smoke, while moving out of the kill zone.
   b. Rangers in disabled vehicles in the kill zone immediately obscure themselves from the enemy with smoke, dismount if possible, seek covered positions, and return fire.
   c. Vehicle gunners and Rangers outside of the kill zone identify the enemy positions; place well aimed suppressive fire on the enemy, and shift fire as Rangers assault the objective.
   d. The unit leader calls for and adjusts indirect fire and request Close Air Support IAW METT-TC.
   e. Rangers in the kill zone assault through the ambush and destroy the enemy.
   f. The unit leader reports the contact to higher headquarters.

Supporting Products

The Infantry Rifle Platoon and Squad **(FM 3-21.8)**
The SBCT Infantry Rifle Platoon and Squad **(FM 3-21.9)**
Warrior Ethos and Soldier Combat Skills **(FM 3-21.75)**

KNOCK OUT BUNKER (07-3-D9406)

CONDITIONS (CUE) —The unit moves tactically, conducting operations. The enemy initiates contact from concealed bunker network. All or part of the unit is receiving accurate enemy direct fire. Some iterations of this task should be performed in MOPP4. The unit receives an order to knock out an enemy bunker from which it is receiving fire.

STANDARDS—The unit destroys the designated bunker(s) by killing, capturing, or forcing the withdrawal of enemy personnel in the bunker(s). The unit maintains a sufficient fighting force to repel an enemy counterattack and continue operations.

PERFORMANCE MEASURES

1. **Deploy**
   a. The squad/team in contact establishes a base of fire.
   b. Weapons squad leader positions machine gun(s) to reinforce rifle squad in contact.
   c. Platoon sergeant moves to support by fire position and assumes control of the position's fires. (The weapon squad leader repositions another machine gun, as needed, based on METT-TC).
   d. The squad in contact gains and maintains fire superiority by—
      (1) Destroying or suppressing enemy crew served weapons.
      (2) Continuing suppressive fires at the lowest possible level.
      (3) [Platoon FO] calling for and adjusting indirect fires as directed by the platoon leader.
      (4) Suppressing the bunker and any supporting positions.
      (5) [Squad] employing shoulder-launched munitions as required.
   e. The squad obscures the enemy position with smoke.
   f. The squad establishes security to rear and flanks of support-by-fire position.
2. **Report**
   a. Submits contact reports.
   b. Submits SALUTE report to commander.
   c. Submits SITREPs as needed.
3. **Evaluate and Develop the Situation**
   a. The platoon leader, his RTO, and the platoon FO move forward to link up with the squad leader of the squad in contact.
   b. The platoon leader evaluates the situation by identifying enemy's composition, disposition and capabilities—
      (1) Identifies enemy disposition: number and locations of enemy bunkers, levels of mutual support and overlapping fires between positions, connecting trenches, and protective obstacles.
      (2) Identifies enemy composition and strength: the number of enemy automatic weapons, the presence of any vehicles, and the employment of indirect fires are indicators of enemy strength.
      (3) Identifies enemy capability to defend, reinforce, attack, and withdraw.
   c. Platoon leader develops the situation by determining where he can move to a position of advantage. He looks for—
      (1) A vulnerable flank or blind spot to at least one bunker.
      (2) A covered and concealed flanking route to the flank of the bunker.
4. **Develop a COA**
   a. The platoon leader determines—
      (1) Which bunker poses the greatest threat.
      (2) Where the adjoining bunkers are located.
      (3) Whether to breach protective obstacles.
   b. Platoon leader determines where to place support positions.
   c. Platoon leader determines size and make up of assault squad.
5. **Execute the COA (Figure 8-13).**
   a. Platoon leader directs the supporting element to suppress bunker as needed—
      (1) Platoon sergeant repositions a squad, a fire team, or a machine gun team to isolate the bunker and to continue suppressive fires.

(2) Forward observer shifts fires.

b. Assault squad leader executes knock out bunker drill—

(1) The assaulting squad, platoon leader, and RTO move along the covered and concealed route to an assault position and avoid masking the fires of the fire element.

(2) Rangers constantly watch for other bunkers or enemy positions in support of bunkers.

(3) The supporting element shifts or ceases fire (direct fire and indirect fire).

(4) Assault squad executes knock out bunker drill. On reaching the last covered and concealed position—

(a) Buddy Team #1 members (Team Leader and Automatic Rifleman) remain where they can cover Buddy Team #2 (grenadier and rifleman).

(b) Platoon Leader/ Squad Leader shifts supporting fires as required.

(5) Buddy team #2 moves to a blind spot near the bunker.

(a) One Ranger takes up a covered position near the exit.

(b) Another Ranger cooks off a grenade, announces, "Frag out," and throws it through an aperture.

(c) After the grenade detonates, the Ranger covering the exit enters first, and then the team clears the bunker.

(6) Buddy team #1 moves to join buddy team #2.

(7) The team leader:

(a) Inspects the bunker.

(b) Marks the bunker IAW unit SOP.

(c) Signals squad leader that bunker is clear.

c. The platoon leader—

(1) Directs the supporting squad to move up and knock out the next bunker. OR

(2) Directs the assaulting squad to continue and knock out the next bunker.

(3) Rotates squads as necessary.

d. The platoon/ squad leader accounts for Rangers, provides a SITREP to higher headquarters, reorganizes as necessary, and continues the mission.

Figure 8-13

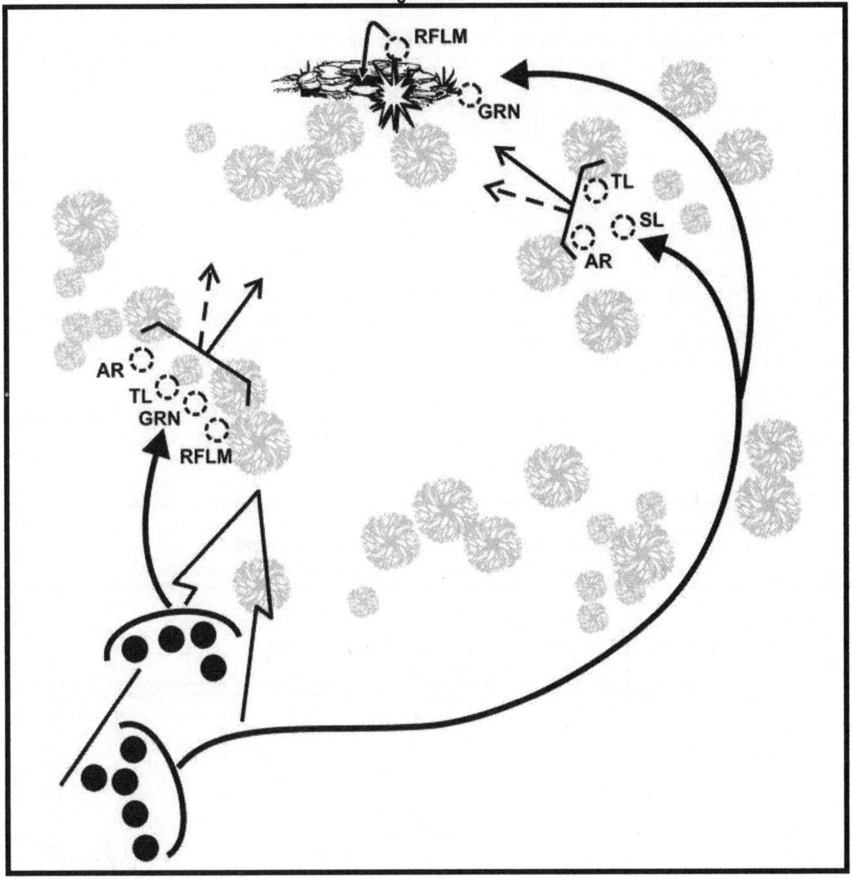

Supporting Products

The Infantry Rifle Platoon and Squad **(FM 3-21.8)**

ENTER AND CLEAR A ROOM (07-4-D9509)

CONDITIONS (CUE)—The element is conducting operations as part of a larger unit and your four-Ranger team has been given the mission to clear a room. Enemy personnel are believed to be in building. Noncombatants may be present in the building and are possibly intermixed with the enemy personnel. Support and security elements are positioned at the initial foothold and outside the building. Some iterations of this drill should be performed in MOPP4.
This drill begins on the order of the unit leader or on the command of the clearing team leader.

STANDARDS—The team secures and clears the room by killing or capturing the enemy, while minimizing friendly casualties, noncombatant casualties, and collateral damage; team complies with Rules of Engagement (ROE). The team maintains a sufficient fighting force to repel an enemy counterattack and continue operations.

PERFORMANCE MEASURES

1. The element leader occupies a position to best control the security and clearing teams.
    a. Element leader directs a team to secure corridors or hallways outside the room with appropriate firepower.
    b. The team leader (normally Ranger #2) takes a position to best control the clearing team outside the room.
    c. The element leader gives the signal to clear the room.

NOTE: If the element is conducting high-intensity combat operations, and it is using grenades, then it must comply with the Rules of Engagement (ROE) and consider the building structure. A Ranger of the clearing team cooks off at least one grenade (fragmentation), throws the grenade into the room, and then announces, "Frag out."

WARNING
When using grenades, consider the building structure.
Rangers can be injured from fragments if walls and floors are thin or damaged.

2. The clearing team enters and clears the room.
    a. The first two Rangers enter the room almost simultaneously **(Figure 8-14)**.

Figure 8-14

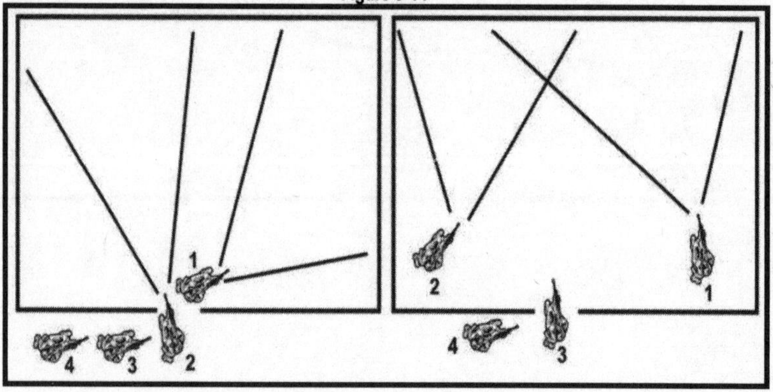

(1) The first Ranger enters the room and moves left or right along the path of least resistance to one of two corners. He assumes a position of domination facing into the room. During movement he eliminates all immediate threats.

(2) The second Ranger (normally the team leader) enters the room immediately after the first Ranger. He moves in the opposite direction of the first Ranger to his point of domination. During movement he eliminates all immediate threats in his sector.

NOTE: During high-intensity combat, the Rangers enter immediately after the grenade detonates. Both Rangers enter firing aimed bursts into their sectors engaging all threats or hostile targets to cover their entry.

NOTE: If the first or second Ranger discovers that the room is small or a short room (such as a closet or bathroom), he announces, "Short room" or "Short." The clearing team leader informs the third and fourth Rangers whether or not to stay outside the room or to enter.

b. The third Ranger moves opposite direction of the second Ranger, scanning and clearing his sector as he assumes his point of domination **(Figure 8-15)**.

Figure 8-15

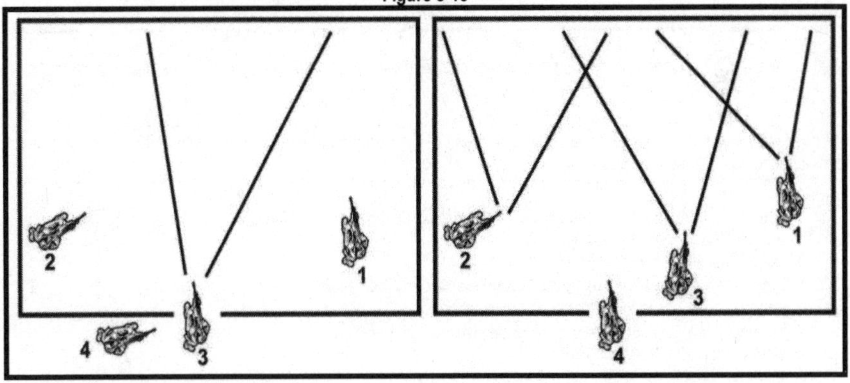

c. The fourth Ranger moves opposite of the third Ranger to a position that dominates his sector **(Figure 8-16).**

Figure 8-16

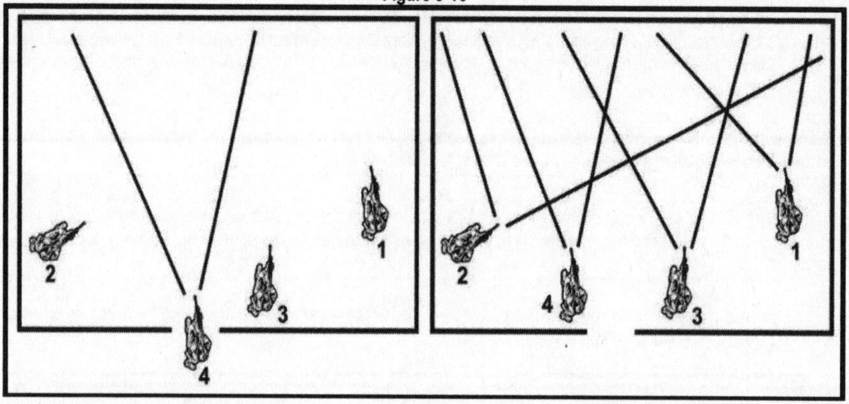

d. All Rangers engage enemy combatants with precision aimed fire, and identify noncombatants to avoid collateral damage.

NOTE: If necessary or on order, number one and two Rangers of the clearing team may move deeper into the room while overwatched by the other team members.

e. The team leader announces to the element leader when the room is "Clear."

3. The element leader enters the room.
   a. Quickly assesses room and threat.
   b. Determines if squad has fire power to continue clearing their assigned sector.
   c. Reports to the unit leader that the first room is clear.
   d. Requests needed sustainment to continue clearing his sector.
   e. Marks entry point IAW unit SOP.
4. The element consolidates and reorganizes as necessary.

Supporting Products

The Infantry Rifle Platoon and Squad **(FM 3-21.8)**
Combined Arms Operations in Urban Terrain **(FM 3-06.11)**

ENTER A TRENCH TO SECURE A FOOTHOLD (07-3-D9410)

CONDITIONS (CUE) — The platoon is moving tactically and receives effective fire from an enemy trench. The platoon is ordered to secure a foothold in the trench. The platoon has only organic weapons support available.
The unit leader initiates drill by giving the order for the assault element to secure a foothold in the trench.

STANDARDS—The platoon leader/squad leader quickly identifies the entry point. Platoon/squad secures the entry point, enters the trench, and secures an area large enough for the follow on unit. The platoon maintains a sufficient fighting force to repel enemy counterattack and continue the mission.

PERFORMANCE MEASURES

1. A platoon's section/squad executes actions on contact to eliminate or suppress fires from the trench.
2. The section/ squad in contact—
   a. **Deploys—**
      (1) Returns fire.
      (2) Seeks cover.
      (3) Establishes fire superiority.
      (4) Establishes local security.
      (5) [Platoon sergeant] repositions other squads to focus supporting fires and increase observation.
   b. **Reports—**
      (1) Squad leader uses SALUTE format to report location of hostile fire to platoon leader from base of fire position using the SALUTE format.
      (2) Platoon leader sends contact report followed by a size, activity, location, unit, time, and equipment (SALUTE) report to commander.
3. Evaluate and develop the situation.
   a. The platoon leader evaluates the situation using the SITREPs from the squad in contact and his personal observations. His evaluation should at least include—
      (1) Number of enemy weapons or volume of fire.
      (2) Presence of vehicles.
      (3) Employment of indirect fires.
   b. The platoon leader quickly develops the situation by—
      (1) Conducting a quick reconnaissance to determine enemy flanks.
      (2) Locating mutually supporting positions.
      (3) Locating any obstacles that impede the assault or provide some type of cover or concealment.
      (4) Determining whether the enemy force is inferior or superior.
      (5) Analyzing reports from squad leaders, teams in contact, or adjacent units.
4. **Choose a COA**
   a. [Platoon leader] decides to enter the trench and selects his entry point.
   b. [Platoon leader] selects a covered and concealed route to his entry point.
   c. [Platoon leader] directs maneuver element to secure near side of entry point and reduce the obstacle to gain a foothold.
   d. [Platoon sergeant] repositions the remaining squad to provide additional observation and supporting fires.
5. **Execute COA.** Use suppress, obscure, secure, reduce, assault (SOSRA) to set conditions for the assault **(Figure 8-17).**

Figure 8-17

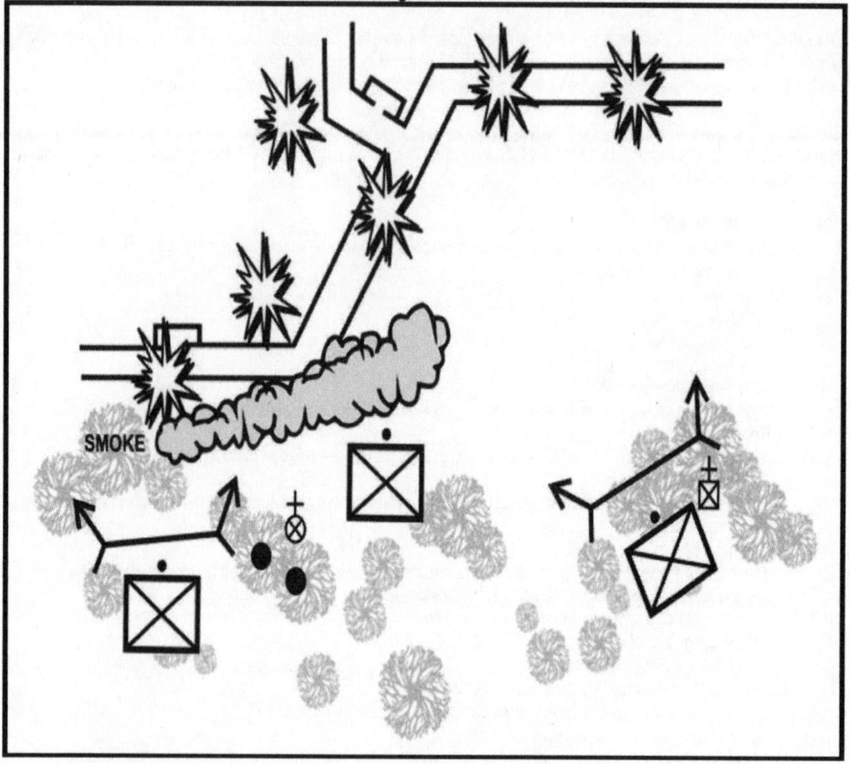

a. Suppress and obscure.
   (1) [Platoon leader or FO] Call for and adjust indirect fire in support of assault.
   (2) [Platoon sergeant] Direct base of fire squad to cover maneuvering squad.
   (3) [Platoon] Obscure maneuver element's movement with smoke, if available.
b. Secure the near side and reduce the obstacle.
   (1) [Maneuver squad] Clear entry point.
   (2) [Squad leader] Move the assaulting squad to last covered and concealed position short of the entry point.
   (3) [Squad leader] Designate entry point.
   (4) [Base of fire squad] Shift fires from entry point and continue to suppress adjacent enemy positions.
   (5) [Squad leader] Use one team to suppress the entry point; position the assaulting team at the entry point.
c. [Platoon leader] Direct FO to shift indirect fires to isolate the OBJ and direct the base-of-fire squads to shifts fire as the assault squad advances.
d. [Platoon] Secure the far side and establish a foothold **(Figure 8-18)**.

Figure 8-18

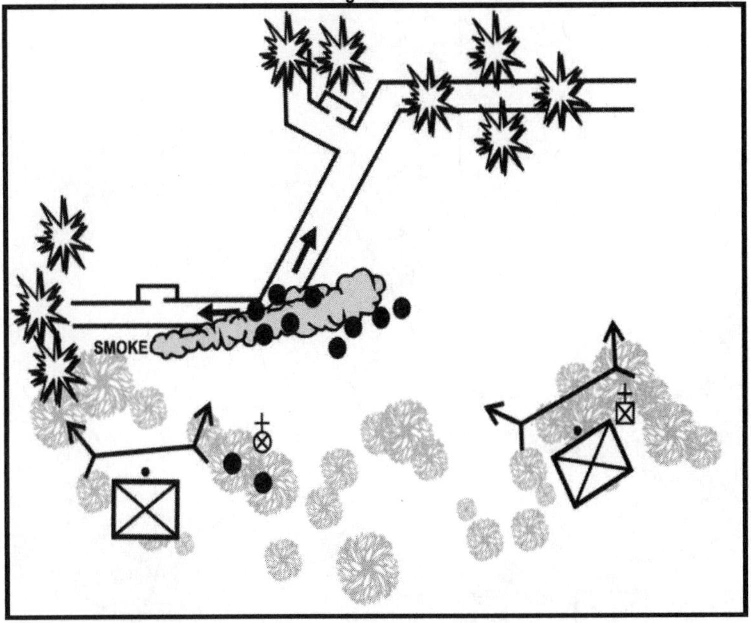

(1) The next two Rangers position themselves against the edge of the trench to roll right and left of the entry point to clear far side of obstacle and establish foothold. They engage all identified or likely enemy positions with rapid, short bursts of automatic fire and scan the trench for concealed enemy positions. The rest of the squad provides immediate security outside the trench.

(2) The team clears enough room for the squad or to the first trench junction and announces, "Clear."

(3) Squad leader marks entry point in accordance with platoon SOP, then sends next team in to increase the size of the foothold by announcing, "Next team in."

(4) Team moves into trench and secures assigned area **(Figure 8-19)**.

Figure 8-19

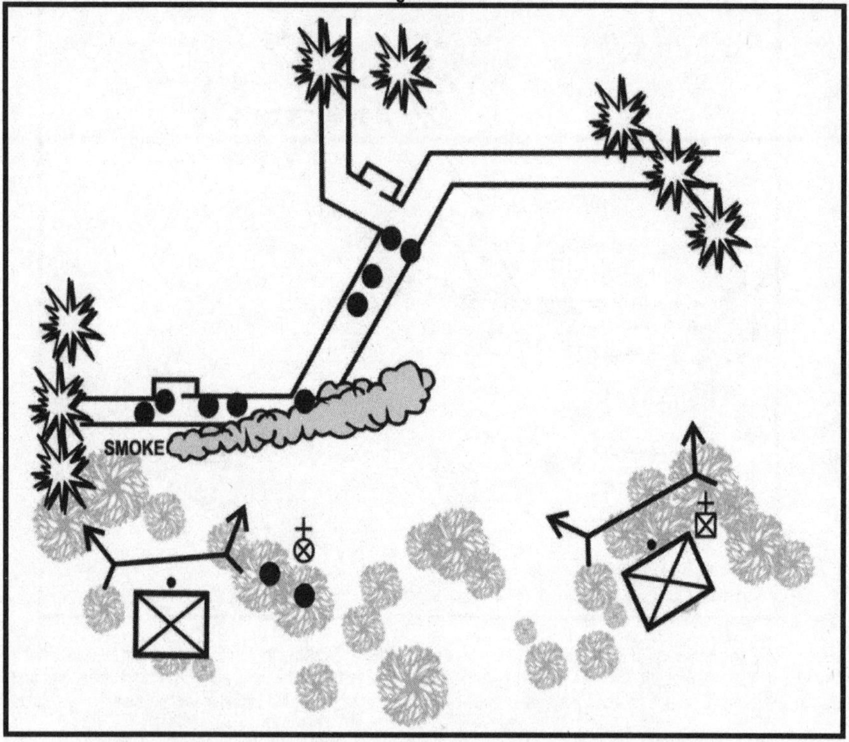

(5) Squad leader reports to platoon leader that the foothold is secure.

(6) Platoon leader moves to the maneuver squad leader to assess the situation.

(7) Platoon sergeant moves forward to control supporting squads outside the trench.

(8) The platoon leader sends necessary teams to clear an area large enough for the platoon, and then reports to the commander that the foothold is secure. He also requests additional support, if needed to continue clearing the trench.

6. The platoon/ squad leaders account for Rangers, provide a SITREP to higher headquarters, reorganize as necessary, and continue the mission.

Supporting Products

The Infantry Rifle Platoon and Squad **(FM 3-21.8)**

BREACH A MINED WIRE OBSTACLE (07-3-D9412)

CONDITIONS (CUE) —The platoon encounters a mined wire obstacle that prevents the platoon from moving forward. The platoon cannot bypass the obstacle. The enemy begins engaging the platoon from positions on the far side of the obstacle. This drill begins when the platoon's lead element encounters a mined wire obstacle, and the unit leader orders an element to breach the obstacle.

STANDARDS—The platoon breaches the obstacle and moves all personnel and equipment quickly through the breach. The platoon moves the support element and follow on forces through the breach and maintains a sufficient fighting force to secure the far side of the breach.

PERFORMANCE MEASURES

1. A platoon's section/squad executes actions on contact to reduce effective fires from the far side of the obstacle.
2. The section/ squad in contact—
   a. **Deploys**
      (1) Returns fire.
      (2) Seeks cover.
      (3) Establishes fire superiority.
      (4) Establishes local security.
      (5) Platoon sergeant repositions other squads to focus supporting fires and increase observation.
   b. **Reports**
      (1) Squad leader reports location of hostile fire to platoon leader from base of fire position using the SALUTE format.
      (2) Platoon leader sends contact report followed by a size, activity, location, unit, time, and equipment (SALUTE) report to commander.
3. Evaluate and develop the situation.
   a. The platoon leader quickly evaluates the situation using the SITREPs from the squad in contact, and using his personal observations. At a minimum, his evaluation should include—
      (1) Number of enemy weapons or volume of fire.
      (2) Presence of vehicles.
      (3) Employment of indirect fires.
   b. The platoon leader quickly develops the situation:
      (1) Conducts a quick reconnaissance to determine enemy flanks.
      (2) Locates mutually supporting positions.
      (3) Locates any obstacles that impede the assault or provides some type of cover or concealment.
      (4) Determines whether the force is inferior or superior.
      (5) Analyzes reports from squad leaders, teams in contact, or adjacent units.
4. The platoon leader directs the vehicles (if available) and the squad in contact to support the movement of another squad to the breach point.
   a. Indicates the route to the base of fire position.
   b. Indicates the enemy position to be suppressed.
   c. Indicates the breach point and the route the rest of the platoon will take.
   d. Gives instructions for lifting and shifting fires.
5. On the platoon leader's signal, the base of fire squad—
   a. Destroys or suppresses enemy weapons that are firing effectively against the platoon.
   b. Obscures the enemy position with smoke.
   c. Continues to maintain fire superiority while conserving ammunition and minimizing forces in contact.
6. After the breach, the platoon leader designates one squad as the breach squad and the remaining squad as the assault squad. (The assault squad may add its fires to the base of fire squad. Normally, it follows the covered and concealed route of the breach squad and assaults through immediately after the breach is made.)
7. The base of fire squad moves to the breach point and establishes a base of fire.

8. The platoon sergeant moves forward to the base of fire squad with the second machine gun and assumes control of the squad.

9. The platoon leader leads the breach and assault squads along the covered and concealed route.

10. The platoon FO calls for and adjusts indirect fires as directed by the platoon leader to support the breach and assault.

11. The breach squad executes actions to breach the obstacle (footpath).

a. The squad leader directs one fire team to support the movement of the other fire team to the breach point.

b. The squad leader designates the breach point.

c. The base of fire team continues to provide suppressive fires and to isolate the breach point.

d. The breaching fire team, with the squad leader, moves to the breach point using the covered and concealed route.

(1) The squad leader and breaching fire team leader employ smoke grenades to obscure the breach point. The platoon base of fire element shifts direct fires away from the breach point and continues to suppress adjacent enemy positions.

(2) The breaching fire team leader positions himself and the automatic rifleman on one flank of the breach point to provide close in security.

(3) The grenadier and rifleman (or the antiarmor specialist and automatic rifleman) of the breaching fire team probe for mines and cut the wire obstacle, marking their path as they proceed. (Bangalore is preferred, if available.)

(4) Once the obstacle is breached, the breaching fire team leader and the automatic rifleman move to the far side of the obstacle using covered and concealed positions. They signal the squad leader when they are in position and ready to support.

e. The squad leader signals the base of fire team leader to move his fire team up and through the breach. The squad leader then moves through the obstacle and joins the breaching fire team, leaving the grenadier (or antiarmor specialist) and rifleman of the supporting fire team on the near side of the breach to guide the rest of the platoon through.

f. Using the same covered and concealed route as the breaching fire team, the base of fire team moves through the breach and to a covered and concealed position on the far side.

12. The breach squad leader reports the situation to the platoon leader and posts guides at the breach point.

13. The platoon leader leads the assault squad through the breach in the obstacle and positions it on the far side to support the movement of the remainder of the platoon or to assault the enemy position covering the obstacle.

14. The breaching squad continues to widen the breach to allow vehicles to pass through.

15. The platoon leader provides a SITREP to the company commander, and directs his breaching squad to move through the obstacle. The platoon leader appoints guides to guide the company through the breach point **(Figure 8-20).**

Figure 8-20

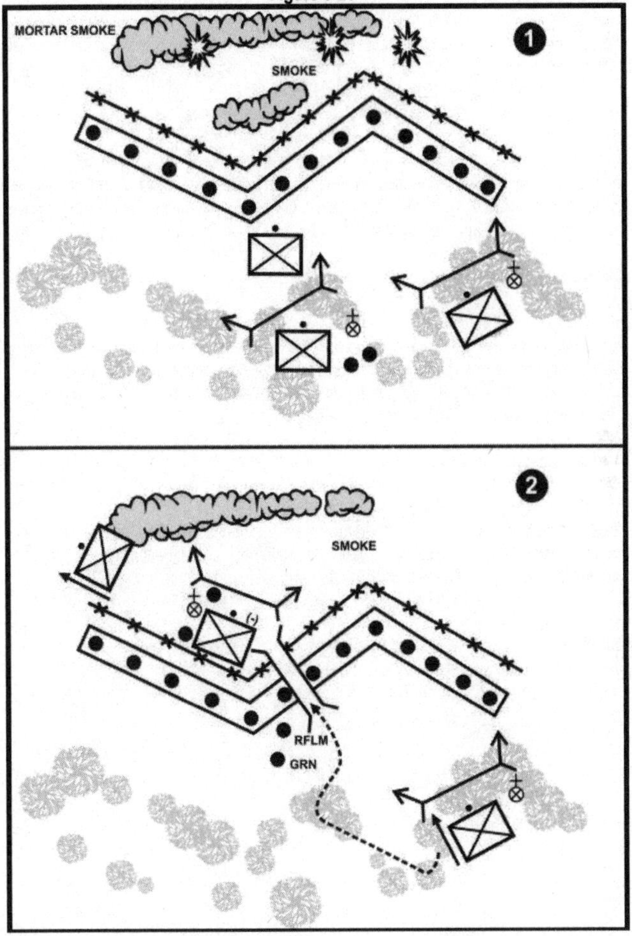

Supporting Products

The Infantry Rifle Platoon and Squad **(FM 3-21.8)**

REACT TO INDIRECT FIRE (07-3-D9504)

CONDITIONS (CUE) ***(Dismounted)***— The unit is moving, conducting operations. Any Ranger gives the alert, "Incoming," or a round impacts nearby. ***(Mounted)***—The platoon/section is stationary or moving, conducting operations. The alert, "Incoming," comes over the radio or intercom, or rounds impact nearby.
This drill begins when any member announces *"Incoming,"* or when a round impacts.

STANDARDS ***(Dismounted)***— Rangers immediately seek the best available cover. Unit moves out of area to the designated rally point after the impacts. ***(Mounted)***—If moving when they receive the alert, drivers immediately move their vehicles out of the impact area in the direction and distance ordered. If stationary when they receive the alert, drivers start their vehicles and move in the direction and distance ordered. The unit leader reports the contact to higher headquarters.

PERFORMANCE MEASURES

1. **Dismounted**
   - a. Any Ranger announces, ***"Incoming!"***
   - b. Rangers immediately assume the prone position or move to immediate available cover during initial impacts.
   - c. The unit leader orders the unit to move to a rally point by giving a direction and distance.
   - d. After the impacts, Rangers move rapidly in the direction and distance to the designated rally point.
   - e. The unit leader reports the contact to higher headquarters.
2. Mounted
   - a. Any Ranger announces, ***"Incoming!"***
   - b. Vehicle commanders repeat the alert over the radio.
   - c. The leader gives the direction and linkup location over the radio.
   - d. Rangers close all hatches if applicable to the vehicle type; gunners stay below turret shields or get down into vehicle.
   - e. Drivers move rapidly out of the impact area in the direction ordered by the leader.
   - f. The unit leader reports the contact to higher headquarters.

Supporting Products

The Infantry Rifle Platoon and Squad **(FM 3-21.8)**
The SBCT Infantry Rifle Platoon and Squad **(FM 3-21.9)**
Warrior Ethos and Soldier Combat Skills **(FM 3-21.75)**

Chapter 9
MILITARY MOUNTAINEERING

In the mountains, commanders face the challenge of maintaining their units' combat effectiveness and efficiency. To meet this challenge, commander's conduct training that provides Rangers with the mountaineering skills necessary to apply combat power in a rugged mountain environment, and they develop leaders capable of applying doctrine to the distinct characteristics of mountain warfare.

9 1. TRAINING. Military mountaineering training provides units tactical mobility in mountainous terrain that would otherwise be inaccessible. Rangers are trained in the fundamental mobility and climbing skills necessary to move units safely and efficiently in mountainous terrain. Rangers conducting Combat Operations in a mountainous environment should receive extensive training to prepare them for the rigor of mountain operations. Some of the areas are as follows:

- Characteristics of the mountain environment.
- Care and use of basic mountaineering equipment.
- Mountain bivouac techniques.
- Mountain communications.
- Mountain travel and walking techniques.
- Mountain navigation, hazard recognition and route selection.
- Rope management and knots.
- Natural and artificial anchors.
- Belay and rappel techniques.
- Installation construction and use such as rope bridges.
- Rock climbing fundamentals.
- Rope bridges and lowering systems.
- Individual movement on snow and ice.
- Mountain stream crossings (to include water survival techniques).

9-2. DISMOUNTED MOBILITY. Movement in class four and five terrain demands specialized skills and equipment. Before Rangers can move in such terrain, a technical mountaineering team might have to secure the high ground. Some basic SOPs for executing combat missions will work in this scenario with small modifications. PL will develop the plan, issue the plan to the squad leaders; the squad leaders only need to disseminate the PL's plan to their units with the details of who conducts what tasks.

9-3. TASK ORGANIZATION. Your platoon will be organized into four different elements, all of which are necessary for mission accomplishment.

a. **Headquarters.** HQ provides overall command and control of the mission:
   - Patrol leader (PL)
   - Patrol leader's RTO
   - Assistant patrol leader (APL)
   - APL's RTO
   - Forward observer (FO)
   - Medic

b. **Support (Evacuation Team).** This team is responsible for care and transportation of the casualty. The Platoon medic, though part of the HQ's element, moves with the Support Squad leader.

c. **Assault (R/ S and Installation Teams).** These will each consist of one squad. Assault has the primary missions of route reconnaissance, navigation, and installation construction.

d. **Security.** This will consist of one squad. Primary focus is to provide security for the platoon at all obstacles and installations as well as serve as back up support/ evacuation.

NOTE: As the PL, you must ensure that the duties are rotated periodically within the platoon to keep any one squad from becoming fatigued.

e. **Equipment.** Once task organization has been completed, the platoon must organize and consolidate all rescue and mountaineering equipment.

9-4. RESCUE EQUIPMENT. In the mountains, Rangers use a rescue stretcher system to transport casualties. The easy-gliding polyethylene stretcher travels easily while supporting and protecting the patient. For spine injuries, add a spinal immobilizer. For shoulder or other injuries, add a short or long backboard. The stretcher also holds a scoop stretcher or other immobilization add ons.

a. **Temperatures.** The rescue stretcher is made of easy gliding polyethylene. Breakage occurs at 120 degrees (Fahrenheit), and melting occurs at 449 degrees (Fahrenheit).

b. **Size.**

- Rolled in storage bag: 9" x 36"
- Flat: 3' X 8'

c. **Weight.** 19 lbs with all accessories.

d. **Strength.**

- Horizontal lift slings, tensile strength: 10,000 lbs
- Vertical lift Slings: 5,000 lbs (depending on the size of the rope.

e. **Components.**

- Stretcher
- Nylon backpack
- Horizontal lift slings
- Vertical lift sling (3/ 8" rope)
- Locking steel carabiner
- Tow strap
- Four webbing handles

f. **Loading Procedure.**

(1) Unpack and unroll the rescue litter.
(2) Bend litter in half backwards to make it and it will lay flat. And lay out flat.
(3) Place a patient in the litter.

g. **Logroll Method.**

(1) Place litter next to patient.
(2) Roll patient on to side and slide litter as far under his as possible.
(3) Roll patient on to litter and carefully slide patient into center of litter.
(4) Secure patient to litter.

h. **Slide Method.**

(1) Place foot end of stretcher at the head of the patient.
(2) One person straddles the stretcher and supports the patient head, neck and shoulders.
(3) Two people grab straps and pull stretcher under patient while slightly lifting patient head and shoulders.

i. **Fastening Straps and Buckles.**

(1) Lift sides of stretcher and fastens straps to buckles directly across from them.
(2) Feed foot straps through unused buckles at the foot of the stretcher and fasten to buckles.

j. **Rigging For Horizontal Lift.** When rigging for a horizontal lift remember the head strap is 4" shorter than the foot strap.

(1) Insert one end of head strap through slot at head end and route under stretcher and then through slot on opposite side.
(2) Repeat at the foot end with foot strap.
(3) Equalize weight on all straps and insert steel carabiner through sewn loops on all 4 straps.

(4) Ensure you remove horizontal lift straps if the stretcher is to be dragged to prevent damage to the straps.

k. **Rigging For Vertical Lift.**

(1) Create a fixed loop in the middle of the rope by tying a double figure knot.

(2) Pass tails through grommet on either side of the head and snug knot against stretcher.

(3) Feed ropes through grommets along the sides, pass through the handles and through the grommets at the foot end of the stretcher and secure with a square knot.

(4) Route the pigtails through the lower carrying handles outside to in, and secure ends with a square knot with two overhand safeties.

l. **Ascending Vertical Terrain with a Casualty.**

(1) Package a casualty in a stretcher for carrying and dragging

(2) Package a casualty in a stretcher for helicopter evacuation horizontally and vertically

(3) Task organization for a platoon for moving a casualty (carrying squad, security squads, machine guns, and key leaders). Emphasize that the PL focuses on the entire tactical situation and controlling the platoon and having a rotation of the carrying squad if they have to move the casualty over long distances. The PSG focuses on controlling the CASEVAC.

(4) Establish the primary anchor (sling rope and 2 opposite and opposed carabiners) and the secondary anchor for the 6 to 8 wrap Prusik safety.

(5) Having teams moving ahead to set up anchors to expedite moving the casualty up multiple pitches

m. **Descending Vertical Terrain with a Casualty.**

(1) Lower the casualty on a Munter Hitch with a 6 to 8 wrap Prusik safety

(2) Everyone else uses a retrievable rappel with the hasty or body rappel to descend

(3) Have teams move down and establish anchors to expedite the lowering if you have multiple pitches

9 5. MOUNTAINEERING EQUIPMENT. Mountaineering equipment refers to all the parts and pieces that allow the trained Ranger to accomplish many tasks in the mountains. The importance of this gear to the mountaineer is no less than that of the rifle to the infantryman.

a. Ropes and Cords. Ropes and cords are the most important pieces of mountaineering equipment. They secure climbers and equipment on steep ascents and descents. They are also used to install rope and hauling equipment. From WWII until the 1980's, the US military mostly used ***7/16-inch nylon laid rope***, often referred to as green line for all mountaineering operations. Since the introduction of kernmantle ropes, ropes designed for more specific purposes are replacing the old all–purpose green line. Kernmantle ropes are constructed similar to parachute cord. It consists of a smooth sheath, surrounding a braided or woven core. Laid ropes are still in use today however, should never be used in situations where rope failure could result in injury or loss of equipment. There are two classifications of kernmantle ropes; static and dynamic.

(1) ***Dynamic Ropes.*** Ropes used for climbing are classified as dynamic ropes. These rope stretch or elongate ***8 to 12 percent*** once subjected to weight or impact. This stretching is critical in reducing the impact force on the climber, anchors, and/or belayer during a fall by softening the catch. ***11mm X 150m*** is generally considered the standard for military use however more specialized ropes in different length and diameters are available.

(2) ***Static Ropes.*** Static ropes are used in situations where rope stretch is undesired, and when the rope is subjected to heavy static weight. Static ropes should never be used while climbing, since even a fall of a few feet could generate enough impact force to injure climber and belayer, and/or cause anchor failure. Static ropes are usually used when constructing rope bridges, fixed rope installations, vertical haul lines, and so on.

(3) ***Sling Ropes and Cordelettes.*** A short section of static rope or static cord is called a ***"sling rope"*** or ***"cordelette."*** These are critical pieces of personal equipment in mountaineering operations. Diameter usually ranges from ***7mm to 8mm***, and up to 21 feet long. ***8mm X 15 feet is the minimum Ranger standard.***

(4) ***Care of Rope.*** Rope that is used daily should be used no longer than one year. Occasionally used rope can be used generally up to five years if properly cared for.

- Inspect ropes thoroughly before, during and after use for cuts, frays, abrasions, mildew, and soft or worn spots.
- Never step on a rope or drag it on the ground unnecessarily.
- Avoid running rope over sharp or rough edges (pad if necessary).

• Keep ropes away from oil, acids and other corrosive substances.
• Avoid running ropes across one another under tension (nylon to nylon contact will damage ropes).
• Do not leave ropes knotted or under tension longer than necessary.
• Clean in cool water, loosely coil and hang to dry out of direct sunlight. Ultraviolet light rays harm synthetic fibers. When wet, hang rope to drip dry on a rounded wooden peg, at room temperature (do not apply heat).

(5) ***Webbing and Slings.*** Loops of tubular webbing or cord, called slings or runners, are the simplest pieces of equipment and some of the most useful. The uses for these simple pieces are endless, and they are a critical link between the climber, the rope, carabiners, and anchors. Runners are predominately made from either 9/16 inch or 1 inch tubular webbing and are either tied or sewn by a manufacturer.

b. **Carabiners.** The carabiner is one of the most versatile pieces of equipment available in the mountains. This simple piece of gear is the critical connection between the climber, his rope, and the protection attaching him to the mountain. Carabiners must be strong enough to hold hard falls, yet light enough for the climber to easily carry a quantity of them. Today's high tech metal alloys allow carabiners to meet both of these requirements. Steel carabiners are still widely used in the military but are being replaced by lighter and stronger materials. Basic carabiner construction affords the user several different shapes.

c. Protection. Protection is the generic term used to describe a piece of equipment (natural or artificial) that is used to construct an anchor. Protection is used with a climber, belayer, and climbing rope to form the lifeline of the climbing team. The rope connects two climbers, and the protection connects them to the rock or to ice. **Figure 9-1** shows removable artificial protection, and **Figure 9-2** shows fixed (usually permanent) artificial protection.

Figure 9-1. EXAMPLES OF TRADITIONAL (REMOVABLE) PROTECTION USED ON ROCK

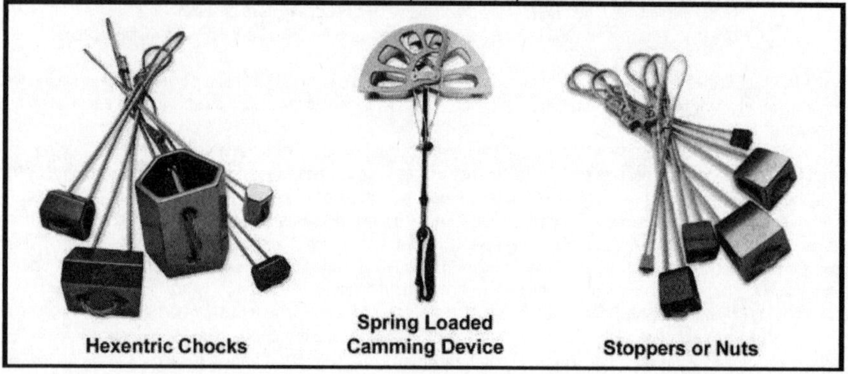

Figure 9-2. EXAMPLES OF FIXED (PERMANENT/SEMIPERMANENT) PROTECTION USED ON ROCK

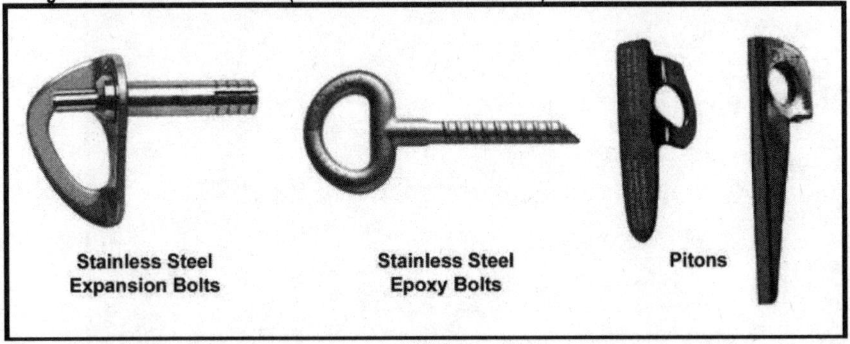

9-6. ANCHORS. Anchors are the base, for all installations and roped mountaineering techniques. Anchors must be strong enough to support the entire weight of the load or impact placed upon them. Several pieces of artificial or natural protection may be incorporated together to make one multi point anchor. Anchors are classified as Artificial or Natural.

a. **Artificial Anchors.** Artificial anchors are constructed using all manmade material. The most common anchors incorporate traditional or fixed protection **(Figure 9-3)**.

b. **Natural Anchors.** Natural anchors are usually very strong and often simple to construct using minimal equipment. Trees, shrubs and boulders are the most common. All natural anchors simply require a method of attaching a rope. Regardless of the type of natural anchor used, the anchor must be strong enough to support the entire weight of the load.

(1) ***Trees.*** These are probably the most widely used of all anchors. In rocky terrain, trees usually have a very shallow root system. Check this by pushing or tugging on the tree to see how well it is rooted. Anchor as low as possible to prevent excess leverage on the tree. Use padding on soft, sap producing trees to keep sap off ropes and slings.

(2) ***Rock Projections and Boulders.*** You can use these, but they must be heavy enough, and have a stable enough base to support the load.

(3) ***Bushes and Shrubs.*** If no other suitable anchor is available, route a rope around the bases of several bushes. As with trees, place the anchoring rope as low as possible to reduce leverage on the anchor. Make sure all vegetation is healthy and well rooted to the ground.

(4) ***Tensionless Anchor.*** This is used to anchor rope on high load installations such as bridging. The wraps of the rope around the anchor **(Figure 9-4)** absorb the tension of the installation and keep the tension off the knot and carabiner. Tie it with a minimum of four wraps around the anchor; however a smooth anchor (small tree, pipe, or rail) may require several more wraps. Wrap the rope from top to bottom. Place a fixed loop into the end of the rope and attached loosely back onto the rope with a carabiner.

Figure 9-4. TENSIONLESS NATURAL ANCHOR

Figure 9-3. CONSTRUCTING A 3 POINT, PRE EQUALIZED ANCHOR USING FIXED ARTIFICIAL PROTECTION

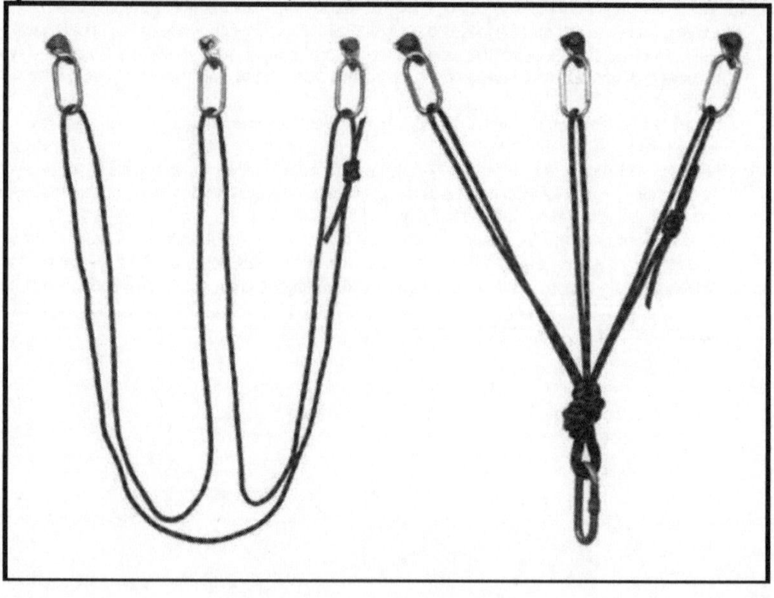

9- 7. KNOTS

a. **Square Knot.** This joins two ropes of equal diameter (**Figure 9-5**): Two interlocking bites, running ends exit on same side of standing portion of rope. Each tail is secured with an overhand knot on the standing end. When you dress the knot, leave at least a 4 inch tail on the working end.

Figure 9-5. SQUARE KNOT

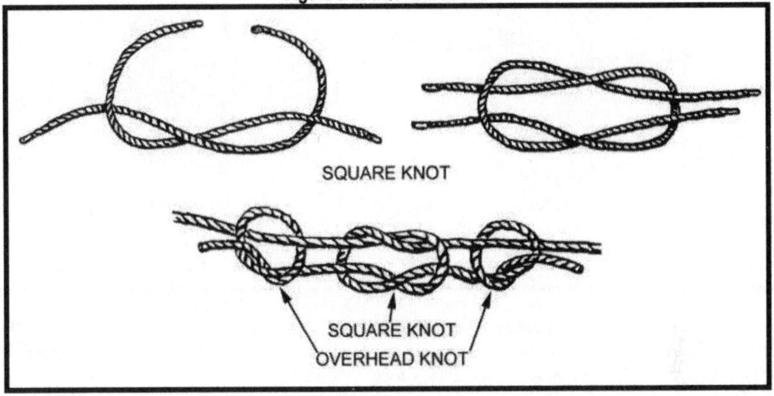

b. **Round Turn with Two Half Hitches.** This is a constant tension anchor knot (**Figure 9-6**). The rope forms a complete turn around the anchor point (thus the name "round turn"), with both ropes parallel and touching, but not crossing. Both half hitches are tightly dressed against the round turn, with the locking bar on top. When you dress the knot, leave at least a 4 inch tail on the working end.

Figure 9-6. ROUND TURN WITH TWO HALF HITCHES

c. **End–of–the–Rope Clove Hitch**. This is an intermediate anchor knot (**Figure 9-7**) that requires constant tension. Make two turns around the anchor (1). A locking bar runs diagonally from one side to the other. Leave no more than one rope width between turns of rope (2). Locking bar is opposite direction of pull. When you dress the knot, leave at least a 4-inch tail on the working end.

Figure 9-7. END–OF–THE–ROPE CLOVE HITCH

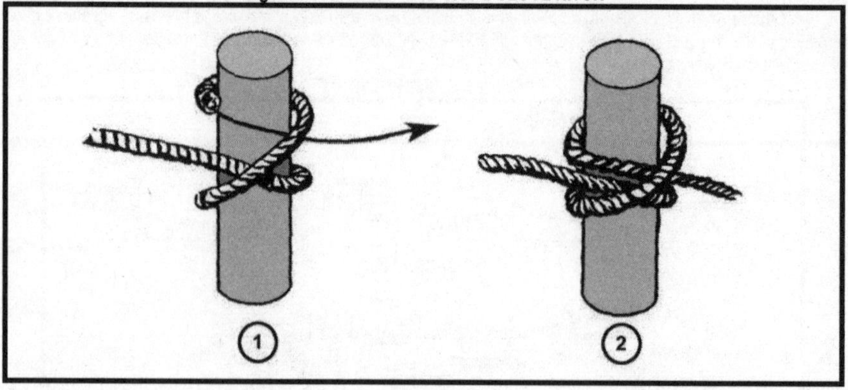

d. **Middle–of–the–Rope Clove Hitch.** This knot **(Figure 9-8)** secures the middle of a rope to an anchor. The knot forms two turns around the anchor (1, 2). A locking bar runs diagonally from one side to the other. Leave no more than one rope width between turns (3). Ensure the locking bar is opposite the direction of pull.

Figure 9-8. MIDDLE–OF–THE–ROPE CLOVE HITCH

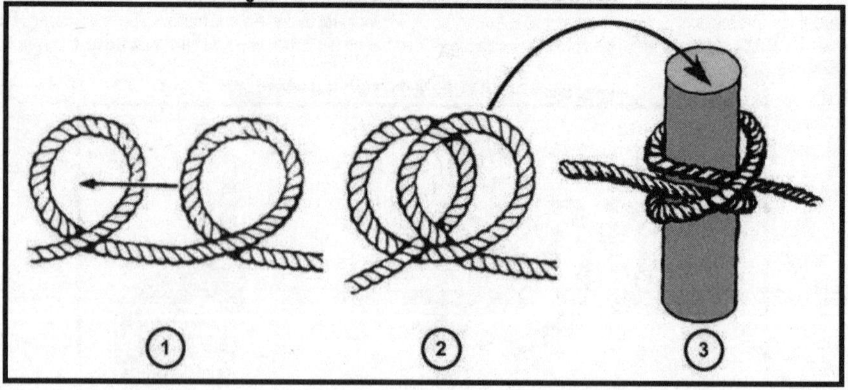

e. **Rappel Seat.** The rappel seat **(Figure 9-9**) is a rope harness used in rappelling and climbing. It can be tied for use with the left or right hand (1). Leg straps do not cross, and are centered on buttocks and tight (2). Leg straps form locking half hitches on rope around waist. Square knot properly tied on right hip (3) and finished with two overhand knots. Tails must be even, within 6 inches (4). Carabiner properly inserted around all ropes with opening gate opening up and away (5). Carabiner will not come in contact with square knot or overhand knot. Rappel seat is tight enough not to allow a fist to be inserted between the rappeller's body and the harness.

Figure 9-9. RAPPEL SEAT

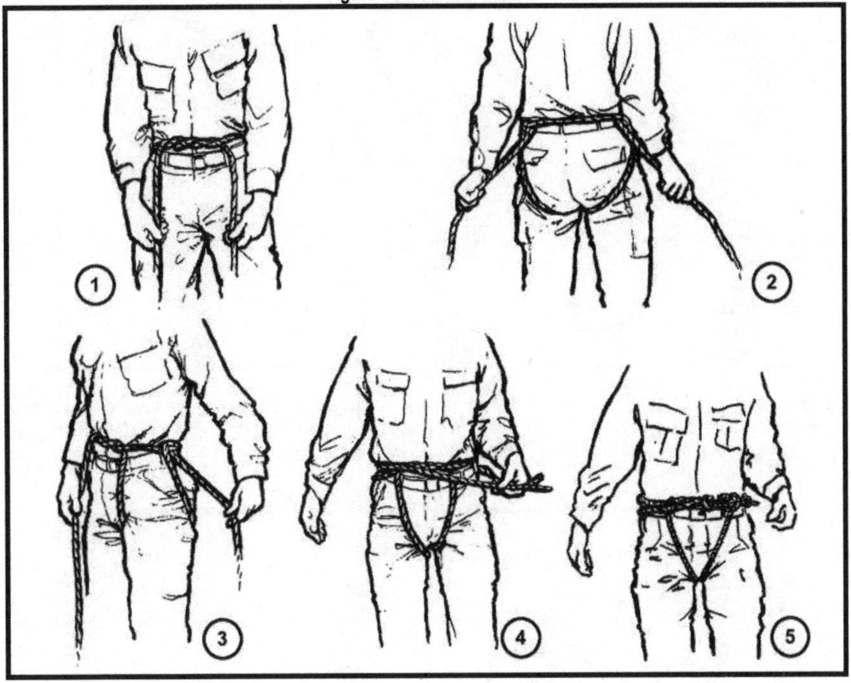

f. **Double Figure 8.** Use a Figure 8 loop knot **(Figure 9-10)** to form a fixed loop in the end of the rope. It can be tied at the end of the rope or anywhere along the length of the rope. Figure 8 loop knots are formed by two ropes parallel to each other in the shape of a Figure 8, no twists are in the Figure 8. Fixed loops are large enough to insert a carabiner. When you dress the knot, leave at least a 4 inch tail on the working end.

Figure 9-10. DOUBLE FIGURE 8 LOOP KNOT

g. **Rerouted Figure 8 Knot.** This anchor knot also attaches a climber to a climbing rope. Form a Figure 8 in the rope, and pass the working end around an anchor. Reroute the end back through to form a double Figure 8 **(Figure 9-11)**. Tie the knot with no twists. When you dress the knot, leave at least a 4 inch tail on the working end.

Figure 9-11. REROUTED FIGURE 8 KNOT

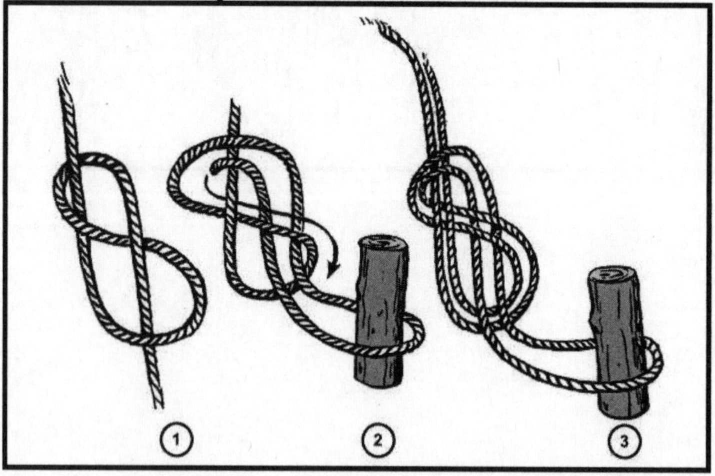

h. **Figure 8 Slip Knot.** The Figure 8 slip is used to form an adjustable bight in the middle of a rope. Knot is in the shape of a Figure 8. Both ropes of the bight pass through the same loop of the Figure 8. The bight is adjustable by means of a sliding section **(Figure 9-12).**

Figure 9-12. FIGURE 8 SLIP KNOT

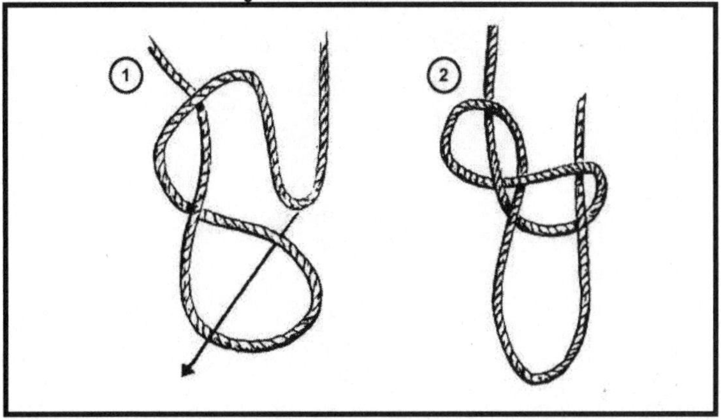

i. **End–of–the–Rope Prusik.** This knot **(Figure 9-13)** attaches a movable rope to a fixed rope. The knot has two round turns, with a locking bar perpendicular to the standing end of the rope. Tie a bow line within 6 inches of the locking bar. When you dress the knot, leave at least a 4 inch tail on the working end.

Figure 9-13. END–OF–THE–ROPE PRUSIK

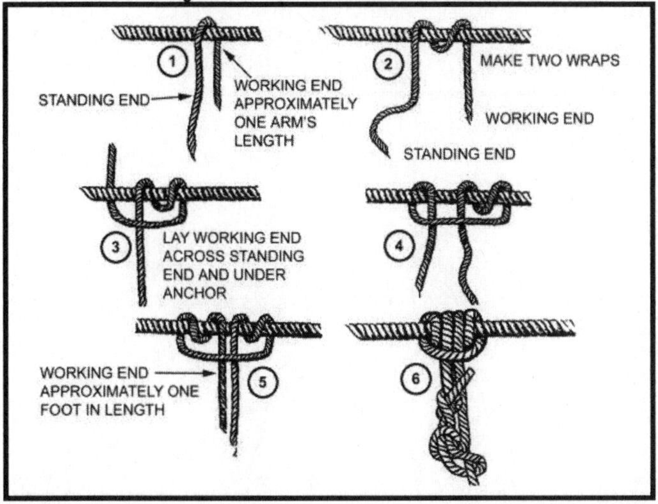

j. **Middle–of–the–Rope Prusik**. The Middle–of–the–Rope Prusik **(Figure 9-14)** attaches a movable rope to a fixed rope anywhere along the length of the fixed rope. To make this knot, make two round turns with a locking bar perpendicular to the standing end. Ensure the wraps do not cross and that the overhand knot is within 6 inches from the horizontal locking bar. Ensure the knot does not move freely on the fixed rope.

Figure 9-14. MIDDLE–OF–THE–ROPE PRUSIK

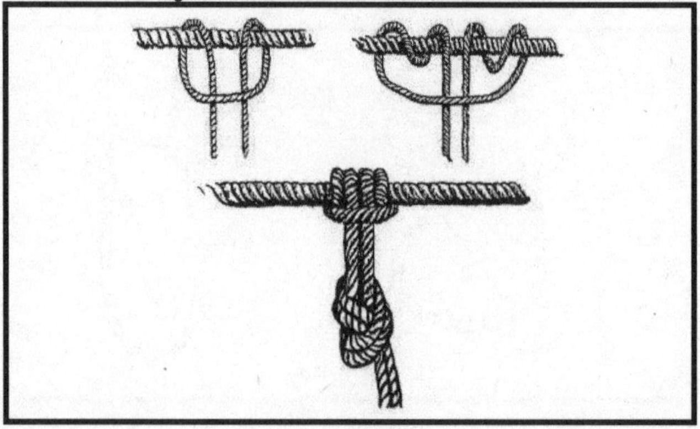

9-8. BELAYS. Belaying is any action taken to arrest a climber that has fallen, or to control the rate of descent of a load from a higher to lower elevation. The belayer also helps manage a climber's rope or the rate of the climber's or rappeller's descent by controlling the amount of rope that is taken out or in. The belayer must be anchored in a stable position to prevent him from being pulled out of position, and losing control of the rope. Two types of belays are body and mechanical.

a. **Body Belay.** This belay **(Figure 9-15)** uses the belayer's body to apply friction. The belayer routes the rope around his body. He must be careful, because his body might have to bear the entire weight of the load.

Figure 9-15. BODY BELAY

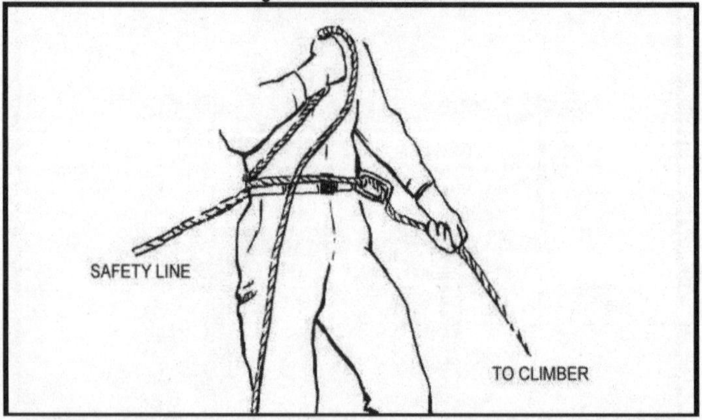

b. **Mechanical Belay**. This belay **(Figure 9-16)** uses mechanical devices to help the belayer control the rope, as in rappelling. A variety of mountaineering devices are used to construct a mechanical belay.

Figure 9-16. MECHANICAL BELAY

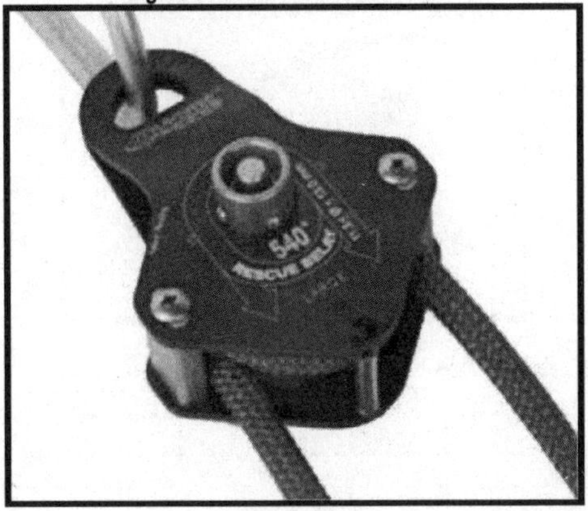

(1) ***Munter Hitch.*** One of the most often used belays; the Munter Hitch **(Figure 9-17)** requires very little equipment. The rope is routed through a locking pear-shaped carabiner, then back on itself. The belayer controls the rate of descent by manipulating the working end back on itself with his brake hand.

Figure 9-17. MUNTER HITCH

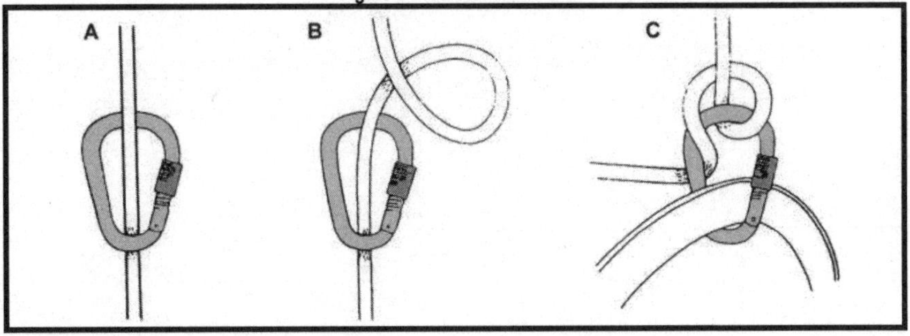

(2) ***Air Traffic Controller.*** The ATC (air traffic controller) is a locking mechanical belay device **(Figure 9-18)**. It locks down on itself once tension is applied in opposite directions. This requires the belayer to apply very little force with his brake hand to control the rate of descent or to arrest a climber's fall.

Figure 9 -8. AIR TRAFFIC CONTROLLER

9-9. CLIMBING COMMANDS. Table 9-1 shows the sequence of commands used by climber and belayer.

Table 9-1. SEQUENCE OF CLIMBING COMMANDS

Command	***Given By***	***Meaning***
BELAY ON, CLIMB	Belayer	Belay is on and climber may climb.
CLIMBING	Climber	Climber is climbing.
UP-ROPE	Climber	Belayer, remove excess slack in the rope
BRAKE	Climber	Belayer, immediately apply brake.
FALLING	Climber	Climber is falling, immediately apply brake, prepare to arrest the fall.
TENSION	Climber	Belayer, removes all slack from climbing rope until rope is tight, apply brake, hold position.
SLACK	Climber	Belayer, allows climber to pull slack into the climbing rope (belayer may have to assist).
ROCK	Anyone	Command given to alert everyone of an object falling near them. Belayer immediately applies brake.
POINT	Climber	Alert belayer that the direction of pull on the climbing rope has changed in the event of a fall.
STAND-BY	Climber or Belayer	Hold position, stand by, I am not ready.
DO YOU HAVE ME?	Climber	Informal command to belayer to prepare for a fall or prepare to lower me.
I HAVE YOU	Belay	The brake is on and I am prepared for you to fall, or to lower you.
OFF-BELAY	Climber	Alert belayer that claimer is safetied in or it is safe to come off belay.
3-METERS	Belayer	Alert climber to the amount of rope between climber and belayer (May be given in feet or meters)
BELAY-OFF	Belayer	I am off belay

9-10. ROPE INSTALLATIONS. Rope installations may be constructed by teams to help units negotiate natural and man-made obstacles. Installation teams consist of a squad-sized element with 2 to 4 trained mountaineers. Installation teams deploy early and prepare the AO for safe, rapid movement by constructing various types of mountaineering installations. Following construction of an installation, the squad, or part of it, remains on site to secure and monitor the system, assist with the control of forces across it, and adjust or repair it during use. After the unit passes, the installation team may disassemble the system and deploy to another area.

a. **Fixed-Rope Installations.** A fixed rope is anchored in place to help Rangers move over difficult terrain. Its simplest form is a rope tied off on the top of steep terrain. As terrain becomes steeper or more difficult, fixed rope systems may require intermediate anchors along the route. Planning considerations follow:

- Does the installation allow you to bypass the obstacle?
- *(Tactical)* Can obstacle be secured from construction through negotiation, to disassembly?
- Is it in a safe and suitable location? Is it easy to negotiate? Does it avoid obstacles?
- Are natural and artificial anchors available?
- Is the area safe from falling rock and ice?

b. **Vertical Hauling Line**. This installation **(Figure 9-19)** is used to haul men and equipment up vertical or near vertical slopes. It is often used with the fixed rope.

(1) ***Planning Considerations.***

- Does the installation allow you to bypass the obstacle?
- *(Tactical)* Can you secure the installation from construction through negotiation to disassembly?
- Does it have good loading and off loading platforms? Are the platforms natural and easily accessible? Do they provide a safe working area?
- Does it allow sufficient clearance for load? Is there enough space between the slope and the apex of the A frame to allow easy loading and off loading of troops and equipment?
- Does it have an A frame for artificial height?
- Does it allow you to haul line in order to move personnel and equipment up and down slope?
- Does the A frame have a pulley or locking carabiner to ease friction on hauling line?
- Does it have a knotted hand line to help Rangers up the installation?
- Does it allow for Rangers top and bottom to monitor safe operation?

(2) ***Equipment.***

- Three 120 foot (37-meter) static ropes.
- Three 15 foot sling ropes for constructing A frame.
- Two A frame poles, 7 to 9 feet long, 4 to 6 inches in diameter (load dependent).
- Nine carabiners.
- One pulley with steel locking carabiner.

Figure 9-19. VERTICAL HAULING LINE

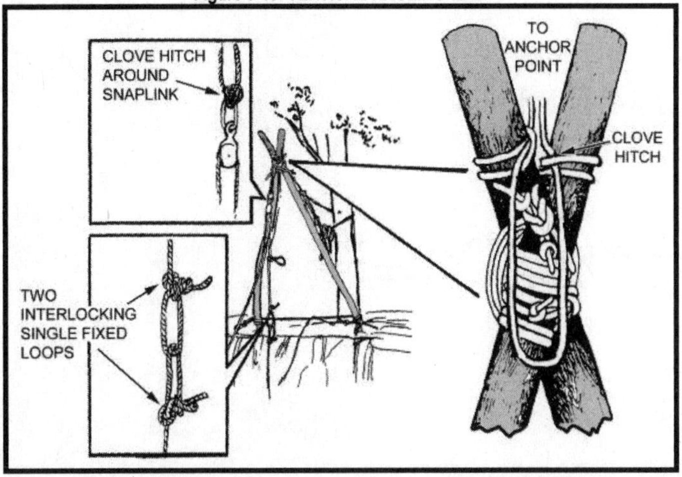

c. **Bridging.** Rope bridges are employed in mountainous terrain to bridge linear obstacles such as streams or rivers where the force of flowing water may be too great or temperatures are too cold to conduct a wet crossing.

(1) ***Construction.*** The rope bridge is constructed using static ropes. The max span that can be bridged is half the length of the rope for a dry crossing, three-fourths for a wet crossing. The ropes are anchored with an anchor knot on the far side of the obstacle, and tied off at the near end with a transport-tightening system. Rope bridge planning considerations follow:

- Does the installation allow you to bypass the obstacle?
- *(Tactical)* Can you secure the installation from construction through negotiation to disassembly?
- Is it in the most suitable location such as a bend in the river? Is it easily secured?
- Does it have near and far side anchors?
- Does it have good loading and off loading platforms?

(2) **Equipment (1 Rope Bridge).**

- One sling rope per Ranger.
- One steel locking carabiner.
- Two steel ovals.
- Two 120 foot static ropes.

(3) ***Construction Steps.*** The first Ranger swims the rope to the far side and ties a tensionless anchor **(Figure 9-4)**, between knee and chest level, with at least 6 to 8 wraps. The BTC ties a transport-tightening system **(Figure 9-20)** to the near side anchor point. Then, he ties a Figure 8 slipknot and incorporates a locking half hitch around the adjustable bight. Insert two steel oval carabiners into the bight so the gates are opposite and opposed. The rope is then routed around the near side anchor point at waist level and dropped into the steel oval carabiners.

(a) A three Ranger pulling team moves forward from the platoon. No more than three are used to tighten the rope. Using more can cause over tightening of the rope, bringing it near failure.

(b) Once the rope bridge is tight enough, the bridge team secures the transport tightening system **(Figure 9-20)** using two half hitches, without losing more than 4 inches of tension.

(c) Personnel cross using either the *Commando Crawl* **(Figure 9-21)**, *Rappel Seat* **(Figure 9-22)**, or *Monkey Crawl* **(Figure 9-23)** method.

Figure 9-20. TRANSPORT-TIGHTENING SYSTEM

Figure 9-21. COMMANDO CRAWL METHOD

Figure 9-22. RAPPEL SEAT (TYROLEAN TRAVERSE) METHOD

Figure 9-23. MONKEY CRAWL METHOD

(4) ***Bridge Recovery.*** Once all except two troops have crossed the rope bridge, the bridge team commander (BTC) chooses either the wet or dry method to dismantle the rope bridge. If the BTC chooses the dry method, he should have anchored his tightening system with the transport knot.

(a) The BTC back stacks all of the slack coming out of the transport knot, ties a fixed loop, and places a carabiner into the fixed loop.

(b) The next to last Ranger to cross should attach the carabiner to his rappel seat or harness, and then move across the bridge using the Tyrolean traverse method.

(c) The BTC removes all knots from the system. The far side remains anchored. The rope should now only pass around the near side anchor.

(d) A three-Ranger pull team, assembled on the far side, takes the end brought across by the next to last Ranger, pulls and holds the rope tight again.

(e) The BTC attaches himself to the rope bridge and moves across.

(f) Once across, the BTC breaks down the far side anchor, removes the knots, and then pulls the rope across. If it is a wet crossing, any method can be used to anchor the tightening system.

(g) All personnel cross except the BTC or the strongest swimmer.

(h) The BTC then removes all knots from the system.

(i) The BTC ties a fixed loop, inserts a carabiner, and attaches it to his rappel seat or harness. He then manages the rope as the slack is pulled to the far side.

(j) The BTC then moves across the obstacle while being belayed from the far side.

d. **Suspension Traverse.** The suspension traverse is used to move personnel and equipment over rivers, ravines, and chasms, or up or down a vertical obstacle. By combining the transport-tightening system used during the rope bridge, an A Frame used for the vertical haul Line (**Figure 9-24**), and belay techniques device, units can make a suspension traverse (**Figure 9-25** and

Figure 9-26). Installing a suspension traverse can be slow and equipment intensive. Everyone must be well-trained and rehearsed in the procedures.

(1) ***Construction.*** The suspension traverse is constructed with static ropes. The max span that can be bridged is generally 75 percent length of the shortest rope. Planning considerations include those for rope bridge and vertical haul line.

(2) ***Equipment.***

- Three static installation ropes.
- Seven sling ropes.
- Nine carabiners.
- One heavy duty double pulley.
- One locking carabiner.
- One canvas pad.

Figure 9-24. ANCHORING THE TRAVERSE ROPE TO THE A FRAME

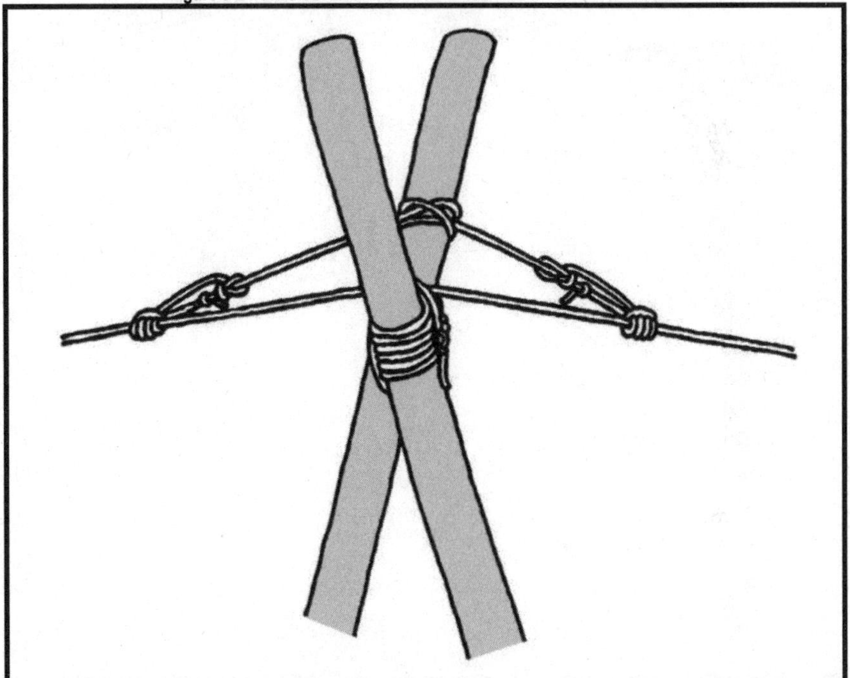

Figure 9-25. CARRYING ROPE FOR USE ON A SUSPENSION TRAVERSE

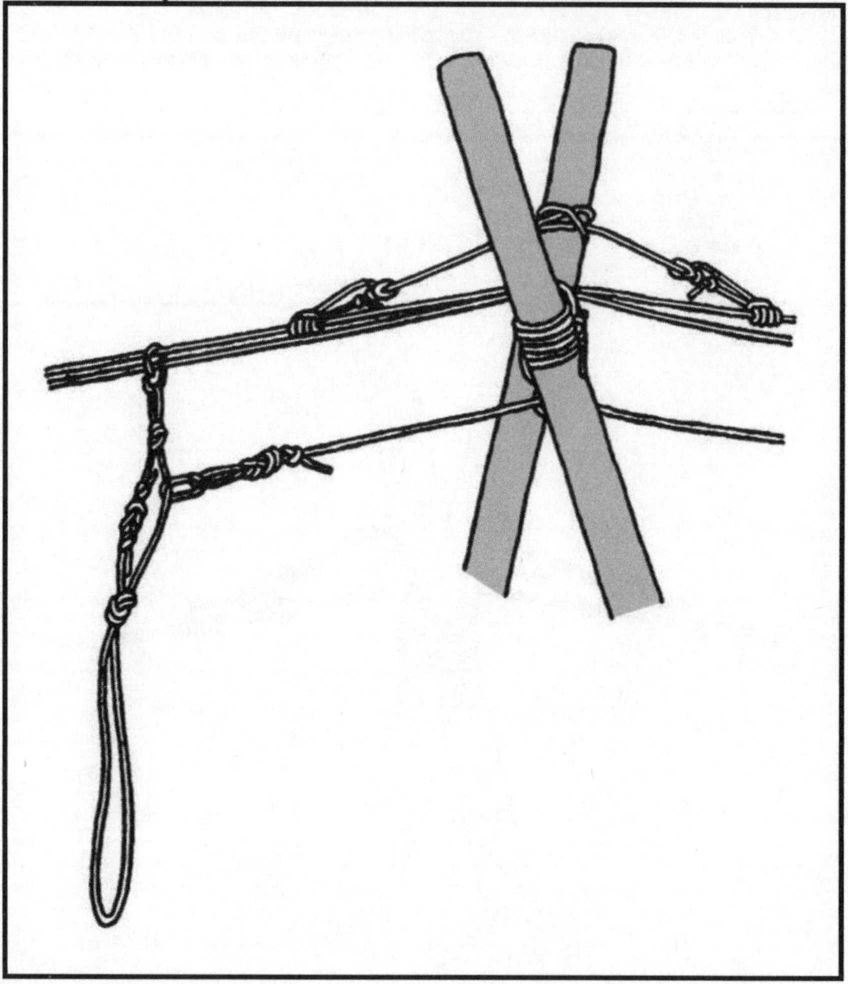

Figure 9-26. SUSPENSION TRAVERSE

9-11. RAPPELLING. Rappelling is a quick method of descent, but it can be extremely dangerous. Dangers include failure of the anchor or other equipment, and individual error. Anchors used in mountainous environments should be chosen carefully. Great care must be taken to load the anchor slowly and to avoid placing too much stress on the anchor. To ensure this, bounding rappels are prohibited–only walk-down rappels are permitted.

a. **Hasty and Body Rappels.** These quick and easy rappels **(Figure 9-27** and **Figure 9-28)** should only be used on moderate pitches–never on vertical or overhanging terrain. Gloves are used with both to prevent rope burns.

Figure 9-27. HASTY RAPPEL

Figure 9-28. BODY RAPPEL

b. **Seat Hip Rappel.** This rappel uses either a Figure 8 descender or a carabiner wrap descender **(Figure 9-29)**. Whichever is used, it is inserted in a sling rope seat, then fastened to the rappeller. This gives the Ranger enough friction for a fast, controlled descent **(Figure 9-30)**.

Figure 9-29. FIGURE 8 DESCENDER

Figure 9-30. CARABINER WRAP DESCENDER (INSET)

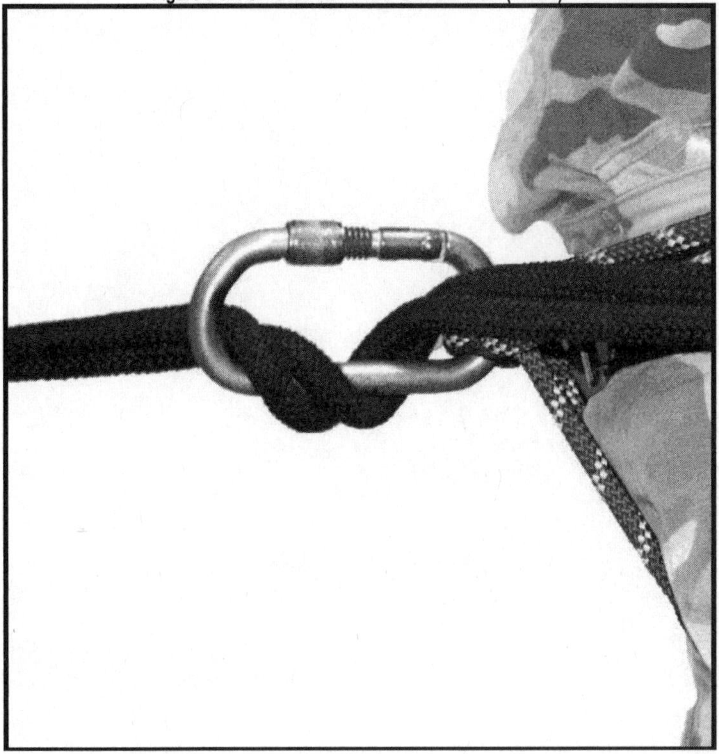

Figure 9-31. SEAT HIP RAPPEL (SHOWN WITH CARABINER WRAP DESCENDER)

c. **Site Selection.** The selection of the rappel point depends on factors such as mission, cover, route, anchor points, and edge composition (loose or jagged rocks). Personnel working near the edge must tie in, and the rappel point must have–

- Smooth route, free of loose rock and debris.
- Good primary and secondary anchors.
- Anchor point above rappeller's departure point.
- Equal tension between all anchor points.
- Suitable loading and off-loading platforms
- Double rope, if possible.
- Long enough ropes to reach the off-loading platform.

Chapter 10
MACHINE GUN EMPLOYMENT

Machine guns are a Ranger platoon's most effective weapons against a dismounted enemy force. Machine guns allow the Ranger unit to engage enemy forces from a greater range and with greater accuracy than individual weapons. A leader's ability to properly employ available machine guns and achieve fire superiority is often the deciding factor on the battlefield.

10-1. SPECIFICATIONS. Table 10-1 shows references and specifications for various machine guns.

Table 10-1. SPECIFICATIONS

WEAPON	M249	M240B	M2	MK 19
Field Manual	**3-22.68**	**3-22.68**	**3-22.65**	**3-22.27**
TM	**9-1005-201-10**	**9-1005-313-10**	**9-1005-213-10**	**9-1010-230-10**
Description	5.56-mm gas-operated automatic	7.62-mm gas-operated medium	.50-caliber recoil-operated heavy	40-mm air- cooled, blowback-operated automatic GL
Weight	16.41 lbs (gun w/barrel) 16 lbs (tripod)	27.6 lbs (gun w/ barrel) 20 lbs (tripod)	128 lbs (gun w/ barrel and tripod)	140.6 lbs (gun w/ barrel and tripod)
Length	104 cm	110.5 cm	156 cm	109.5 cm
Maximum Range	3,600 m	3,725 m	6,764 m	2,212 m
Maximum Effective Range	Bipod/ point: 600 m Bipod/ area: 800 m Tripod/ area: 1,000 m Grazing: 600 m	Bipod/point: 600m Tripod/point: 800m Bipod/area: 800m Tripod/area:1,100m Suppression: 1,800 m Grazing: 600 m	Point: 1,500 m (single shot) Area: 1,830 m Grazing: 700 m	Point: 1,500 m Area: 2,212 m
Tracer Burnout	900 m	900 m	1,800 m	-
Sustained Rate of Fire Rounds/ burst Interval Barrel change	50 RPM 6 to 9 rounds 4 to 5 sec Every 10 min	100 RPM 6 to 9 rounds 4 to 5 sec Every 10 min	40 RPM 6 to 9 rounds 10 to 15 sec End of day or if damaged	40 RPM
Rapid Rate of Fire Rounds/ burst Interval Minutes to barrel change	100 RPM 6 - 9 rounds 2 - 3 sec 2 minutes	200 RPM 10 - 13 rounds 2 - 3 sec 2 minutes	40 RPM 6 - 9 rounds 5 - 10 sec Change barrel end of day or if damaged	60 RPM
Cyclic Rate of Fire/ Minutes to barrel change	850 RPM, continuous burst/ min	650 - 950 RPM, continuous burst/ min	450 - 550 RPM, continuous burst	325 – 375 RPM, continuous burst

10-2. DEFINITIONS. Table 10-2 defines machine gun terms and **Figure 10-1** shows some of them.

Table 10-2. MACHINE GUN TERMS

Line of Sight	The imaginary line drawn from the firer's eye through the sights to the point of aim.
Burst of Fire	A number of successive rounds fired with the same elevation and point of aim when the trigger is held to the rear. The number of rounds in a burst can vary depending on the type of fire employed.
Trajectory	The curved path of the projectile in its flight from the muzzle of the weapon to its impact. As the range to the target increases, so does the curve of trajectory.
Maximum Ordinate	The height of the highest point above the line of sight the trajectory reaches between the muzzle of the weapon and the base of the target. It always occurs at a point about two-thirds of the distance from weapon to target and increases with range.
Cone of Fire	The pattern formed by the different trajectories in each burst as they travel downrange. Vibration of the weapon and variations in ammunition and atmospheric conditions all contribute to the trajectories that make up the cone of fire.
Beaten Zone	The elliptical pattern formed when the rounds in the cone of fire strike the ground or target. The size and shape of the beaten zone change as a function of the range to and slope of the target, but is normally oval or cigar shaped and the density of the rounds decreases toward the edges. Gunners and automatic riflemen should engage targets to take maximum effect of the beaten zone. Due to the right-hand twist of the barrel, the simplest way to do this is to aim at the left base of the target.
Sector of Fire	An area to be covered by fire that is assigned to an individual, a weapon, or a unit. Gunners are normally assigned a primary and a secondary sector of fire.
Primary Sector of Fire	The primary sector of fire is assigned to the gun team to cover the most likely avenue of enemy approach from all types of defensive positions.
Secondary Sector of Fire	The secondary sector of fire is assigned to the gun team to cover the second most likely avenue of enemy approach. It is fired from the same gun position as the primary sector of fire.
Final Protective Fire (FPF)	An immediately available, prearranged barrier of fire to stop enemy movement across defensive lines or areas.
Final Protective Line (FPL)	A predetermined line along which grazing fire is placed to stop an enemy assault. If an FPL is assigned, the machine gun is sighted along it except when other targets are being engaged. An FPL becomes the machine gun's part of the unit's final protective fires. An FPL is fixed in direction and elevation. However, a small shift for search must be employed to prevent the enemy from crawling under the FPL and to compensate for irregularities in the terrain or the sinking of the tripod legs into soft soil during firing. Fire must be delivered during all conditions of visibility.
Principal Direction of Fire (PDF)	Assigned to a gunner to cover an area that has good fields of fire, or that has a likely dismounted avenue of approach. A PDF also provides mutual support to an adjacent unit. If no FPL has been assigned, then sight machine guns using the PDF. If a PDF is assigned and other targets are not being engaged, then machine guns remain on the PDF. It is used only if an FPL is not assigned; it then becomes the machine gun's part of the unit's final protective fires.

Figure 10-1. TRAJECTORY AND MAXIMUM ORDINATE

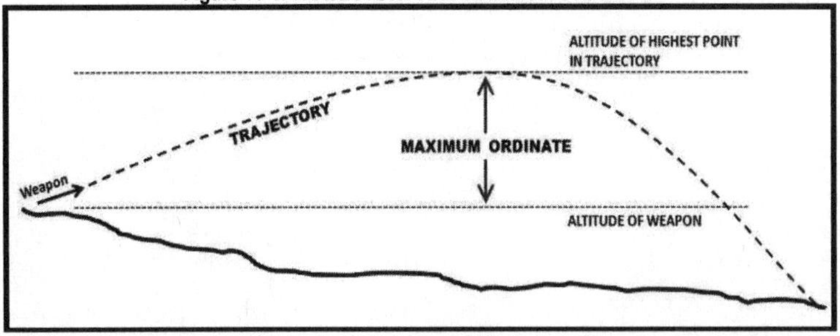

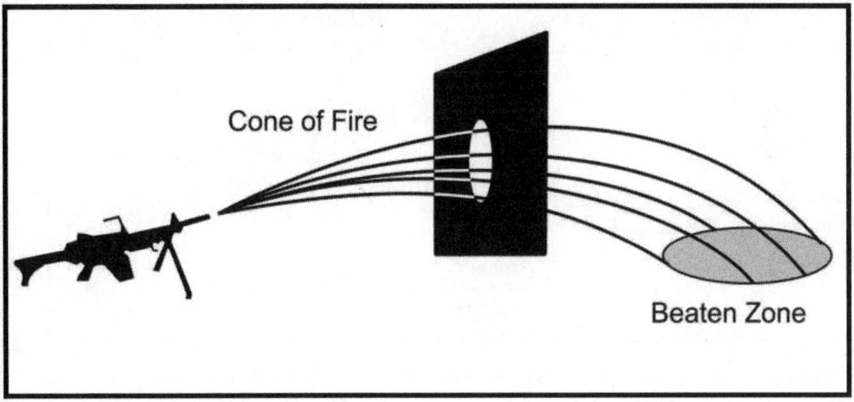

10-3. CLASSES OF AUTOMATIC WEAPONS FIRE. The U.S. Army classifies automatic weapons fires with respect to ground, target, and weapon.

a. **Respect to Ground. (See Table 10-3** and **Figure 10-2)**

Table 10-3. CLASSES OF FIRE - RESPECT TO THE GROUND

Grazing Fires	Automatic weapons graze fire when the center of the cone of fire fails to rise more than 1 meter aboveground. Grazing fire is employed in the ***final protective line*** (FPL) in defense. It is possible only when the terrain is level or uniformly sloping. Any dead space encountered along the FPL must be covered by indirect fire, such as from an M203. When firing over level or uniformly sloping terrain, the machine gun M240B and M249 can attain a maximum of 600 meters of grazing fire. The M2 can attain a maximum of 700 meters.
Plunging Fires	Plunging fire occurs when there is little or no danger space from the muzzle of the weapon to the beaten zone. It occurs when weapons fire at long range, from high to low ground, into abruptly rising ground, or across uneven terrain, resulting in a loss of grazing fire at any point along the trajectory.

Table 10-2 CLASSES OF FIRE - RESPECT TO THE GROUND

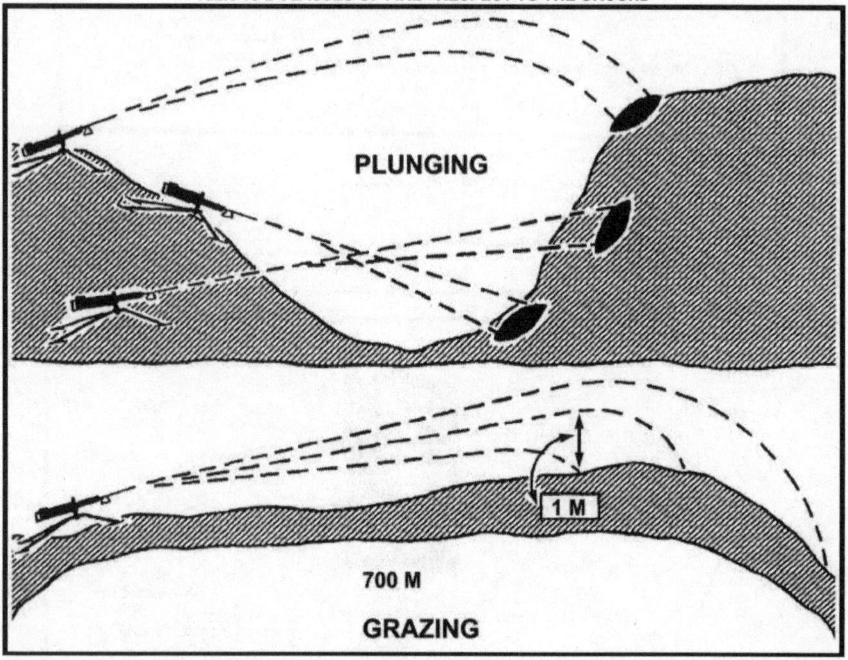

b. **Respect to Target.** Leaders and gunners should strive at all times to position their gun teams where they can best take advantage of the machine gun's beaten zone with respect to an enemy target. Channeling the enemy by use of terrain or obstacles so they approach a friendly machine gun position from the front in a column formation is one example. In this situation, the machine gun would employ enfilade fire on the enemy column, and the effects of the machine gun's beaten zone would be much greater than if it engaged the same enemy column from the flank. **Table 10-4** defines and compares the four classifications of fire with respect to the target, and **Figure 10-3A** and **Figure 10-3B** show them.

Table 10-4 CLASSES OF FIRE - RESPECT TO TARGET

ENFILADE FIRE **Best COF/ T**	• Occurs when long axes of *beaten zone* and target coincide/ nearly coincide. • Can be *frontal fire* on column or *flanking fire* on line. • ***Most desirable class of fire with respect to the target***, because it… • Makes maximum use of the *beaten zone.* • Leaders and gunners should always try to position guns for *enfilade fire.*
FRONTAL FIRE **Column…Yes** **Line……..No**	• Occurs when the long axis of the *beaten zone is at a right angle to the front of the target.* • **Highly desirable against a column.** • Becomes *enfilade fire* as *beaten zone* coincides with long axis of target. • ***Less desirable against a line***, because most of the *beaten zone* normally falls below or after the enemy target.
FLANKING FIRE **Column…No** **Line……..Yes**	• Delivered directly against the flank of the target. • ***Most desirable against a line.*** • Becomes *enfilade fire* as *beaten zone* will coincide with the long axis of the target. • ***Least desirable against a column***, because most of the *beaten zone* normally falls before or after the enemy target.
OBLIQUE FIRE	• Gunners and automatic riflemen. • Occurs when long axis of *beaten zone* is at any angle other than a right angle to the front of the target.

Table 10-3A CLASSES OF FIRE - RESPECT TO TARGET

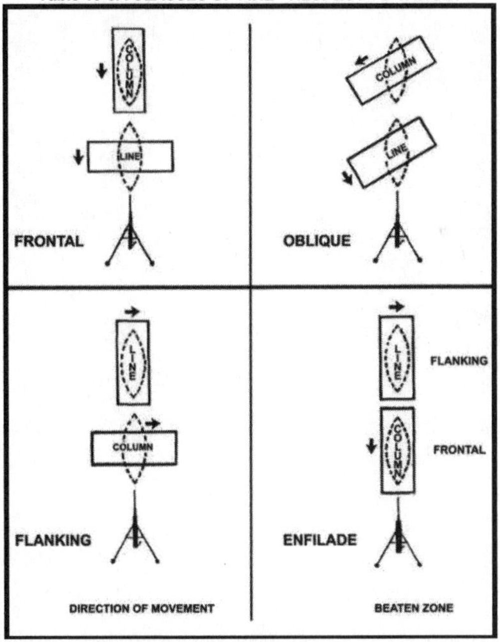

Table 10-3B CLASSES OF FIRE - RESPECT TO TARGET

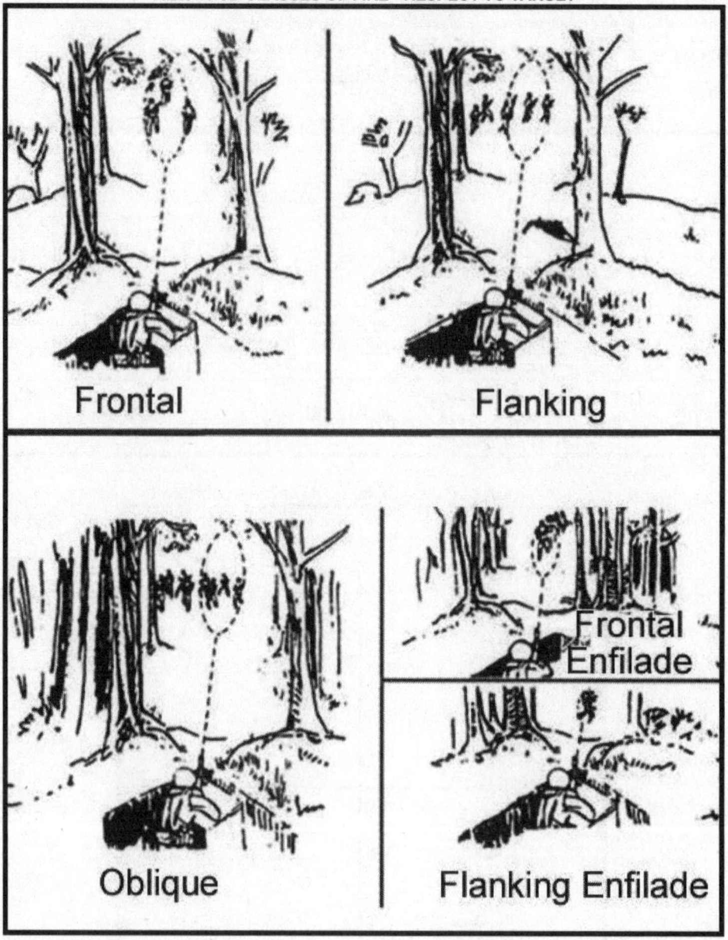

c. **Respect to the (Machine) Gun.** Fires with respect to the weapon include fixed, traversing, searching, traversing and searching, swinging traverse, and free gun fires. (See **Table 10-5** and **Figure 10-4.**)

Table 10-5. CLASSES OF FIRE - RESPECT TO GUN

Fixed	Fixed fire is delivered against a stationary point target when the depth and width of the beaten zone will cover the target with little or no manipulation needed. After the initial burst, the gunners will follow any change or movement of the target without command.
Traversing	Traversing disperses fires in width by successive changes in direction, but not elevation. It is delivered against a wide target with minimal depth. When engaging a wide target requiring traversing fire, the gunner should select successive aiming points throughout the target area. These aiming points should be close enough together to ensure adequate target coverage. However, they do not need to be so close that they waste ammunition by concentrating a heavy volume of fire in a small area.
Searching	Searching distributes fires in depth by successive changes in elevation. It is employed against a deep target or a target that has depth and minimal width, requiring changes in only the elevation of the gun. The amount of elevation change depends upon the range and slope of the ground.
Traversing and Searching	This class of fire is a combination in which successive changes in direction *and* elevation result in the distribution of fires both in width and depth. It is employed against a target whose long axis is oblique to the direction of fire.
Swinging Traverse	Swinging traverse fire is employed against targets that require major changes in direction but little or no change in elevation. Targets may be dense, wide, in close formations moving slowly toward or away from the gun, or vehicles or mounted troops moving across the front. If tripod mounted, the traversing slide lock lever is loosened enough to permit the gunner to swing the gun laterally. When firing swinging traverse, the weapon is normally fired at the cyclic rate of fire. Swinging traverse consumes a lot of ammunition and does not have a beaten zone because each round seeks its own area of impact.
Free Gun	Free gun fire is delivered against moving targets that must be rapidly engaged with fast changes in both direction and elevation. Examples are aerial targets, vehicles, mounted troops, or infantry in relatively close formations moving rapidly toward or away from the gun position. When firing free gun, the weapon is normally fired at the cyclic rate of fire. Free gun fire consumes a lot of ammunition and does not have a beaten zone because each round seeks its own area of impact.

Table 10-4. CLASSES OF FIRE - RESPECT TO GUN

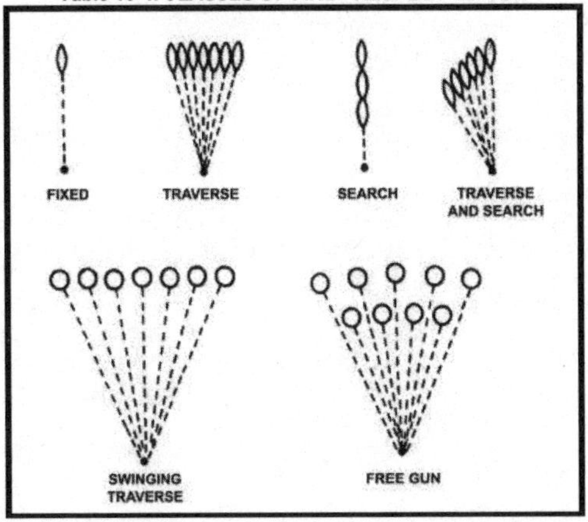

10-4. OFFENSE. Successful offensive operations depend on effective employment of fire and movement. They are both essential, and they depend on each other. For example, without the support of covering fires, maneuvering in the presence of enemy fire can produce huge losses. Covering fires, especially those that provide fire superiority, allow maneuvering in the offense. However, fire superiority alone rarely wins battles. The primary objective of the offense is to advance, occupy, and hold the enemy position.

a. **Medium Machine Guns**. In the offense, the platoon leader can establish his base of fire element with the M240B, the M249 light machine gun, or a combination of the weapons. When the platoon scheme of maneuver is to conduct the assault with the Infantry squads, the platoon sergeant or weapons squad leader may position this element and control its fires. The M240B machine gun is more stable and accurate at greater ranges, but takes longer to maneuver, on the tripod than on the bipod. Machine gunners can–

- Target key enemy weapons until the enemy's assault element masks the machine gunners' fires.
- Suppress the enemy's ability to return accurate fire
- Hamper the maneuver of the enemy's assault element.
- Fix the enemy in position
- Isolate the enemy by cutting off his avenues of reinforcement.
- Shift fire to the flank opposite the one being assaulted and continue targeting any automatic weapons providing enemy support
- Engage enemy counterattack, if any.
- Cover the gap created between the forward element of the friendly assaulting force and terrain covered by indirect fires when the direct fires are lifted and shifted.
- On signal, displace (along with the base of fire element) to join the assault element on the objective.

b. **MK 19 and M2.** As part of the base-of-fire element, the MK 19 and M2 can help the friendly assault element. They can do this by suppressing enemy bunkers and lightly armored vehicles. Even if their fire is too light to destroy enemy vehicles, well aimed suppressive fire can keep the enemy buttoned up and unable to place effective fire on friendly assault elements. The MK 19 and M2 are particularly effective in preventing lightly armored enemy vehicles from escaping or reinforcing. Both vehicle-mounted weapons can fire from a long range standoff position, or be moved forward with the assault element.

c. **Base of Fire.** Machine gun fire from a support by fire (SBF) position must be the minimum possible to keep the enemy from returning effective fire. Ammunition must be conserved so the guns do not run out of ammunition. The weapon squad leader positions and controls the fires of all machine guns in the element. Machine gun targets include key enemy weapons or groups of enemy targets either on the objective or attempting to reinforce or counterattack. The nature of the terrain, desire to achieve some standoff, and the other factors of METT-TC prompt the leader to the correct tactical positioning of the base of fire element. There are distinct phases of rates of fire employed by the base of fire element:

- Initial heavy volume (rapid rate) to gain fire superiority.
- Slower rate to conserve ammunition (sustained rate) while still preventing effective return fire as the assault moves forward.
- Increased rate as the assault nears the objective.
- Lift and shift to targets of opportunity.
- Machine guns in the SBF role should be set in and assigned a primary and alternate sector of fire as well as a primary and alternate position.
- Machine guns are suppressive fire weapons used to suppress known and suspected enemy positions. Therefore, gunners cannot be allowed to empty all of their ammunition into one bunker simply because that is all they can identify at the time.
- Shift and shut down the weapon squad gun teams one at a time, not all at once. M203 and mortar or other indirect fire can be used to suppress while the machine guns are moved to where they can shoot.
- Leaders must take into account the SDZ of the machine guns when planning and executing the lift and or shift of the SBF guns. The effectiveness of the enemy on the objective will play a large role in how much risk should be taken with respect to the lifting or shifting of fires.
- Once the SBF line is masked by the assault element, fires are shifted and or lifted to prevent enemy withdrawal or reinforcement.

d. **Maneuver Element.** Under certain terrain conditions, and for proper control, machine guns may join the maneuver or assault unit.

(1) When this is the case, they are assigned a cover fire zone or sector. The machine guns seldom accompany the maneuver element. The gun's primary mission is to provide covering fire. The machine guns are only employed with the maneuver element when the area or zone of action assigned to the assault or company is too narrow to permit proper control of the guns. The machine guns are then moved with the unit and readied to employ on order from the leader and in the direction needing the supporting fire.

(2) When machine guns move with the element undertaking the assault, the maneuver element brings the machine guns to provide additional firepower. These weapons are fired from a bipod, in an assault mode, from the hip, or from the underarm position. They target enemy automatic weapons anywhere on the unit's objective.

(3) After destroying the enemy's automatic weapons, if any, the gunners distribute fire over their assigned zones or sectors. The machine gunner in the assault position engages within 300 meters of his target, often at point blank ranges.

(4) If the platoon's organic weapons fail to cover the area or zone of action, the company commander can assign more machine guns and personnel. This might help the platoon accomplish its assigned mission. Each machine gunner is assigned a zone or a sector to cover, and they move with the maneuver element.

e. **Controlled Occupation of the Support-by-Fire Position.** This is one of the key elements in setting up an SBF position. To remain undetected, use stealth and control.

(1) The WSL must move to and establish a release point just short of the SBF position.

(a) Order of movement for the Weapons Squad during movement to their position is the WSL, Gun Team 2 (Gunner, AB, AG), Gun Team 1 (AG, Gunner, AB), and Gun Team 3 (AG, Gunner, AB).

(2) The WSL then moves forward from the release point with the Gun 2 Gunner. The gunner will get into position and remain in bipod mode to provide security.

(3) The WSL will then bring forward the Gun 2 AG and AB.

(a) The AG moves to the left of the gun and emplaces the tripod.

(b) The AB drops off all ammunition at the gun position and then moves to pull flank or rear security.

(4) Once Gun 2 is in place, the WSL brings the Gun 1 AC forward with the tripod (T&E already attached). The WSL emplaces the tripod.

(5) Once the tripod is emplaced, the WSL brings the gunner forward. The gunner places the gun on the tripod. The WSL gets down behind the gun to ensure it covers down on its sector of fire, and that everything is IAW the PL's guidance.

(6) The WSL directs the Gun 1 AB to move forward, drop off the ammunition, and then assume his security position.

(7) Once Gun 1 is emplaced, Gun 3 is occupied exactly the same as Gun 1.

(8) Once Gun 1 and Gun 3 are both on tripods, the Gun 2 gunner also places his gun on the tripod.

(9) The WSL calls the PL to notify him that the SBF position is occupied.

f. **Controlled Withdrawal of the Support-by-Fire Position.** The platoon leader can use this method to cover the withdrawal of the platoon as well as to provide security for the support-by-fire position itself.

(1) Before the platoon moves off an objective, the weapons squad leader shifts the machine guns' sectors of fire to cover the objective.

(2) After he has the guns covering the objective, the weapons squad leader starts breaking down the gun positions, one at a time.

(3) After the main body of the platoon starts to moves off the objective, the gun teams move one at a time into the order of movement, with the last gun breaking down as soon as the platoon is completely off the objective.

(4) The entire weapons squad moves tactically to link up with the rest of the platoon.

10-5. DEFENSE. The platoon's defense centers on its machine guns. The platoon leader positions the rifle squad to protect the machine guns against the assault of a dismounted enemy formation. The machine gun provides the necessary range and volume of fire to cover the squad front in the defense.

a. **Position.**

(1) ***Requirements.*** The main requirement of a suitable machine gun position in the defense is its effectiveness in accomplishing specific missions. The position should be accessible and afford cover and concealment. Machine guns are positioned to protect the front, flanks, and rear of occupied portions of the defensive position, and to be mutually supporting. Attacking troops usually seek easily traveled ground that provides cover from fire. For each machine gun, the leader chooses three positions: primary, alternate, and supplementary. In choosing them, he ensures they cover his sector and have protection on their flanks.

(2) ***Employment.*** The leader positions each machine gun to cover the entire sector or to overlap sectors with the other machine guns. The engagement range may extend from over 1,000 meters where the enemy begins his assault to point blank range. Machine gun targets include enemy automatic weapons and command and control elements.

b. **Distribution of Fire.** Machine gun fire is distributed in width and depth in a defensive position. Machine guns are the backbone or framework of the defense, because the leader can use them–

- To subject the enemy to increasingly devastating fire from the initial phases of his attack, and
- To neutralize any partial enemy successes the enemy might attain by delivering intense fires in support of counterattacks.
- To help the unit hold ground (thanks to its tremendous firepower).

c. **Medium Machine Guns**. In the defense, the medium machine gun provides sustained direct fires that cover the most likely or most dangerous enemy dismounted avenues of approach. It also protects friendly units against the enemy's dismounted close assault. Platoon leader positions his machine guns to concentrate fires in locations where he wants to inflict the most damage to the enemy. He also places them where they can take advantage of grazing enfilade fires, stand off or maximum engagement range, and best observation of the target area. Machine guns provide overlapping and interlocking fires with adjacent units and cover tactical and protective obstacles with traversing or searching fires. When final protective fires are called for, machine guns (aided by M249 fires) place an effective barrier of fixed, direct fire across the platoon front. Leaders position machine guns to—

- Concentrate fires where they want to kill the enemy.
- Fire across the platoon front
- Cover obstacles by direct fire.
- Tie in with adjacent units.

d. **MK 19 and M2.** In the defense, MK 19 and M2 machine guns may be fired from the vehicle mount or dismounted from the vehicle and mounted on a tripod at a defensive fighting position designed for the weapon system.

- These guns provide sustained direct fires that cover the most likely enemy mounted avenue of approach. Their maximum effective range enables them to engage enemy vehicles and equipment at far greater ranges than the platoon's other direct-fire weapons.
  • When mounted on the tripod, the M2 and MK 19 are highly accurate to their maximum effective ranges. Predetermined fires can be planned for likely high payoff targets. The trade off is that these weapon systems are heavy, and thus slow to move.
- These guns are less accurate mounted on vehicles than when fired from the tripod mounted system. However, they are more easily maneuvered to alternate firing locations should the need arise.

10-6. CONTROL OF MACHINE GUNS. Leaders use control measures, coordinating instructions, and fire commands to control the engagements of their machine guns. Rehearsals are key in a leader's ability to control his machine guns.

a. Methods of Fire Control. The noise and confusion of battle may limit the ability of a leader to control his machine guns. Therefore a leader must be able to use a combination of methods that will accomplish the mission. The following are several successful methods for a leader to control fires:

- Oral
- Hand and arm signals
- Prearranged signals
- Personal contact
- Range cards

b. Fire Commands. A fire command is given to deliver effective fire on a target quickly and without confusion. It is essential that the commands delivered by the weapons squad leader are understood and echoed by the assistant gunner/ gun team leader and the gunner. The elements of a fire command follow:

(1) ***Alert.*** Lets the gun crew know that they are about to engage a target
(2) ***Direction.*** Lets the gun team know where to engage
(3) ***Description.*** Lets the gun team know what they are engaging
(4) ***Range.*** If not already set on predestined target, the gun team can adjust the T&E
(5) ***Method of Fire.*** This element includes manipulation and rate of fire. Manipulation dictates the class of fire with respect to the weapon. It is announced as FIXED, TRAVERSE, SEARCH, or TRAVERSE AND SEARCH. Rate controls the volume of fire (Sustained, rapid, and cyclic).
(6) ***Command to Open Fire.*** Initiates the firing of the weapon system

10-7. AMMUNITION PLANNING. Leaders must carefully plan for the rates of fire to be employed by machine guns as they relate to the mission and the amount of ammunition available. The weapons squad leader must fully understand the mission, the amount of available ammunition, and the application of machine gun fire needed to fully support all key events of the mission. Careful planning helps ensure the guns do not run out of ammunition.

a. A mounted platoon might have access to enough machine gun ammunition to support the guns throughout any operation. A dismounted platoon with limited resupply capabilities has to plan for only the basic load to be available. In either case, leaders must take into account key events the guns must support during the mission. They must plan for the rate of machine gun fire needed to support the key events, and the amount of ammunition needed for the scheduled rates of fire.

b. The leader must estimate how much ammunition is needed to support all the machine guns. He adjusts the amount used for each event to ensure enough ammunition remains for all phases of the operation. Examples of planning rates of fire and ammunition requirements for a platoon's machine guns in the attack are shown in **Figure 10-5.**

Figure 10-5. WEAPONS SQUAD TACTICS, TECHNIQUES, AND PROCEDURES

a. Use a starter belt when moving (about 50 to 70 rounds).
b. Ensure ammo and NVDs are in packs, such as an assault pack for mounted and city operations, rucksack for long sustainment missions, and are readily accessible.
c. Carry T&E and tripod together.
d. Mission dependent on when you take the tripod. (Urban operations)
e. Use optics, lasers, NVDs. For example, in urban operations, think about using a reflexive sight, because most of your engagements will be 150m or less. Also, zero your iron sights.
f. Use your terrain analysis to help decide how best to employ your machine guns.
g. A support-by-fire position will not always be at a 90-degree angle to the assault element. Emplace it where the weapon system can best support the platoon and cause the most damage.
h. Senior Ranger on the gun team is the AG for command and control; gunner is next in charge, then the AB.
i. Plan time to perform maintenance on the weapon systems.

Chapter 11
CONVOY OPERATIONS

This chapter outlines a technique for conducting vehicle convoy operations. Convoy operations present a challenge to the Ranger leader. Trucks and other combat vehicles produce a large signature on the battlefield and increase your unit's value as a target. Vehicle movement is restricted to roads and terrain that they can traverse; therefore, a sound plan must be implemented to minimize the possibility of compromise (**FM 3-21.8** and **FM 55-30**).

11-1. PLANNING. When conducting a vehicle convoy as part of your operation, it is important to incorporate the convoy as a leader uses the eight steps of the troop leading procedures. The following information should be included when conducting a mission analysis using METT-TC:

a. **Mission.** The PL will extract the following information from the Company OPORD.
- Vehicle support (number and type of vehicles, ACL)
- Weather: road conditions
- Vehicle pickup and drop off location/ marking
- Vehicle movement timeline (pick up time, movement time and so on.)
- Vehicle routes (primary and alternate, checkpoints)

b. **Enemy.**
- Known or suspected enemy locations in the AO or along planned routes.
- Potential locations for enemy ambush or IED emplacement.
- Recent enemy activities or reactions to convoy operations.

c. **Terrain.**
- Identify potential pick up and drop off locations.
- Evaluate routes, pick up and drop off locations using OAKOC.
- Consider weather/ road conditions.

d. **Troops.**
- Number of PAX per vehicle
- Chalks and Chalk Leaders identified
- Tactical cross load
- Linkup and marking teams identified
- Pick up and drop off security plan.

e. **Time.** Backwards planning sequence:
- Ground tactical plan
- Unload plan
- Ground movement plan
- Loading plan
- Staging plan

NOTE: Allocate time for movement/ recon/ establishment of security.

f. **Civilians.**
- Known civilian locations along route
- ROE/ Actions with civilians and civilian vehicles during movement

11-2. FIVE PHASES OF TRUCK MOVEMENT. Each phase must support the ground tactical plan, which specifies actions in the objective area to accomplish the commander's intent for the assigned mission, be it a raid, ambush, recon or other follow on missions.

a. **Staging Plan.**

(1) Establish security of entrucking/ pickup point.

(2) Employ markings/ recognition signals (day/ night).
(3) Link up.
(4) Conduct final friendly unit coordination with Convoy/ Truck Commander.
(5) Disseminate information and any changes to subordinate leaders.

b. **Loading Plan–Task Organization/ Tactical Cross-Loading.** Each Ranger is assigned to a vehicle ensuring tactical cross load of weapon systems and key personnel.

- Truck #, Key Leader, Key Weapon Systems, Additional Personnel, Communications.
- Location of PL.
- Location of PSG and Medic.
- Location of WSL.
- Location of communication (FO/ RTO).

c. **Ground Movement Plan.**
(1) Troops awake and alert pulling active security during movement.
(2) Platoon Leader and vehicle commanders tracking route progress.
(3) Compromise and contingency plan.
(a) React to IED.
(b) React to Ambush.
(c) Vehicle breakdown.

d. **Unloading Plan.**
(1) Dismount vehicles (IAW SOP/ Reverse Load Plan).
(2) Establish security of de trucking point.
(3) PSG accounts for personnel and clears all vehicles for departure.
(4) Establish security halt/ perimeter.
(5) Adjust perimeter as vehicles depart area.

e. **Ground Tactical Plan.** Prepare to continue movement and conduct follow on mission

Convoy Warning Order

1. SITUATION: A brief statement of the enemy and friendly situation. Who, What, Where
2. MISSION: Who, What, When, Where, Why
3. TASK ORGANIZATION: Based on tasking from higher WARNO.
4. INITIAL TIME SCHEDULE:

When	Who	What

5. SPECIAL INSTRUCTIONS: PCC/PCI Guidance, Rehearsals, additional tasks to be accomplished.
6. SERVICE SUPPORT:

Class I: (Rations & Water)

Class III: (POL)

Class V: (Ammunition /Pyrotechnics)

Weapon System	Rounds	Type

Pyrotechnic Device	Number	Location

Class VIII: (Medical)

7. UNIFORM AND EQUIPMENT COMMON TO ALL.

CONVOY BRIEF **(Modified from FM 55-30)**

Movement Order No. ____

References: A. ______________________________ *(Maps, tables and relevant documents)*

B. ______________________________

TASK ORGANIZATION: (Internal organization for convoy – Manifest)

1. **SITUATION:**

a. **Enemy Forces:**

(1) Weather. General forecast.

(2) Light Data (EENT, % Illumination, MR, MS, BMNT)

(3) Discuss Enemy.

Identification of enemy (If known).
Composition / capabilities / strength / equipment
Location
(Hot Spots highlighted on map)

b. **Friendly Forces:**

(1) Operational support provided by higher headquarters.

Helicopter / Gunships
Quick Reaction Forces (QRFs)
MP Escorts / Rat Patrols
Fire Support elements

c. **Attachments:** (From outside the organization)

2. **MISSION: (WHO, WHAT, WHEN, WHERE, & WHY)**

3. **Employment Tactics, Techniques and Procedures**

a. Readiness Level
   i. Ride / Scanning (Observation)
   ii. At the Ready

b. Scanning / Sector of Fire
   i. Driver
      1. Sector of Scan is 9 – 1 clock position
      2. Observation with Mirrors
      3. Sector of Fire is 9 – 11 o'clock
   ii. TC
      4. Sector of Scan is 11 – 3 clock position
      5. Sector of Fire is 1 – 3 clock position
   iii. Other Systems are based off vehicle type and load

b. Target ID
   i. Communicate / Signal

c. Body Positioning
   i. Engage as you Train (Right or Left handed Firing)
   ii. Firing-side Shoulder Down
   iii. Maintain your Body Position
   iv. Weapon to Head

d. Rules of Engagement Concerns – Employ the appropriate systems based off the threat.

e. Point of Aim
   i. Moving Platform – Stationary Target requires aim to the rear (trail) and low dependant upon speed.

CONVOY BRIEF (Modified from FM 55-30) (Continued)

3. Employment Tactics, Techniques and Procedures

e. Point of Aim
   - ii. Moving Platform – Moving Target requires aim directly on and low.
   - iii. Stationary Platform – Moving Target requires aim to the front (lead) and low dependent upon speed.

f. Rate of Fire
   - i. Burst Mode
   - ii. Steady Suppression (ROE)

g. Magazine Awareness
   - i. Serviceability / maintenance
   - ii. Tracer Mix
   - iii. Magazine Storage / Placement
   - iv. Mounted, Reload when time is available

Dismounted, Seek cover prior to the need to Reload

REACT TO AMBUSH (NEAR)

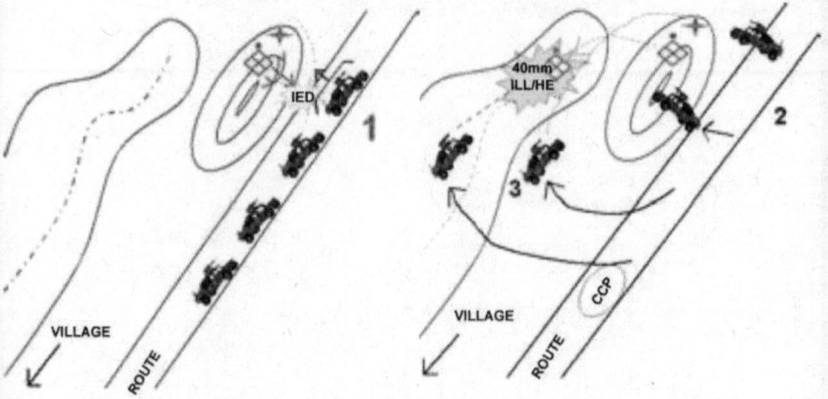

(Enemy Initiates Weapons Fire / RPG / IED)

1. Lead element of convoy reacts to contact and immediately returns fire. Lead fire team fires 40mm HE (at night also white star if NVGs drown out by vehicle headlights - leader calls it. FYI, most travel at night will be with headlights on).

2. Lead elements assault through ambush - fire team dismounts while gun trucks use M249, MK-19, and .50 cal in close support.

3. Trail fire team and trail gun truck move to block most likely enemy escape route. Fire team uses 40mm HE/Illumination if needed. (Use of 40mm becomes less likely as units get closer to urban areas due to collateral damage risk.) Establish CCP upon action completion or enemy neutralized.

REACT TO AMBUSH (FAR)

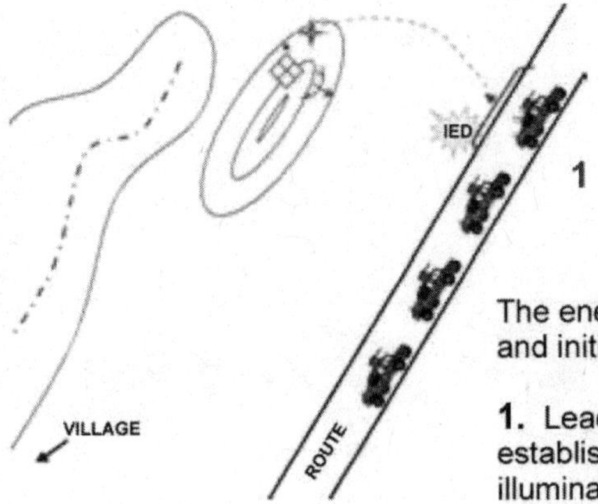

The enemy is outside 100 meters and initiates with IED/RPG.

1. Lead gun truck and lead fire team establish SBF - fire team fires 40mm illumination.

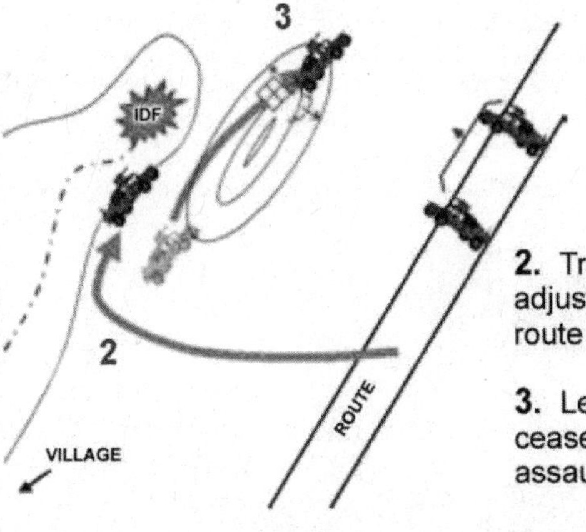

2. Trail element flanks - SL/PL adjusts IDF into enemy escape route (if in range).

3. Lead element shifts fire - arty cease loading - trail element assaults across - enemy destroyed.

CONVOY FORCED TO STOP
(Weapons Fire / RPG / IED / Indirect)

1. Vehicles forced to stop. Activate vehicle turn signal to indicate direction of contact.
2. Vehicle(s) / Personnel not in direct contact, report on internal communication, identifying truck number, type of contact, clock direction and eight digit grid if available.

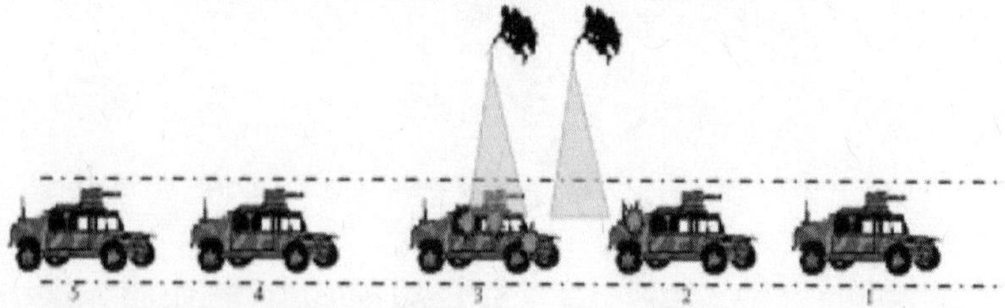

3. Personnel on vehicle(s) forced to stop dismount on the non-contact side, assume covered position(s) and provide initial base of fire.
4. The <u>entire</u> convoy halts; personnel will dismount vehicle(s) on the non-contact side and provide additional base of fire on the enemy if in range. Vehicle(s) not in contact will reposition to attack the attackers. M16's should now be on <u>Semi-Automatic</u> in order to conserve ammunition.
5. PL/CC/SC will move to better assess the situation and position the Gun Truck(s) in order to best suppress the enemy while maintaining <u>standoff</u>. Gun trucks will close with and engage the enemy.
6. Once the PL/CC/SC determines the convoy has either gained fire superiority or defeated the enemy contact, Recovery / CASEVAC operations will begin.
7. If the PL/CC/SC determines the convoy cannot gain fire superiority, leader will then conduct Break Contact procedures

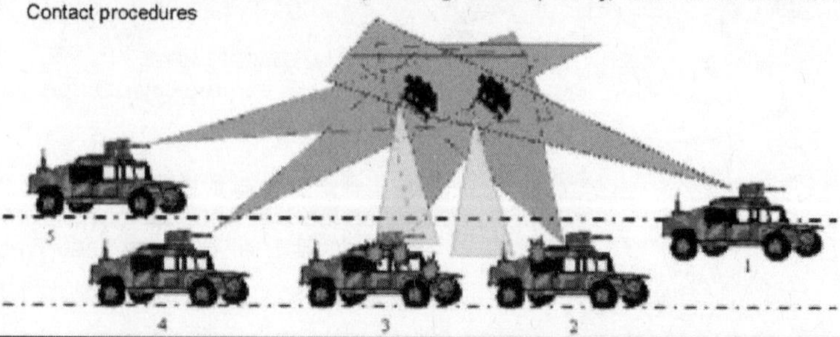

CONVOY FORCED TO STOP - METHOD 2

(Weapons Fire /RPG / IED / Indirect)

1. Vehicle(s) forced to stop. Activate vehicle turn signal to indicate direction of contact.

*** ALL PERSONNEL STAY IN VEHICLES ***

2. All vehicles immediately drive forward out of the kill zone.

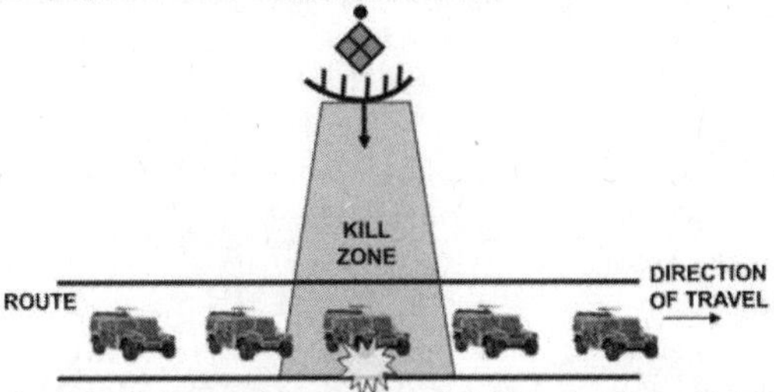

3. The vehicle(s) directly behind the disabled vehicle(s) push the disabled vehicle(s) out of the kill zone.
4. The vehicle(s) that are not disabled will establish a base of fire towards the suspected/known contact.
5. If fire superiority can be gained the PL will use the minimum force necessary to destroy the enemy.
6. If the PL determines the convoy cannot gain fire superiority, the leader will conduct break contact.

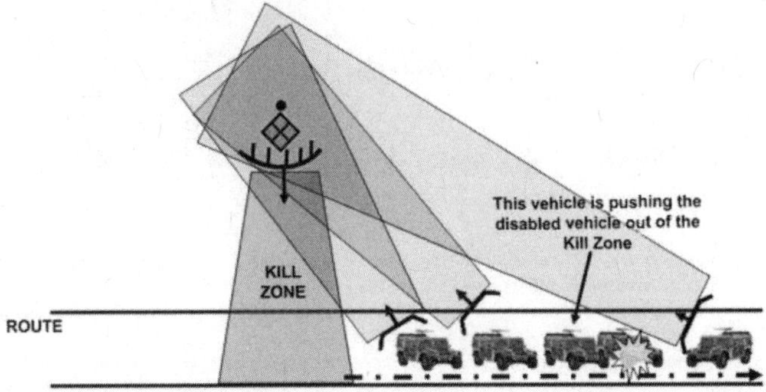

BREAK CONTACT

Always try to close with and destroy the enemy first. This way he cannot come back later on to attack you or another convoy again. However – if you must...

1. The PL/CC/SC determines the convoy cannot gain fire superiority and the decision has been made to Break Contact.
2. The PL/CC/SC designates that either Rally Point "Rear" or "Forward" will be used. If necessary, both Rally Points may be used. Communication systems and appropriate pyrotechnic signals will be used to communicate Break Contact and Rally Point.
3. Personnel will deploy obscuration measures if available. Utilizing cover and concealment, Aid & Litter team(s) will evacuate all casualty(s) under support of Gun Truck and other protective fire(s).
4. Personnel will maintain position and suppression in contact zone and assist Aid & Litter team(s) as necessary.
5. Disabled vehicle(s) will be hauled back or destroyed as directed by leaders. (Thermite or explosives)

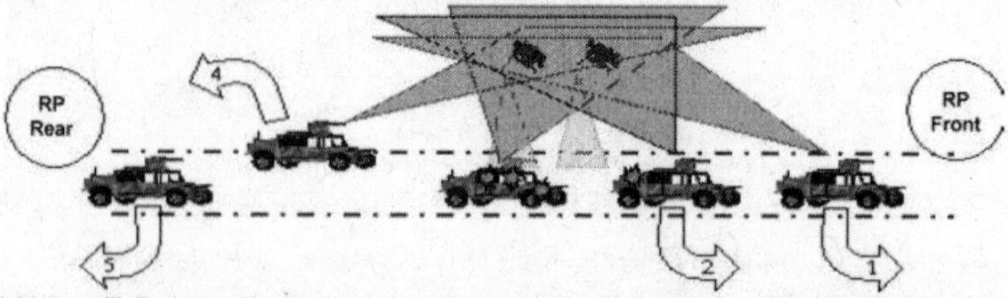

6. Vehicles will displace either backwards or forwards through the convoy lines under control of leaders. The most forward vehicle in the contact zone moves first, then the next most forward vehicle moves second. Vehicles will continue to displace. As vehicles displace, Gun Truck repositions as necessary until contact is broken.

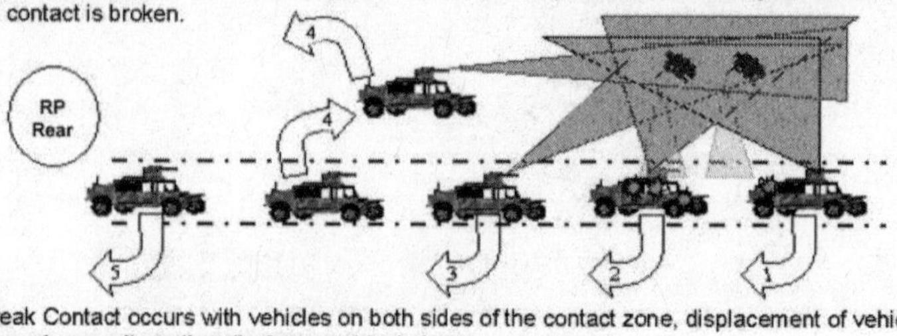

7. If Break Contact occurs with vehicles on both sides of the contact zone, displacement of vehicles will occur using an alternating displacement technique.
8. Upon occupation of the Rally Point, leaders will immediately position vehicles, setting 360 degree security, and conduct Consolidation and Reorganization.
9. If the convoy vehicles get separated when not in contact with the enemy, personnel and vehicles stay together and move to the closest Rally Point or Check Point.

RECOVERY / CASEVAC OPERATIONS

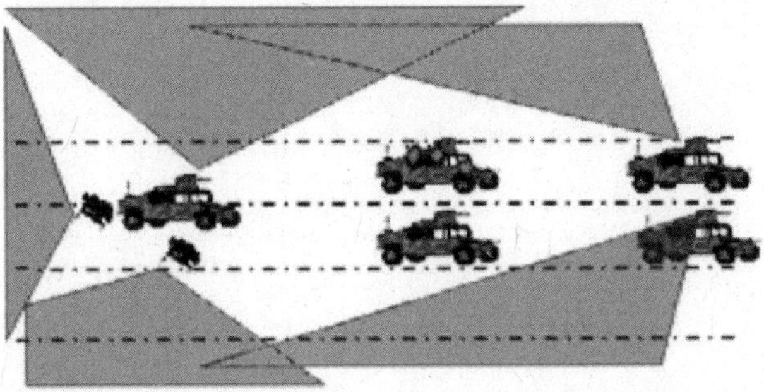

1. Once the leader assesses the enemy threat to be destroyed, neutralized, defeated and the area secured - Recovery / CASEVAC operations will begin. This keeps soldiers focused first on defeating and destroying the threat.

2. CASEVAC:

   a. Aid & Litter team will position on the safe side of the vehicle and extract casualty(s) and personnel.

   b. Treatment of casualty(s) will occur once they are safely removed from the contact area.

3. Vehicle Recovery Procedures:

   a. Recovery team will position on the safe side of the disabled vehicle.

   b. TC will dismount and assess the disabled vehicle.

   c. If determined the vehicle can be safely recovered, TC guides the recovery vehicle into position and conducts a hasty hook-up. TC will operate the disabled vehicle is necessary.

   d. Upon exiting the contact area, complete and correct hook-up procedures will occur.

   e. If assessment results in outside support necessary for recovery, leader will contact higher for guidance.

4. Once recovery operations are complete, the team will displace and conduct link-up with the convoy at the Rally Point.

   Disabled vehicles(s) will be abandoned or destroyed as directed by leaders. (Thermite or explosives).

Chapter 12
URBAN OPERATIONS

Today's security environment demands more from leaders than ever before. Leaders must not only be able to lead Rangers but also influence other people. They must be able to work with members of other Services and governmental agencies. They must win the willing cooperation of multinational partners, both military and civilian. Urban offensive operations pose the greatest risks to Army forces and noncombatants. Yet, ultimately, the military demands self aware and adaptive leaders who can compel enemies to surrender in war and master the circumstances facing them in peace. Victory and success depend on the effectiveness of these leaders' organizations. Developing effective organizations requires hard, realistic, and relevant training.

12-1. AN URBAN PERSPECTIVE. Urban operations include full spectrum operations—offensive, defensive, and stability or civil support—that may be executed, either sequentially or (more likely) simultaneously during an urban operation. For further study, see **FM 3-06, FM 3-06.11, FM 3-21.8,** *75th Ranger Regiment Advanced UO SOP*, and **Ranger Training Circular 350-1-2**, where asterisks (*) represent collective guidance and tactics acquired from lessons learned throughout several units.

12-2. STRATEGIC IMPORTANCE OF URBAN AREAS. Several factors attract armies to combat in urban areas:

- Use the defensive advantages of urban environment
- Develop allegiance and support of populace
- Adapt urban resources for operational or strategic purposes: infrastructure, capabilities, and other resources.
- Draw the enemy in
- Play on area's symbolic importance
- Use the area's geographical advantages
  - Dominance of a region
  - Avenue of approach

12-3. MODERN ARMY URBAN OPERATIONS. Conflict is not the goal. Conflict is just the inevitable means of transition from a (perceived) unsatisfactory state of affairs to a better one. To achieve a successful transition, leaders must carefully orchestrate military and civilian capabilities.

12-4. TASK ORGANIZATION. Task-organizing subordinate units for urban operations depend largely on the nature of the operation. Some units, however, are always part of the task organization to ensure the success of UO. Infantry, SPECOPS, CA, aviation, military police, PSYOP, military intelligence, and engineers are units required for all urban operations across full spectrum operations. Other type forces—such as armor, artillery, and chemical—have essential roles in specific types of urban operations, and may apply less to other operations. When using armor, consider the destructive power of the weapon system and its limitations.

12-5. FULL SPECTRUM OPERATIONS. Military forces conduct full spectrum operations within urban areas. Commanders conduct full spectrum operations abroad by executing offensive, defensive, and stability urban operations as part of a joint, interagency, and multinational effort. The situation will mandate that one type of operation—offense, defense, stability, or civil support—dominates the urban operation. Commanders will often find themselves executing offensive, defensive, stability, or civil support operations at the same time. In fact, waiting until all combat operations are concluded before beginning stability or civil support operations often results in lost, sometimes irretrievable, opportunities. The dominant type of operation will vary between different urban areas even in the same campaign.

a. **Offense.**

(1) ***Characteristics.***

- Surprise
- Concentration
- Tempo
- Audacity

(2) **Organization.**

- Decisive
- Shaping
- Sustaining

(3) ***Movement to Contact.*** In an urban area where the threat situation is vague, Army forces often conduct a movement to contact to establish or regain threat contact and develop the situation. A movement to contact in an urban area occurs as both sides try to establish their influence or control over a contested urban area. The situation determines whether the movement to contact or its specific technique, the search and attack, is appropriate.

(4) ***Attack.*** The attack is the most common and likely offensive operation that Army forces conduct in an urban environment. Commanders conducting major operations and commanders of large tactical units usually execute deliberate attacks. Hasty attacks normally occur at company level or bellow as units use their initiative to take advantage of tactical opportunities.

(5) ***Exploitation.*** Exploitation follows a successful attack to disrupt the threat in depth. The exploitation focuses on the urban area as well as on the remnants of the threat. A successful exploitation to seize an urban area works efficiently because the attack preempts the defense and denies the threat the full advantages of urban terrain. Commanders conducting exploitation must acknowledge the vulnerability of their forces to counterattack and ambush in urban areas. An urban area provides ideal cover and concealment to hide threat reserves, reinforcements, or reorganized forces. Constrictions of routes into and through the urban area make exploitation forces a potentially dense target and limit maneuver options.

(6) ***Pursuit.*** The pursuit is designed to destroy threat forces attempting to escape. It focuses on the threat and not on urban areas. The agility of Army aviation forces for attack, reconnaissance, and transportation is essential to execute a successful pursuit around and through urban areas. Against a large conventional enemy in a major urban area with a large civil population present, offensive operations might require a greater commitment of Army resources than in other environments.

b. **Defense.** Defensive UO are generally conducted as a shaping operation within a larger major operation. These temporary operations often set conditions for successful offensive operations, stability operations, or civil support operations. Commanders often conduct defensive UO within other types of operations to protect essential facilities in the urban area, protect flanks against counterattack, prevent the breakout of isolated enemies, or protect valuable supply bases or vulnerable convoy routes. In UO, essential facilities will likely include urban ports and airfields required by sister services to support ground operations. There are five general characteristics of a successful UO defense: preparation, security, disruption, massing effects, and flexibility.

c. **Stability.** Stability operations in an urban environment require offensive and defensive operations, combined with other tasks unique to each stability operation. Military forces conduct urban stability operations for various reasons, including noncombatant evacuation operations and peace operations. Urban stability operations will require an offensive capability to destroy any military capability that overtly threatens its objectives. Various stability operations will also require the distribution of food or aid and the protection or assistance of agencies conducting economic or humanitarian activities.

d. **Civil Support.** Army civil support operations in an urban environment aid other U.S. agencies and organizations in mitigating the consequences of natural and man–made disasters. In response to the disaster, civil support operations require the equipment, personnel, or organizational abilities of Army forces rather than the Army's combat capabilities.

12-6. PREPARATIONS FOR FUTURE URBAN OPERATIONS. To operate successfully in a complex urban environment requires a thorough understanding of the urban environment and rigorous, realistic UO training. Training should cover full spectrum operations to include appropriate tactics, techniques, and procedures. Training should also replicate the following:

a. The psychological impact of intense, close combat against a well trained, relentless, and adaptive enemy.

b. The effects of noncombatants, including governmental and nongovernmental organizations and agencies in close proximity to Army forces. This necessitates—

- An in depth understanding of culture and its effects on perceptions.
- An understanding of civil administration and governance.
- The ability to mediate and negotiate with civilians including the ability to effectively communicate through an interpreter.
- The development and use of flexible, effective, and understandable rules of engagement.

c. A complex intelligence environment requiring lower echelon units to collect and forward essential information to higher echelons for rapid synthesis into timely and useable intelligence for all levels of command. The multifaceted urban environment requires a bottom fed approach to developing intelligence.

d. The communications challenges imposed by the environment as well as the need to transmit large volumes of information and data.

e. The medical and logistic problems associated with operations in an urban area including constant threat interdiction against lines of communications and sustainment bases.

12-7. CONDUCT OF LIVE, VIRTUAL, AND CONSTRUCTIVE TRAINING. Force preparedness mandates integrating the actual use of urban terrain, exercises at urban training sites, simulations, or any combination into tactical and operational level intra and inter service training.

12-8. RANGERS – URBAN WARRIORS. In a complex urban environment, every Ranger—regardless of branch or military occupational specialty—must be committed and prepared to close with and kill or capture threat forces in an urban environment. Every Ranger must also be prepared to effectively interact with the urban area's noncombatant population and assist in his unit's intelligence collection efforts. In UO, every Ranger will likely be required to—

a. Perform advanced rifle marksmanship to include advanced firing positions, short range marksmanship, and night firing techniques (unassisted and with the use of optics).

- Operate unit's crew served weapons.
- Conduct urban reconnaissance and combat patrolling.
- Enter and clear buildings and rooms as part of an urban attack or cordon and search operation.
- Tactical site exploitation (SSE)
- Defend an urban area.
- Act as a member of a combat convoy (including specific drivers training).
- Recover own vehicles.
- Control civil disturbances.
- Navigate in an urban area.
- Prepare for follow on missions.
- Identify explosives, bombs, booby traps, materials used, and methods for marking and clearing them.
- Link up with battlespace owner.

b. While not all inclusive and necessarily urban specific, other critical individual and collective UO tasks (often modified for the urban environment) might include—

- Conduct troop leading procedures.
- React to contact, ambush, snipers, indirect fire, and improvised explosive devices.
- Set up personnel or vehicle checkpoint, or blocking positions around TGT location.
- Establish overwatch positions and support-by-fire positions such as sniper positions.
- Simultaneous clearing of top and bottom floors of the building.
- Assign climbing and roof-clearing teams for overwatch or sniper support.
- Teach how to use LRS, scout, and sniper teams effectively.
- Secure a disabled vehicle or downed aircraft.
- Call for indirect fire and close air support.
- Create and employ explosive charges.
- Handle detainees and enemy prisoners of war. Know how to extract HVTs.
- Treat and evacuate casualties.
- Accurately report information.
- Understand the society and culture specific to the area of operations.
- Use basic commands and phrases in the region's dominant language.
- Conduct tactical questioning.
- Interact with the media.
- Conduct thorough after action reviews.

12-9. PRINCIPLES

a. **Surprise.** Strike the enemy when, where, or how he is unprepared. Surprise is key to success. It gives the assaulting element the advantage.

b. **Security.** Never let the enemy gain an unexpected advantage. Maintaining security while moving through an ever changing urban environment is an art. It requires all members of the clearing team to remain at a high level of security and to maintain total situational awareness. Transitioning security of a sector from one team member to another needs to occur smoothly after cross-talk coordination between team members. The mission is only complete once you leave the urban environment. An urban environment offers less security than does "open" terrain. The key to survivability is a constant state of situational awareness.

- Maintain during all phases of the operation.
- Secure the four dimensional battlefield (height, depth, width, subterranean).
- Always maintain 360 degree security (include elevated and subterranean areas).

c. **Simplicity.** Prepare clear, uncomplicated plans, and provide subordinates with concise orders to ensure thorough understanding.

- Always keep plans simple.
- Ensure everyone understands the mission and the commander's intent.
- Plan and prepare for the worst.

d. **Speed.** Rate of military action.

- Move in a careful hurry.
- Smooth is fast and fast is smooth.
- Never move faster than you can accurately engage targets.
- Exercise tactical patience; let the situation develop, stay several steps ahead so the situation does not turn bad.

e. **Violence of Action.** Eliminate the enemy with sudden, explosive force.

- Combined with speed gives surprise.
- Prevents enemy reaction.
- Both physical and mental.

12-10. METT–TC. To effectively plan combat operations in urban environments, leaders must use troop-leading procedures and conduct a thorough analysis using METT–TC factors. The following lists specific guidance for planning urban operations. For more, see **Chapter 2**:

a. **Mission.** Know correct task organization to accomplish the mission (offense, defense, or stability and support operations).

b. **Enemy.**

(1) ***Disposition.*** Analyze the arrayal of enemy forces in and around your objective, known and suspected. Example: Known or suspected locations of minefields, obstacles, and strong points.

(2) ***Composition and Strength.*** Analyze the enemy's task organization, troops available, suspected strength, and amount of support from local civilian populace based on intelligence estimates. Is the enemy a conventional or unconventional force?

(3) ***Morale.*** Analyze the enemy's current operational status based on friendly intelligence estimates. For example, is the enemy well supplied? Has he recently won against friendly forces or taken many casualties? What is the current weather?

(4) ***Capabilities.*** Determine what the enemy can employ against your forces, for example, what weapon systems does he have? Does he have snipers? What IED, artillery assets, engineer assets, air defense assets, NBC threats, thermal/ NVD capabilities, close air support, armor threat, and so on. Be able to discriminate between threats and nonthreats such as suicide vests.

(5) ***Probable Course(s) of Action.*** Based on friendly intelligence estimates, determine how the enemy will fight within his AO (in and around yours). Know the enemy AO TTPs such as trip wires, pressure plate IEDs, or snipers. Analyze historical data from attacks, where, what, how and time of day.

c. **Terrain.** Leaders conduct a detailed terrain analysis of each urban setting, considering the types of BUAs and composition of existing structures. They use OCOKA when analyzing terrain, in and around the AO.

d. **Observation and Fields of Fire.** Always be prepared to conduct UO under limited visibility conditions.

e. **Cover and Concealment.** Thoroughly analyze areas inside and on the edge of urban areas. Identify routes to objectives to give assault forces the best possible cover and concealment. Take advantage of limited visibility, which allows forces to move undetected to their final assault/ breaching positions. Use overwatch elements and secondary entry teams for security while initial entry/ breaching teams move forward. When in the final assault position, forces should move as rapidly as tactically possible to access structures, which afford cover and concealment.

(1) It is human nature to stick together and to seek safety, but you must try to avoid bunching up at entry points, funnels, walls, or indoors. Maintain a safe but securable distance between teams and squads. This helps ensure that one grenade cannot take out the whole team at once.

(2) Learn to properly use obscurants, and use "tactical patience" to fully take advantage of these effects.

(3) Practice noise and light discipline. Avoid unnecessary voice communications, learn the proper use of white light, and limit contact with surfaces that could draw the enemy's attention.

f. **Obstacles.** Many man-made and natural obstacles exist on the periphery, as well as within the urban environment. Conduct a detailed reconnaissance of routes and objectives, including subterranean complexes, and consider route adjustments and special equipment needs. Ensure routes are clear (not blocked). Avoid roads that run along or thru market places, these roads can become blocked easily.

g. **Key Terrain.** Analyze which buildings, intersections, bridges, LZ/ PZ, airports, and elevated areas that provide a tactical advantage to you or the enemy. The leader must also identify critical infrastructure within his area of operations, which would provide the enemy with a tactical advantage on the battlefield. These may include, but are not limited to, communication centers, medical facilities, governmental facilities, and facilities with psychological significance.

h. **Avenues of Approach.** Consider roads, intersections, inland waterways, and subterranean constructions (subways, sewers, and basements). Leaders should classify areas as go, slow go, or no go based on the navigability of the approach. *Always have alternate infiltration and exfiltration routes. Keep in mind that a wall can be breached as an emergency exfiltration route.

> **NOTE:** Military maps may provide too little detail for urban terrain analysis, and they may or may not show the underground water and sewer systems, subways, and mass transit routes. You might have to obtain some current aerial photographs.

i. **Troops.** Analyze your forces using their disposition, composition, strength, morale, capabilities, and so on. Leaders must also consider the type and size of the objective to plan effective use of troops available.

j. **Time.** Operations in an urban environment have a slower pace and tempo. Leaders must consider the amount of time required to secure, clear, or seize the urban objective and stress and fatigue Rangers will encounter. Additional time must also be allowed for area analysis efforts; these may include, but are not limited to

- Maps and urban plans–recon and analysis. Collect historical data from other units as well as indigenous forces.
- Hydrological data analysis.
- Line of sight surveys.
- Long range surveillance and scout reconnaissance.
- Is artillery supporting you and someone else at the same time?
- How long does it take to shift a 155-mm howitzer and prep the gun?
- How close is Armor to your target? Will their presence compromise your mission? How long will it take them to move to a location?
- If Armor assets are not previously coordinated, how long will it take to get them?
- What is your priority level for getting Armor assets?
- How close is Armor to your target?
- Will their presence compromise your mission?
- How long will it take them to move to a location?
- Engineers. How much prep, survey and emplacement time of charges do they need?

k. **Civilians.** Authorities such as the National Command establish the Rules of Engagement. Commanders at all levels may provide further guidance for dealing with civilians in the AO. Leaders must daily remind subordinates of the latest ROE/to subordinates, and immediately inform them of any changes to the ROE. Rangers must have the discipline to identify the enemy from noncombatants and to ensure civilians understand and follow all directed commands.

> **NOTE:** Civilians may not speak English, may be hiding (especially small children), or may be dazed from a breach. Do not give them the means to resist. Rehearse how clearing/ search teams will react to these variables. Never compromise the safety of your Rangers. Consider having the TERP use a marking system to separate MAMs from women and children. Have designated dirty and clean rooms as well as a tactical questioning area.

12-11. CLOSE QUARTERS COMBAT. Due to the nature of a CQC encounter, engagements will be very close (within 10 meters) and very fast (targets exposed for only a few seconds). Most close quarter's engagements are won by who hits first and puts the enemy down. It is more important to knock an enemy down as soon as possible than it is to kill him. In order to win a close quarters engagement, Rangers must make quick, accurate shots by mere reflex. This is accomplished by reflexive fire training. Remember, no matter how proficient you are, always fire until the enemy goes down. All reflexive fire training is conducted with the eyes open.

> **NOTE:** Research has determined that only three out of ten people actually fire their weapons when confronted by an enemy during room cleaning operations. Close quarters combat success for the Ranger begins with the Ranger being psychologically prepared for the close quarter's battle. The foundation for this preparedness begins with the Ranger's proficiency in basic rifle marksmanship. Survival in the urban environment does not depend on advanced skills and technologies. *Rangers must be proficient in the basics.*

12-12. REHEARSALS. Similar to the conduct of other military operations, leaders need to designate time for rehearsals. UOs require a variety of individual, collective, and special tasks that are not associated with operations on less complex terrain. These tasks require additional rehearsal time for clearing, breaching, obstacle reduction, casualty evacuation, and support teams. Additionally, time must be identified for rehearsals with combined arms elements. These may include, but are not limited to

a. **Stance.** Feet are shoulder width apart, toes pointed straight to the front (direction of movement). The firing side foot is slightly staggered to the rear of the non firing foot. Knees are slightly bent and the upper body leans slightly forward. Shoulders are not rolled or slouched. Weapon is held with the butt stock in the pocket of the shoulder maintaining firm rearward pressure into the shoulder. This will allow for more accurate shot placement on multiple targets. The firing side elbow is kept in against the body and the hand should be forward on the weapon not on the magazine well. This allows for better control of the weapon. The stance should be modified to ensure the Ranger maintains a comfortable boxer stance.

b. **Low Carry Technique.** The butt stock of the weapon is placed in the pocket of the shoulder. The barrel is pointed down so the front sight post and day optic are just outside of Ranger's field of vision. The head is always up identifying targets. This technique is safest and is recommended for use by the clearing team once inside the room.

c. **High Carry Technique.** The butt stock of the weapon is held in the armpit. The barrel is pointed slightly up with the front sight post in the peripheral vision of the individual. To assume the proper firing position, push out on the pistol grip, thrust the weapon forward, and pull the weapon straight back into the pocket of the shoulder. This technique is best suited for the line up outside the door. Exercise caution with this technique always maintaining situational awareness, particularly in a multi floored building.

> **NOTE:** Muzzle awareness is critical to the successful execution of close quarter's operations. Rangers must never, at any time, point their weapons at or cross the bodies of their fellow Rangers. Rangers should also avoid (always) exposing the muzzle of their weapons around corners; this is referred to as "flagging."

d. **Malfunction.** If a Ranger has a malfunction with his weapon during any CQC training, he will take a knee to conduct immediate action. Once the malfunction is cleared, there is no need to immediately stand up to engage targets. Rangers can save precious seconds by continuing to engage from one knee. Whenever other members of the team see a Ranger down, they must automatically clear his sector of fire. Before rising to his feet, the Ranger warns his team members of his movement and only rises after he has checked his rear to make sure no one is shooting over him, and after they acknowledge him. If a malfunction occurs after he has committed to a doorway, the Ranger must enter the room far enough to allow those following him to enter, and then move away from the door. This drill must be continually practiced until it is second nature.

e. **Approach to a Building or Breach Point.** One of the trademarks of Ranger operations is the use of limited visibility conditions. Whenever possible, breaching and entry operations should be executed during hours and conditions of limited visibility. Rangers should always take advantage of all available cover and concealment when approaching breach and entry points. When natural or manmade cover and concealment is not available, Rangers should employ obscurants to conceal their approach. There are times when Rangers will want to employ obscurants to enhance existing cover and concealment. Members of the breach/ entry team should be numbered for identification, communication, and control purposes.

- The Ranger #1 should always be the most experienced and mature member of the team, *other than the team leader. The Ranger #1 is responsible for frontal and entry and breach point security.
- The Ranger #2 is directly behind the Ranger #1 in the order of movement, and he moves through the breach point in the opposite direction from the Ranger #1.
- The Ranger #3 will simply go opposite the Ranger #2 inside the room at least 1 meter from the door.
- The Ranger #4 moves the opposite the Ranger #3 and is responsible for rear security (and is normally the last Ranger into the room). An additional duty of the Ranger #4 is breaching.
- The team leader is responsible for initiating all voice and physical commands. He must exercise situational awareness at all times with respect to the task, friendly force, and enemy activity. He must be in a position to maintain control his team.
- The possibility of civilians in the building or rooms, Rangers may decide to only enter with precision weapons such as M 4s not M 249s to avoid civilian casualties.

> **NOTE:** Consider how much firepower each Ranger delivers. Where do you put the SAW gunner in the order? Weigh firepower against quick, accurate shots. If the # 4 Ranger has breaching responsibilities, it should not be the SAW gunner, because this would reduce your firepower.

f. **Actions Outside the Point of Entry.** Entry point position and individual weapon positions are important. The clearing team members should stand 1 to 2 feet from the entry point, ready to enter. They should orient their weapons so the team can provide its own 360–degree security at all times. Team members must signal to each other that they are ready at the point of entry. This is best accomplished by sending up a "squeeze" or rocking motion. If a tap method is used, an inadvertent bump may be misunderstood as a tap.

g. **Entering Building/ Clearing a Room.** See Battle Drill 6, Chapter 8.

h. **Locking Down the Room.**

- Control the situation within the room.
- Use clear, concise arm and hand signals. Voice commands should be kept to a minimum to reduce the amount of confusion and to prevent the enemy who might be in the next room from discerning what is going on. This enhances the opportunity for surprise and allows the assault force the opportunity to detect any approaching force.
- Physically and psychologically dominate.
- *Assess the situation. In a less hostile situation it may be better to slow clear instead of dominating the room with brute force. This will keep noncombatants calm and more manageable.
- Establish security and report status.
- Cursory search of the room to include the ceiling (three-dimensional fight).
- Identify the dead using reflexive response techniques (eye thump method or kick to the groin for males).
- Search the room for PIR, precious cargo as per the mission and time available.
- Evacuate personnel.
- Mark room clear using chemical lights, engineer tape, chalk, paint, VS 17 panels, and so on.

12-13. TTPS FOR MARKING BUILDINGS AND ROOMS. Units have long identified a need to mark specific buildings and rooms during UO. Sometimes rooms need to be marked as having been cleared, or buildings need to be marked as containing friendly forces. The US Army Infantry School is currently testing a remote marking device that can be used to mark doors from as far away as across a wide street. In the past, units have tried several different field expedient marking devices; some with more success than others. Chalk has been the most common. It is light and easily obtained but less visible than other markings. Other techniques include spray paint and paintball guns.

a. **NATO Standard Marking SOP.** The North Atlantic Treaty Organization (NATO) has developed a standard marking SOP for use during urban combat. It uses a combination of colors, shapes, and symbols. These markings can be fabricated from any material available. (**Figure 12-1** shows examples.)

b. **Spray Paint.** Canned spray paint is easily obtained and comes in a wide assortment of colors including florescent shades that are highly visible in daylight. Spray paint cannot be removed. Cans are bulky and noisy, and hard to carry with other combat equipment. Paint is invisible during darkness and it does not show up well through thermal sights.

c. **Paintball Guns.** Commercial paintball guns have been purchased by some units and issued to small unit leaders. Some models can be carried in standard military holsters. They can mark a building or door from about 30 meters. The ammunition and propellant gas is hard to obtain. The ammunition is fragile and, if it gets wet, will often jam the gun. Colors are not very bright and, like spray paint, cannot be seen at night or through thermal sights.

Figure 12-1. EXAMPLE NATO STANDARD MARKINGS

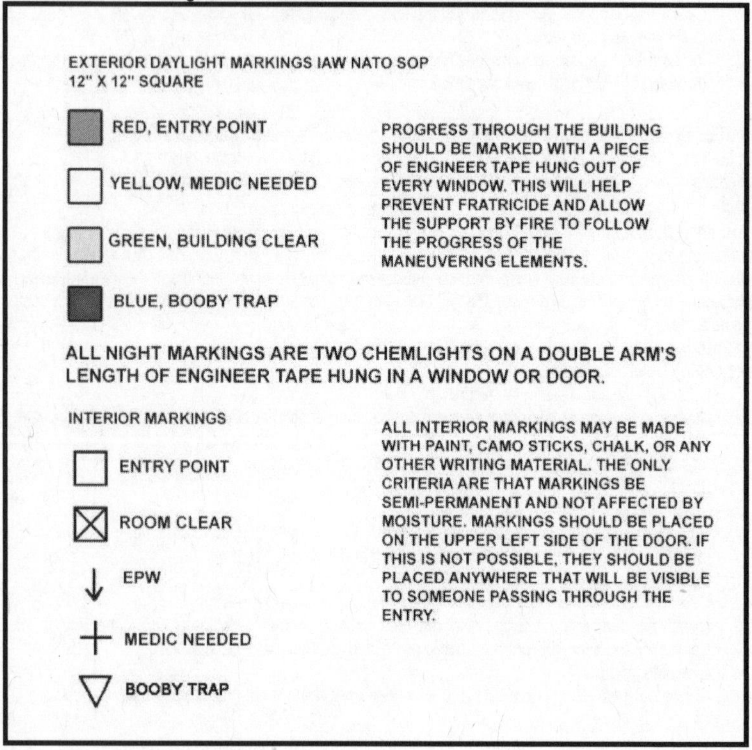

NOTE: Avoid permanently marking buildings and rooms, as this may not only cause collateral damage, but will most likely also deteriorate relationships built with local nationals.

d. **Chemlights and 100 Mile-an-Hour Tape.** These are available and reliable.

- RED – CCP
- GREEN – room clear
- ORANGE – UXO
- BLUE – clean room
- IR – breach point

e. **Wolf Tail**. A simple, effective, easy to make, lightweight device called a "wolf tail" can be fabricated to mark buildings, doorways, and windows. A unit has changed its tactical SOP to require that each Infantryman carry one of these devices in his ACU cargo pocket. Wolf tails, when used IAW a simple signaling plan understood by all members of the unit, can aid in command and control, reduce the chances of fratricide, and speed up casualty collection during urban combat.

Chapter 13
WATERBORNE OPERATIONS

This chapter discusses rope bridges, poncho rafts, and other watercraft.

13-1. ROPE BRIDGE. The Ranger patrol seldom has ready made bridges, so they must know how to employ field expedient stream-crossing techniques.

a. **Personnel.**

(1) Ranger #1 – Lead safety swimmer and far side lifeguard.
(2) Ranger #2 – Rope puller; swims water obstacle pulling 150 foot rope, ties off rope on far side anchor point.
(3) Ranger #3 – Near-side lifeguard is the last Ranger to cross water obstacle.
(4) Ranger #4 – Bridge team commander (BTC; most knowledgeable person on team).
(5) Rangers #5 and # 6 – Rope tighteners.

b. **Wet Crossing (One-Rope Bridge).** Special equipment:

- Two carabiners for each piece of heavy equipment.
- Two more carabiners for each 120 feet of rope.
- One 14 foot utility rope per person (swimmer safety line).
- Two carabiners per person.
- One waterproof bag per RTO.
- Two B 7 life preservers.
- Three flotation work vests.
- Two 150 foot nylon ropes.

c. **Planning Considerations.** A stream-crossing annex is prepared with the unit's OPORD. Special organization is accomplished at this time. For a platoon-sized patrol, a squad is normally given the task of providing the bridge team. The squad leader designates the most technically proficient Ranger in the squad as the bridge team commander (BTC).

d. **Rehearsals and Inspections.**

(1) Rehearse the entire stream crossing, emphasizing—

- Security and actions on enemy contact.
- Actual construction of the rope bridge on dry land within the 8 minute time standard.
- Individual preparation.
- Order of crossing.
- All signals and control measures.
- Reorganization.

(2) Conduct rehearsals as realistically as possible.
(3) Ensure personnel are proficient in the mechanics of a stream-crossing operation.
(4) Inspect for equipment completeness, correct rigging and preparation, personnel knowledge, and understanding of the operation.
(5) During the preparation phase, Ranger #4 (BTC) rehearses the bridge team, accounts for all equipment in the bridge kit, and ensures the 150 foot rope is coiled.

e. **Execution Phase.**

(1) Establish and conduct a bridge stream crossing:

(a) [Leader] Halts short of the river, establishes local security, and reconnoiters the area for the presence of the enemy and for crossing site suitability or necessity. He directs the BTC to construct the bridge.
(b) The BTC constructs a one rope bridge and selects near side and (visibility permitting) far side anchor points. To anchor himself to the bridge, he ties a swimmer's safety line around his waist and secures it with an overhand knot. He ties the free running end of the bow line into an overhand knot, and attaches a carabiner to the loop in the knot. He ensures the bow line is just long enough to place the carabiner at arm's length. This ensures he remains within reach of the rope bridge, should he lose his grip.
(c) The bridge team begins to establish the rope bridge while unit members begin individual preparation.

(d) Each Ranger puts a carabiner in his end of the bow line and in the front sight assembly of every M 4, M16, or M203. M240 gunners put a carabiner through the front sight assembly and rear swivel of their M240 MG. RTOs (and others with heavy rucksacks) place an additional carabiner on the top center of their rucksack frames.

(e) Team establishes security upstream and downstream, while unit leader briefs the BTC on anchor points. The leader counts the Rangers across.

(f) The BTC enforces noise and light discipline, and maintains security.

(2) The bridge team is responsible for constructing the rope bridge as follows:

(a) Ranger #1 (Lead Safety Swimmer and Far Side Lifeguard) grounds his rucksack (with carabiner through top of frame) to the rear of the near side anchor point. He wears equipment in the following order (body out). He carries a safety line to assume duties of far side lifeguard:

- Waterborne uniform (top zipped up, neck collar fastened, and pants unbloused)
- B7 life preserver or engineer work vest
- FLC
- Weapon (across the back)

(b) Ranger #1 enters the water upstream from Ranger #2 and stays an arms length away from Ranger #2 on the upstream side. Ranger #1 identifies the far side anchor point upon exiting the water. Once Ranger #2 has exited the water, he moves to his far side lifeguard position downstream of the rope bridge, with knotted safety line on wrist, FLC/ weapon grounded, and work vest held in throwing hand. He continues to wear the B7.

(c) Ranger #2 (Rope Puller) in waterborne uniform (same as Ranger #1) wears his equipment in the following order. He grounds his rucksack (with Ranger through top of frame) to the rear of the near side anchor point. His duties are to swim across the water obstacle pulling the rope. He ties off the rope on the anchor point identified by Ranger #1 with a round turn and two half hitches with a quick release. The direction of the round turn is the same direction as the flow of water (current) to facilitate exit off the rope bridge:

- Weapon (across the back)
- Swimmer safety line
- Work vest
- FLC

(d) Ranger #3 positions himself on the downstream side of the bridge before Rangers #1 and #2 enter the water.

(e) Ranger #3 (Near Side Lifeguard) wears the same type (waterborne) uniform as the far side lifeguard. He grounds his rucksack (with carabiner through top of frame) on rear of near side anchor point. After the PSG crosses and verifies the headcount, Ranger #1 unties the quick release at the near side anchor point. Ranger #3 reties his safety line into an Australian rappel seat, hooks the end-of-line bow line into his carabiner, and connects his carabiner to the one on the end of line bow line. Ranger #3 is the last pulled across the water obstacle. Before crossing the water obstacle, he dons his equipment in the following order:

- B7
- Work vest
- FLC
- Weapon

(f) Ranger #4 (BTC). He wears the standard waterborne uniform with FLC and sling rope tied in safety line (around the waist bow line with end-of-line bow line no more than one arms length). He is responsible for construction of the rope bridge and organization of bridge team. He is also responsible for back-feeding the rope and tying the end-of-line bow lines. He designates the near side anchor point, ties the Figure 8 slip of the transport-tightening system, and hooks all personnel to the rope bridge. He ensures that the transport-tightening knot is on the upstream side of the rope bridge. He ensures that all individuals are in the waterborne uniform, hooked into the rope facing the current with the safety line routed through the trailing shoulder of the individual's FLC and rucksack. He ensures that the weapon is hooked onto the rope. He controls the flow of traffic on the bridge. He is responsible for crossing with Ranger #1's rucksack. He is generally the next to the last Ranger to cross (follows PSG, who is keeping a head count).

(g) Rangers #5 and #6 (Rope Tighteners) wear the waterborne uniform with FLC and safety line. They tighten the transport-tightening knot. They also take the rucksacks of Ranger #2 and #3 across. Once they reach the far side, Rangers #5 and #6 pull the last Ranger (#3) across.

(h) Rangers #4, #5, and #6 transport the rucksacks of Rangers #1, #2, and #3 across. To do so, they hook the rucksacks into the rope by running the carabiner through the top of the frames, then pulling the rucksacks across. They attach their own weapons between themselves and the rucksack they are pulling across the bridge.

(i) BTC rehearses the bridge team during the planning sessions, and then directs the construction and emplacement. The unit leader selects the crossing site, which complements the tactical plan.

- Ranger #3 positions himself downstream of crossing site.
- Ranger #1 enters water upstream of #2. He stays one arm's length from Ranger #2 and is prepared to render any assistance to Ranger #2. They stay together to help compensate for the current. BTC feeds rope out of the rucksack positioned on the downstream side of the near side anchor point.

(j) Ranger #1 exits and identifies the far side anchor point (if BTC cannot identify it for Ranger #2). Ranger #2 exits on the upstream side of the far side anchor point. The rope is now routed to facilitate movement on and off the bridge.

(k) Radios and heavy equipment are waterproofed and rigged. Rangers don waterborne uniform and tie safety lines. PSG moves to anchor point and maintains accountability by counting heads.

(l) Ranger #2 signals the BTC that the rope is temporarily attached to the far side anchor point, and the BTC pulls out excess slack and ties the transport-tightening system using a Figure 8 slip. The BTC signals Ranger #2 to pull the knot 12 to 15 feet from the near side anchor point. After this, Ranger #2 ties round turns 18 to 24 inches off the water with the remaining rope, and secures the rope to itself, with a carabiner. Ranger #2 signals the BTC and the pulling team (Rangers #4, #5, and #6) tightens the bridge, pulling the transport-tightening system as close as possible to the near side anchor point.

(m) Ranger #1 moves downstream and assumes his duties as the far side lifeguard. The bridge team commander ties off the rope with a round turn and two half hitches around the near side anchor point. The BTC places himself on the upstream side of the bridge (facing downstream) and starts hooking individuals into the rope and inspecting them for safety.

> **NOTE:** Any Ranger identified as a weak swimmer crosses alone so the near and far side lifeguards can watch him without distraction.

(n) Ranger #2 moves to the upstream side of the rope bridge, assists personnel off the rope on the far side, and keeps the head count going. Rangers #5 and #6 cross with the rucksacks of Rangers #1 and #2.

(o) The BTC maintains the flow of traffic, ensuring that no more than three Rangers are on the bridge at any one time (one hooking up, one near the center, and one being unhooked). Once the PSG has accounted for everyone on the near side, he withdraws left and right (L/ R) security and sends them across. PSG follows security across. Ranger #3 hooks the BTC (with #3's rucksack) onto the rope. Once the BTC crosses, Ranger #3 unhooks the near side anchor point and the BTC unties the far side anchor point. Ranger #3 ties an Australian rappel seat with carabiner to the front. He hooks onto the carabiner that is in the end of the line bow line on the 120 foot rope, and then signals Rangers #4 to #6 to take in slack. Ranger #3 extends his arms in front of his head, slightly upstream, to fend off debris, and then he is pulled across by the #4, #5, and #6. Except for Rangers #1 through #3, everyone wears a rucksack across. Rangers #4 through #6 hook the rucksacks of Rangers #1/ 2/ 3 onto the bridge by the carabiners. All the Rangers cross facing upstream.

(p) The PSG and Ranger #5 verify weapons and equipment between PSG and Ranger #5. After that, personnel reorganize, and then they continue the mission.

(q) Personnel with heavy equipment:

M240. All major groups are tied together with 1/ 4 inch cord. An anchor line bow line runs through the rear swivel, down the left side of gun. Tie a round turn through the trigger guard. Route the cord down the right side and tie off two half hitches around the forearm assembly with a round turn and two half hitches through the front sight posts. Tie off the rest of the working end with an end of the rope bow line about one foot from the front sight post large enough to place leading hand through. The M240 is secured to the bridge by carabiners on the front sight post and rear swivel. The M240 is pulled across by the trailing arm of the M240 gunner.

AN/PRC–119s. These are waterproofed before crossing a one rope bridge. Once far-side FM communications are set up, the near side RTO breaks down and waterproofs his radio, and prepares to cross the bridge. He puts a carabiner in the top center of the rucksack frame (same as for Rangers #1/ 2/ 3). The BTC will hook the rucksack to the rope.

> **NOTE:** Using two carabiners binds the load on the rope. Adjust arm straps all the way out. RTO pulls radio across the rope bridge.

13-2. PONCHO RAFT. Normally a poncho raft is constructed to cross rivers and streams when the current is not swift. A poncho raft is especially useful when the unit is still dry and when the platoon leader wants to keep their equipment dry also.

a. **Equipment Requirements.**

- Two serviceable ponchos.
- Two weapons (poles can be used in lieu of weapons).
- Two rucksacks per team.
- 10 feet of utility cord per team.
- One sling rope per team.

b. **Conditions.** Poncho rafts are used to cross water obstacles when at least one of the following conditions is found:

- The water obstacle is too wide for a 150 foot-long section of rope.
- No sufficient near or far shore anchor points are available to allow rope bridge construction.
- Under no circumstances will poncho rafts be used to cross a water obstacle if current is unusually swift.

c. **Choosing a Crossing Site.** Before a crossing site is used, a thorough reconnaissance of the immediate area is made. Analyzing the situation using METT-TC, the patrol leader chooses a crossing site that offers as much cover and concealment as possible and has entrance and exit points that are as shallow as possible. For speed of movement, it is best to choose a crossing site that has near and far shore banks that are easily traversed by an individual Ranger.

d. **Execution Phase.** To construct a poncho raft–

(1) Pair off the unit/ patrol in order to have the necessary equipment.

(2) Tie off the hood of one poncho and lay out on the ground with the hood up.

(3) Place weapons in the center of the poncho, about 18 inches apart, muzzle to butt.

(4) Place rucksacks and FLC between the weapons, with the two people placing their rucksacks as far apart as possible.

(5) Start to undress, bottom to top, boots first. Take the laces completely out for subsequent use as tie downs if necessary).

(6) Place the boots over muzzle/ butt of weapon toe in.

(7) Continue to undress, folding each item neatly and placing it on top of boots.

(8) Once all of the equipment is placed between the two weapons or poles, snap the poncho together. Lift the snapped portion of the poncho into the air and tightly rolled it down to the equipment. Start at the center and work out to the end of the raft creating pigtails at the end. This is faster and easier with two Rangers working. Fold the pigtailed ends inward and tie them off with a single boot lace.

(9) Lay out the other poncho on the ground with the hood up. In the center of this poncho, place the other poncho, with equipment. Snap, roll, and tie the whole package up as you did before. Tie the third and fourth boot laces (or utility cord) around the raft about one foot from each end for added security. The poncho raft is now complete.

NOTE: The patrol leader must analyze the situation using METT-TC and make a decision on the uniform for crossing the water obstacle such as whether to place weapons inside the poncho raft or slung across the back, and whether to remain dressed or strip down with clothes placed inside raft.

13-3. OTHER WATERCRAFT. Use of inland and coastal waterways may add flexibility, surprise, and speed to tactical operations. Use of these waterways will also increase the load-carrying capacity of normal dismounted units. Watercraft are employed in reconnaissance and assault operations.

a. **Inflation Method.** Inflate watercraft using foot pumps. There are four separate valves inside the buoyancy tubes. There are eight separate airtight compartments. To pump air into the boat, turn all valves into the "orange" or "inflate" section of the valve. Once the assault boat is filled with air, turn all valves onto the "green" or "navigation" section. This will section the assault boat into eight separate compartments.

b. **Characteristics.**

- Maximum payload: 2,756 pounds.
- Crew: 1 coxswain + either 10 paddlers or a 65 HP short shaft outboard motor.
- Overall length: 15 feet, 5 inches.
- Overall width: 6 feet, 3 inches.
- Weight: 265 pounds.

c. **Preparation.**

(1) ***Rubber Boat.***

- Each rubber boat will have a 12-foot bow line secured to the front starboard D ring. This rope will be tied with an anchor line bow line, and the knot will be covered with 100 MPH tape.
- Each rubber boat will have a 15-foot center line tied to the rear floor D ring. The same procedure for securing the bow line will be used for the centerline.
- Each rubber boat will be filled to 240 millibars of air, and checked to ensure that all valve caps are tight, and set in the NAVIGATE position.
- Each rubber boat will have one foot pump, which will be placed in the boat's front pouch or, if no pouches are present, the foot pumps will be placed on the floor.
- Each rubber boat must be inspected using the maintenance chart.

(2) ***Personnel and Equipment.***

- All personnel will wear work vest or kapok (or another suitable positive flotation device).
- FLC is worn over the work vest, unbuckled at the waist.
- Individual weapon is slung across the back, muzzle pointed down and facing toward the inside of the boat.
- Crew-served weapons, radios, ammunition, and other bulky equipment are lashed securely to the boat to prevent loss if the boat should overturn. Machine guns with hot barrels are cooled prior to being lashed inside the boats.
- Radios and batteries are waterproofed.
- Pointed objects are padded to prevent puncturing the boat.

d. **Positions.** Assign each Ranger a specific boat position (**Figure 13-1**).

e. **Duties.**

- Designate a commander for each boat, (normally coxswain)
- Designate a navigator (normally a leader within the platoon)—observer team as necessary.
- Position crew as shown in **Figure 13-2**.

- Duties of the coxswain.
  - Responsible for control of the boat and actions of the crew.
  - Supervises the loading, lashing, and distribution of equipment.
  - Maintains the course and speed of the boat.¬
  - Gives all commands.
- Paddler #2 (Long Count) is responsible for setting the pace.
- Paddler #1 is the Observer, stowing and using the bow line unless another observer is assigned.

f. **Embarkation and Debarkation Procedures.**

(1) When launching, the crew will maintain a firm grip on the boat until they are inside it: similarly, when beaching or debarking, they hold on to the boat until it is completely out of the water. Loading and unloading is done using the bow as the entrance and exit point.

(2) Keep a **low center of mass** when entering and exiting the boat to avoid capsizing. Maintain **3 points of contact** at all times.

(3) The **long count** is a method of loading and unloading by which the boat crew embarks or debarks individually over the bow of the boat. It is used at river banks, on loading ramps, and when deep water prohibits the use of the short count method.

(4) The **short count** is a method of loading or unloading by which the boat crew embarks or debarks in pairs over the sides of boat while the boat is in the water. It is used in shallow water allowing the boat to be quickly carried out of the water. The short count method of organization is primarily used during surf operations.

(5) **Beaching** the boat is a method of debarking the entire crew at once into shallow water and quickly carrying the boat out of the water.

Figure 13-1. BOAT POSITIONS

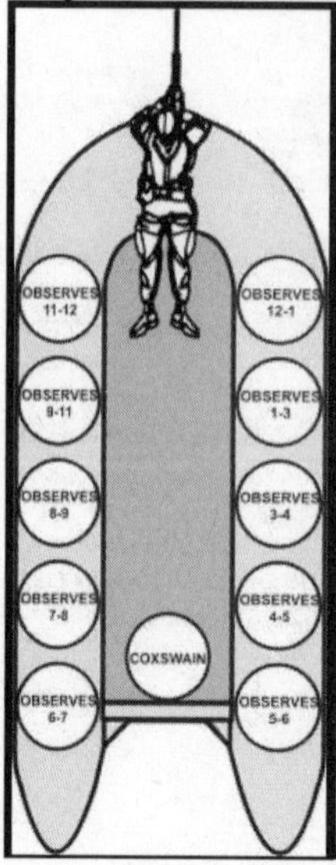

Figure 13-2. CREW POSITIONS, LONG COUNT AND SHORT COUNT

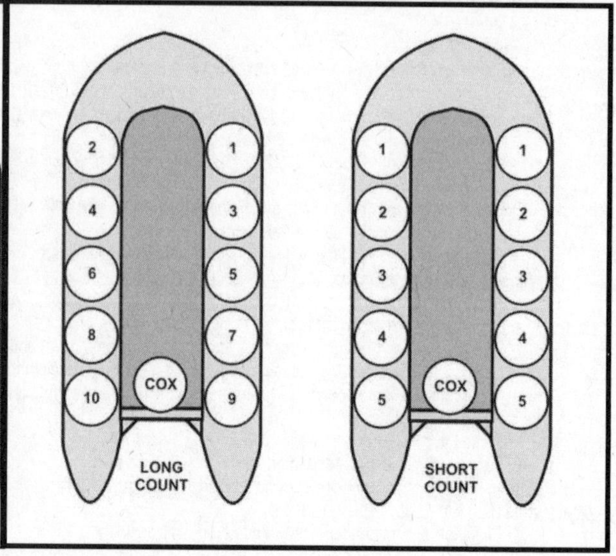

g. **Commands.** Commands are issued by the coxswain to ensure the boat is transported over land and controlled in the water. All crew members learn and react immediately to all commands issued by the coxswain. The various commands are as follows:

(1) *"Short count...count off,"* Crew counts off their position by pairs, such as 1, 2, 3, 4, 5 (Passenger #1, #2, if applicable), coxswain.
(2) *"Long count…count off,"* Crew counts off the position by individual, such as 1, 2, 3, 4, 5, 6, 7, 8, 9, 10, (Passenger #1, #2, if applicable), coxswain.
(3) *"Boat stations,"* Crew takes positions alongside the boat.
(4) *"High carry…move"* (used for long distance move overland).
  (a) On the preparatory command of *"High carry,"* the crew faces the rear of the boat and squats down, grasping carrying handles with the inboard hand.
  (b) On the command *"Move,"* the crew swivels around, lifting the boat to their shoulders so that the crew is standing and facing to the front with the boat on their inboard shoulders.
  (c) Coxswain guides the crew during movement.
(5) *"Low carry…move,"* (used for short distance moves overland).
  (a) On preparatory command of *"Low carry,"* the crew faces the front of the boat, bent at the waist, and grasps the carrying handles with their inboard hands.
  (b) On the command of *"Move,"* the crew stands up straight raising the boat about 6 to 8 inches off the ground.
  (c) Coxswain guides the crew during movement.
(6) *"Lower the boat move,"* Crew lowers the boat gently to the ground using the carrying handles.
(7) *"Give way together,"* Crew paddles to front with #2 setting the pace.
(8) *"Hold,"* Entire crew keeps paddles straight downward motionless in the water stopping the boat.
(9) *"Left side hold" (right),* Left crew holds, right continues with previous command.
(10) *"Back paddle,"* Entire crew paddles backward propelling the boat to the rear.
(11) *"Back paddle left"* (right), left crew back paddles causing the boat to turn left, right crew continues with previous command.
(12) *"Rest paddles,"* crewmembers place paddles on their laps with blades outboard. This command may be given in pairs such as *"#1s, rest paddles"*).

h. **Securing of Landing Site.**

(1) If the patrol is going into an unsecured landing site, it can have a security boat land and reconnoiter the landing site, and then signal to the remaining boats to land. This is the best way.
(2) If the landing site cannot be secured prior to the waterborne force landing, some form of early warning, such as scout swimmers, should be considered. These Rangers swim to shore from the assault boats and signal the boats to land. All signals and actions are rehearsed prior to the actual operation.
(3) The landing site can be secured by force with all the assault boats landing simultaneously in a line formation. While this is the least desirable method of securing a landing site, it is rehearsed in the event the tactical situation requires its use.
(4) Arrival at the debarkation point.
  (a) Unit members disembark according to leaders order (**Figure 13-3**).
  (b) Local security is established.
  (c) Leaders account for personnel and equipment.
  (d) Unit continues movement.
    • Rangers pull security initially with work vest on.
    • Coxswains and two men unlash and de rig rucksacks.
    • Rangers return in buddy teams to secure rucksack and drop off work vest.
    • Boats are camouflaged/ cached if necessary prior to movement.

Figure 13-3. DEBARKATION

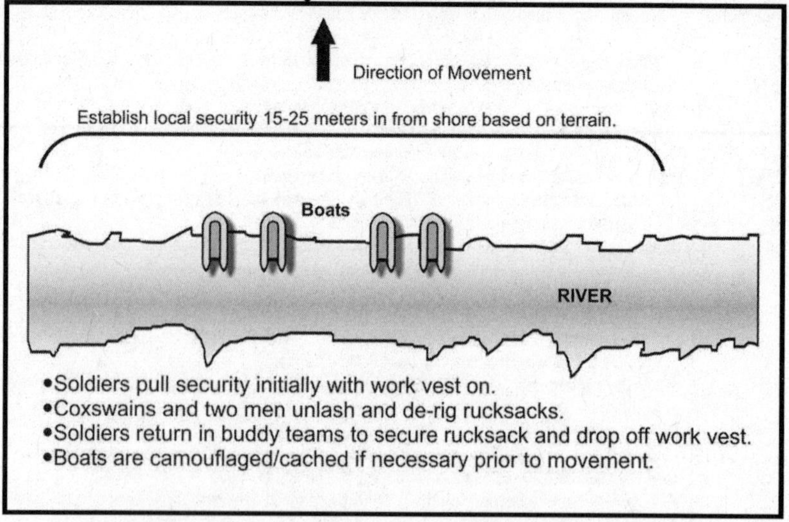

i. **Capsize Drill.** The following commands and procedures are used for capsize drills or to right an overturned boat:

(1) ***Prepare to Capsize.*** This command alerts the crew and they raise paddles above their heads, with the blades pointed outward. Before capsizing, the coxswain will conduct a long count.

(2) ***Pass Paddles.*** All paddles are passed back and collected by Rangers #9/ 10.

(3) ***Capsize the Boat.*** All personnel slide into the water except Rangers #3/ 5/ 7. Ranger #1 secures the bow line. The three men in the water grasp the capsize lines (ensuring the lines are routed under the safety lines) and stand on the buoyancy tubes opposite the capsize lines anchor points. The boat is then turned over by Rangers #3/ 5/ 7 men, who lean back and straighten their legs while pulling back on the capsize lines. As the boat lifts off the water, Ranger #4 grasps the center carrying handle and rides the boat over. Once the boat is over, Ranger #4 helps Rangers #3/ 7 men back onto the boat. Ranger #5 holds onto the center carrying handle and turns the boat over the same way. Ranger #5 rides the boat back over and helps the rest of the crew into the boat. As soon as the boat is capsized, the coxswain commands a long count to ensure that no one sank or was trapped under the boat. Every time the boat is turned over, he conducts another long count.

j. **River Movement.**

(1) ***Characteristics of River.***

(a) Know local conditions prior to embarking on river movement.

(b) A *bend* is a turn in the river course.

(c) A *reach* is a straight portion of river between two curves.

(d) A *slough* (pronounced "sloo") is a dead end branch from a river. They are normally quite deep and can be distinguished from the true river by their lack of current.

(e) *Dead water* is a part of the river, due to erosion and changes in the river course that has no current. Dead water is characterized by excessive snags and debris.

(f) An *island* is usually a pear shaped mass of land in the main current of the river. Upstream portions of islands usually catch debris and are avoided.

(g) The *current* in a narrow part of a reach is normally greater than in the wide portion. The current is greatest on the outside of a curve; sandbars and shallow water are found on the inside of the curve.

(h) *Sandbars* are located at those points where a tributary feeds into the main body of a river or stream.

(j) Because Rangers #1 and #2 are sitting on the front left and right sides of the boat, they observe for obstacles as the boat moves downriver. If either notices an obstacle on either side of the boat, he notifies the coxswain. The coxswain then adjusts steering to avoid the obstacle.

(2) ***Navigation.*** The patrol leader is responsible for navigation. The three acceptable methods of river navigation are

(a) *Checkpoint and General Route.* These methods are used when the drop site is marked by a well defined checkpoint and the waterway is not confused by a lot of branches and tributaries. They are best used during daylight hours and for short distances.

(b) *Navigator Observer Method.* This method is the most accurate means of river navigation and is used effectively in all light conditions.

(c) *Equipment Needed.*

- Compass
- GPS
- Photo map (1st choice)
- Topographical map (2nd choice)
- Poncho (for night use)
- Pencil/ Grease pencil
- Flashlight (for night use)

(d) *Procedure.* Navigator is positioned in center of boat and does not paddle. During hours of darkness, he uses his flashlight under the poncho to check his map. The observer (or Ranger #1) is at the front of the boat.

- The navigator keeps his map and compass oriented at all time.
- The navigator keeps the observer informed of the configuration of the river by announcing bends, sloughs, reaches, and stream junctions as shown on his map.
- The observer compares this information with the bends, sloughs, reaches, and stream junctions he actually sees. When these are confirmed, the navigator confirms the boat's location on his map.
- The navigator also keeps the observer informed of the general azimuths of reaches as shown on his map and the observer confirms these with actual compass readings of the river.
- The navigator announces only one configuration at a time to the observer and does not announce another until it is confirmed and completed.
- A strip map drawn on clear acetate backed by luminous tape may be used. The drawing is to scale or a schematic. It should show all curves and the azimuth and distance of all reaches. It may also show terrain features, stream junctions, and sloughs.

k. **Formations.** Various boat formations are used (day and night) for control, speed, and security (**Figure 13-4**). The choice of which is used depends on the tactical situation and the discretion of the patrol leader. He should use hand and arm signals to control his assault boats. The formations are:

- Wedge
- Line
- File
- Echelon
- Vee

Figure 13-4. FORMATIONS

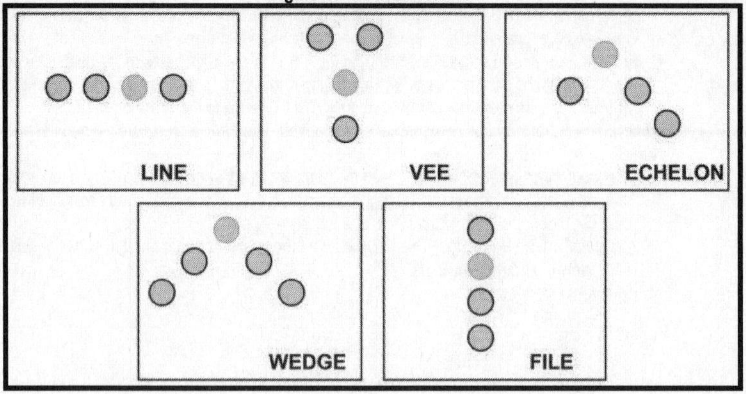

Chapter 14

EVASION AND SURVIVAL

This chapter will cover basic materials dealing with evading the enemy and survival techniques. It covers field-expedient methods of acquiring food and water; land navigation; construction of shelters; and fire-starting techniques. See also **FM 3-05.70** *(Survival)* and **FM 3-55.93** *(Long-Range Surveillance Unit Operations)*.

Section I. EVASION

When you become isolated or separated in a hostile area, either as an individual or as a group, your evasion and survival skills will determine whether or not you return to friendly lines.

14-1. PLANNING CONSIDERATIONS. While certain units formally plan for evasion at the receipt of every mission, every leader should have a plan to facilitate the recovery of his personnel should things turn bad. He can use the following methods and materials for both formal and informal evasion planning:

a. Construct an evasion plan of action (EPA) for each mission, and give a copy to higher headquarters. At a minimum, include the overall plan; routes; a personnel roster; a list of survival equipment and signaling devices you plan to carry; authentication means (letter of the day, number of the day, word, for example), and your evasion intentions, should the situation arise (initial, immediate less than 48 hrs, and extended more than 48 hrs).

b. Prior to initiating movement, review the Air Tasking Order Special Instructions (ATOSPINS), ISOPREP cards, info on operational environment (cultural boundaries, social or political events, and norms and trends), and have everyone "sterilize" their personal effects.

c. You can get other evasion aids through higher headquarters. These include evasion charts (EVCs), which are maps made of waterproof paper. In the margins, they contain survival information such as edible and poisonous plants. Information cards printed in the destination language and in English are useful for emergency communications with locals.

14-2. INITIAL EVASION POINT. This is when you realize that you have become separated from friendly forces. Here is what to do when you reach this point:

- Immediately move to a concealed point
- S.L.L.S. (Do at your earliest opportunity)
- Try to establish communications with friendly forces
- Observe enemy activity
- Evaluate the situation
- Security, camouflage
- Formulate a plan of action
- Employ keyword SURVIVAL
- Inventory supplies (food, water, equipment)

14-3. EVASION MOVEMENT. Do not move from the area just for something to do. Move only if you have to, in which case, consider the following. Traveling alone offers the least possibility of detection, but traveling in groups of 2 or 3 is more desirable:

- Health of evader/ ability to move
- Enemy activity
- Recovery potential

14-4. ROUTES. Plan primary and alternate routes. Consider distance, cover, food, and water. The easiest and shortest route may not be the best.

- Food and water are daily requirements. You can do without food for several days; water, however, is essential.
- Move at night. Use the daylight to observe, plan, and rest in a hide position.
- Linkup only during daylight hours. Place friendly lines under observation.
- Attempt to identify the unit you approach, note their movements and routine.

• After carefully considering your approach route, make voice contact with the unit as soon as possible.

14-5. COMMUNICATIONS. During your evasion, you may be required to relay your location to higher over unsecure means such as cell phone, radio that has either lost or expired, COMSEC or that has been zeroed) you can use one of the following methods. They can be found in ATOSPINS along with authentication means:

Search and Rescue Numeric Encryption Grid. SARNEG is a 10-letter word with no repeating letters; each letter corresponding to a number 0 to 9.

Search and Rescue DOT. SARDOT is a geographic location that is used as a reference to relay your location. It is relayed as an azimuth and distance from you to the SARDOT.

Code Words. Words used to send vital information quickly and in a secure manner so that the meaning cannot be understood by an intercepting element. Words are either pulled from the ATOSPINS, passed down from higher, or generated by the element in planning. Words may have a theme for ease of remembering such as Types of liqueur, animals or sports.

14-6. HIDE SITE. Should be an isolated, covered and concealed site selected during evasion. In the hide site you should take inventory of your situation and accomplish tasks such as camouflage, resting, and planning the next movement. It is seldom used or occupied for more than 24 hrs. When selecting a site consider the following:

• Distance from natural lines of drift (water, roads and trails, ridges, and key terrain)
• Vegetation–thick?
• Concealed from all directions?
• Away from populated or built up areas
• Escape route
• Location where humans do not normally habitat

14-7. HOLE UP AREA. This is an isolated area selected during movement. Use it when your physical condition requires that you stop for food, water, equipment maintenance, and rest. Generally, avoid occupying such a position for more than 72 hrs. When selecting a hole up area, consider–

• Abundance of food and water
• Isolated
• Low population density
• Security at all times
• Movement in or around hole up area is still kept to a minimum
• Decentralize–separate rest, food procurement, food preparation and so on.

14-8. CAMOUFLAGE. While evading, you may need to use field-expedient means to camouflage yourself, your equipment, and your site. Mud, charcoal, berries, vegetation, ash and blood can all be used to camouflage exposed skin and equipment. Natural vegetation can be fixed to clothing and equipment by using vines to secure it or cutting small holes in the rip stop uniform material and feeding vegetation into it. Whenever a site is constructed and camouflaged keep the following memory aid in mind. B L I S S

B LENDS IN
L OW IN SILHOUETTE
I RREGULAR IN SHAPE
S MALL IN SIZE
S ECLUDED

Section II. SURVIVAL

With training, equipment, and the *will to survive*, you can overcome any obstacle you may face. *You will survive*. Understand the emotional states associated with survival. *"Knowing thyself"* is extremely important in a survival situation. It bears directly on how well you cope with serious stresses, anxiety, pain, injury, illness; cold, heat, thirst, hunger, fatigue, sleep deprivation, boredom, loneliness and isolation.

14-9. MEMORY AID. You can overcome and reduce the shock of being isolated behind enemy lines if you keep the key word **S-U-R-V-I-V-A-L** foremost in your mind. Its letters can help guide you in your actions.

S – **S** ize up the situation, the surroundings, your physical condition, and your equipment.

U – **U** ndue haste makes waste; don't be too eager to move. Plan your moves.

R – **R** emember where you are relative to friendly and enemy units and controlled areas; water sources (most important in the desert); and good cover and concealment. This information will help you make intelligent decisions.

V – **V** anquish fear and panic.

I – **I** mprovise/Imagine. You can improve your situation. Learn to adapt what is available for different uses. Use your imagination.

V – **V** alue living. Remember your goal to get out alive. Remain stubborn. Refuse to give in to problems and obstacles. This will give you the mental and physical strength to endure.

A – **A** ct like the natives; watch their daily routines and determine when, where, and how they get their food and water.

L – **L** ive by your wits. Learn basic skills.

14-10. SURVIVAL KITS. Upon finding yourself in a survival situation you will be required to provide for your basic needs; water, food, fire, shelter, medical, signaling and protection. This will be accomplished by using the resources you have on hand and those that may be found or constructed. The more detailed your survival kit the less scavenging or constructing you will have to do. Some examples of individual survival kits follow. All items should be small, portable and most importantly multipurpose.

Level 1 Kit (carried on individual) should consist of at a minimum of a knife, some form of fire starter, such as lighter matches or flint striker, watch, signal mirror and some 550 cord.

Level 2 Kit (carried in FLC or rack) waterproof container, water purification tabs, 2ft sq. aluminum foil, fishing kit (line, hooks and weights) medical supplies, snare wire, signaling devices, compass and survival knife.

Level 3 Kit (carried in assault pack or ruck) water proof container with more of the materials listed in the level 1 and 2 kits plus shelter making materials (poncho, tarp, bungee cords, or space blanket) and a hatchet or saw.

14-11. NAVIGATION. In a survival situation, you might find yourself without a compass. The ability to determine direction can enable you to navigate back to your unit or to a friendly sanctuary. In sunlight, there are two simple ways to determine direction: the shadow tip and the watch.

a. **Shadow Tip.** Use the sun to find approximate true North. Use this in light bright enough to cast shadows. Find a fairly straight stick about 3 feet long, and follow the diagram below (**Figure 14-I**).

Figure 14-1. SHADOW TIP METHOD

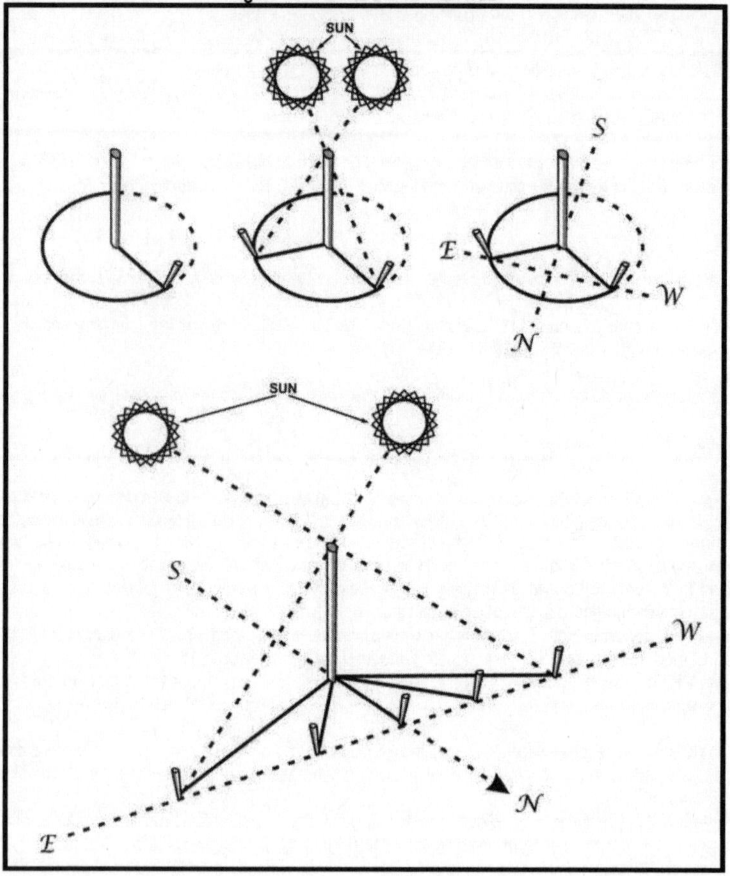

b. **Watch Method.** You can also determine direction using a watch (**Figure 14-2**). The steps you take will depend on whether you are in the Northern Temperate Zone or in the Southern Temperate Zone. The Northern Temperate Zone is located between 23.4 North and 26.6 North. The Southern Temperate Zone is located between 23.4 South and 66.6 South.

Figure 14-2. WATCH METHOD

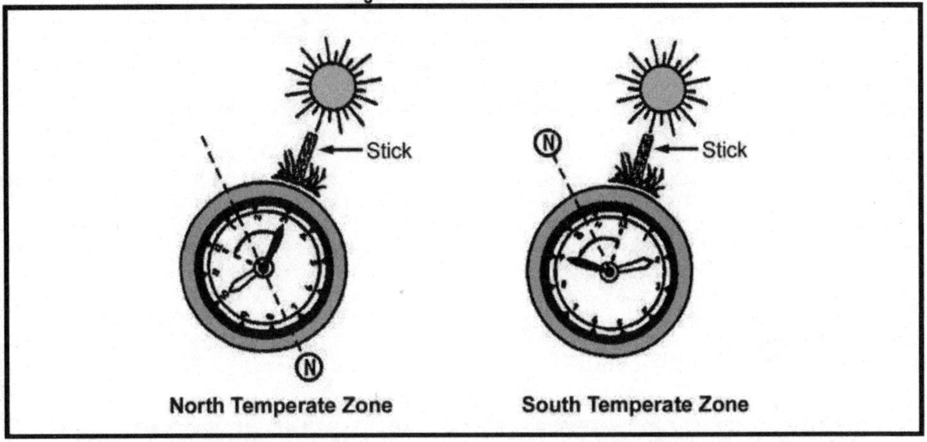

c. **Northern Temperate Zone.** Procedures in the Northern Temperate Zone using a conventional watch are as follows:

(1) Place a small stick in the ground so that it casts a definite shadow.

(2) Place your watch on the ground so that the hour hand points toward and along the shadow of the stick.

(3) Find the point on the watch midway between the hour hand and 12 o'clock and draw an imaginary line from that point through and beyond the center of the watch. This imaginary line is a North South line. You can then tell the other directions.

> **NOTE:** If your watch is set on daylight savings time, then use the midway point between the hour hand and 1 o'clock to draw your imaginary line.

d. **Southern Temperate Zone.** Procedures in the southern temperate zone using a conventional watch are as follows:

(1) Place a small stick in the ground so that it casts a definite shadow.

(2) Place your watch on the ground so that 2 o'clock points to and along the shadow.

(3) Find the midway point between the hour and 12 o'clock and draw an imaginary line from the point through and beyond the center of the watch. This is a North South line.

(4) A hasty shortcut using a conventional watch is simply to point the hour hand at the sun in the Northern temperate zone (or point the 12 at the sun in the Southern temperate zone). Follow the last step of the watch method above to find your directions. This shortcut is less accurate than the regular method, but quicker. Your situation will dictate which method to use.

e. **Moon.** Because the moon has no light of its own, we can only see it when it reflects the sun's light. As it orbits the earth on its 28 day circuit, the shape of the reflected light varies according to its position. We say there is a new moon or no moon when it is on the opposite side of the earth from the sun. Then, as it moves away from the earth's shadow, it begins to reflect light from its right side and waxes to become a full moon before waning, or losing shape, to appear as a sliver on the left side. You can use this information to identify direction. If the moon rises before the sun has set, the illuminated side will be the west. If the moon rises after midnight, the illuminated side will be the east. This obvious discovery provides us with a rough East West reference during the night.

f. **Stars.** Your location in the Northern or Southern Hemisphere determines which constellation you use to determine your North or South direction. Each sky is explained below.

g. **The Northern Sky.** The main constellations to learn are the Ursa Major, also known as the Big Dipper or the Plow, and Cassiopeia, also known as the Lazy W (Figure 14 3). Use them to locate Polaris, also known as the polestar or the North Star. Polaris is considered to remain stationary, as it rotates only 1.08 degrees around the northern celestial pole. The North Star is the last star of the Little Dipper's handle and can be confused with the Big Dipper. However, the Little Dipper is made up of seven rather dim stars and is not easily seen unless you are far away from any town or city lights. Prevent confusion by attempting to use both the Big Dipper and Cassiopeia together. The Big Dipper and Cassiopeia are generally opposite each other and rotate counterclockwise around Polaris, with Polaris in the center. The Big Dipper is a seven star constellation in the shape of a dipper. The two stars forming the outer lip of this dipper are the "pointer stars" because they point to the North Star. Mentally draw a line from the outer bottom star to the outer top star of the Big Dipper's bucket. Extend this line about five times the distance between the pointer stars. You will find the North Star along this line. You may also note that the North Star can always be found at the same approximate vertical angle above the horizon as the northern line of latitude you are located on. For example, if you are at 35 degrees north latitude, Polaris will be easier to find if you scan the sky at 35 degrees off the horizon. This will help to lessen the area of the sky in which to locate the Big Dipper, Cassiopeia, and the North Star .Cassiopeia or the Lazy W has five stars that form a shape like a "W." One side of the "W" appears flattened or "lazy." The North Star can be found by bisecting the angle formed on the lazy side. Extend this line about five times the distance between the bottom of the "W" and the top. The North Star is located between Cassiopeia and the Ursa Major (Big Dipper). After locating the North Star, locate the North Pole or true North by drawing an imaginary line directly to the earth.

Figure 14-3. NORTHERN SKY

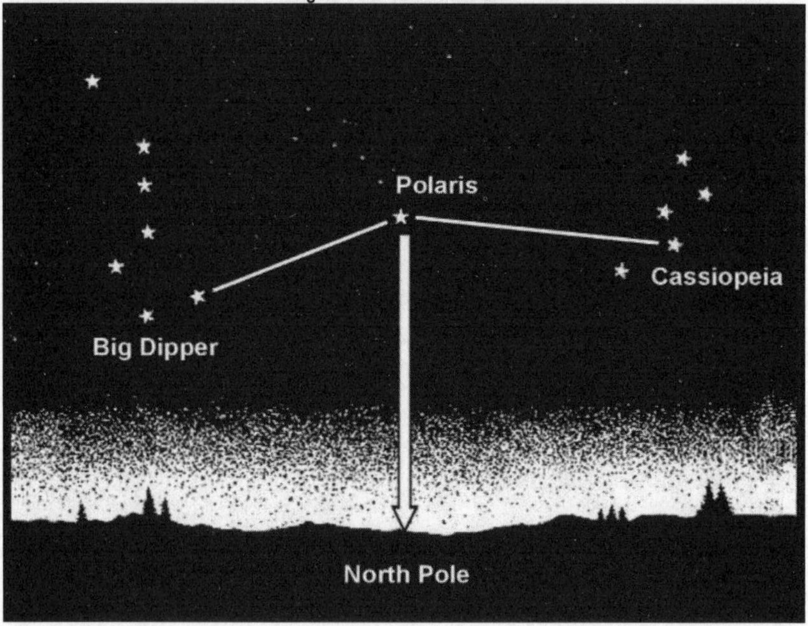

h. **The Southern Sky.** Because there is no single star bright enough to be easily recognized near the South celestial pole, you can use a constellation known as the Southern Cross. You can use it as a signpost to the South. The Southern Cross or Crux has five stars. Its four brightest stars form a cross. The two stars that make up the Cross's long axis are used as a guideline. To determine south, imagine a distance 4.5 to 5 times the distance between these stars and the horizon. The pointer stars to the left of the Southern Cross serve two purposes. First, they provide an additional cue toward South by imagining a line from the stars toward the ground. Second, the pointer stars help accurately identify the true Southern Cross from the False Cross. The intersection of the Southern Cross (**Figure 14-4**) and the two pointer stars is very dark and devoid of stars. This area is called the coal sac. Look down to the horizon from this imaginary point and select a landmark to steer by. In a static survival situation, you can fix this location in daylight if you drive stakes in the ground at night to point the way.

Figure 14-4. SOUTHERN SKY

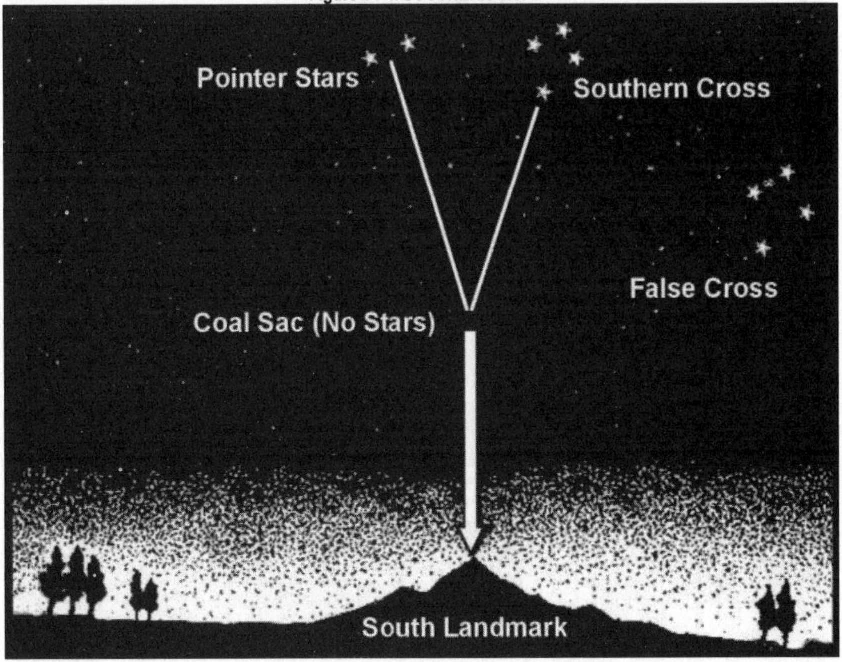

i. **Water.** Water is one of your most urgently needed resources in a survival situation. You can't live long without it, especially in hot areas where you lose so much through sweating. Even in cold areas, you need a minimum of 2 quarts of water a day to maintain efficiency. More than three fourths of your body is composed of fluids. Your body loses fluid as a result of heat, cold, stress, and exertion. The fluid your body loses must be replaced for you to function effectively. So, one of your first objectives is to obtain an adequate supply of water.

(1) ***Purification.*** Purify all water before drinking. Either

- Boil it for at least one minute (plus 1 more minute for each additional 1,000 feet above sea level) or for a maximum of 10 minutes anywhere.
- Use water purification tablets.

- Add eight drops of 2 1/ 2 percent iodine solution to a quart (canteen full) of water. Let it stand for 10 minutes before drinking.
- Collect rain water directly in clean containers or on plants. This is generally safe to drink without purifying. Never drink urine or sea water; the salt content is too high. Use old, bluish sea ice, but newer, grayer ice may be salty. Glacier ice is safe to melt and drink.

(2) ***Desert Environment***. In a desert environment, water has a huge effect on Rangers. If a unit fails to plan properly for water, and resupply is unavailable, then they can run out of water. In the desert, look for four signs of water: animal trails, vegetation, birds, and civilization. Adequate water is critical in a hot environment if a unit is to survive and maintain the physical condition necessary to accomplish the mission. Unit leaders must enforce water discipline and plan for water resupply. The leader can use the following planning considerations for water resupply:

- Units average water consumption.
- Drop sites.
- Aviation support.
- DZ and LZ parties.
- Caches.
- Targets of opportunity (enemy).

(3) ***Survival Water Still***. Dig a below-ground still (Figure 14 5). Select a site where you believe the soil will contain moisture such as a dry stream bed or a spot where rain water has collected. The soil should be easy to dig, and be in sunlight most of the day:

(a) Dig a bowl shaped hole about 3 feet across and 2 feet deep.
(b) Dig a sump in center of the hole. The depth and the perimeter of the sump will depend on the size of the container that you have to set inside of it. The bottom of the sump should allow the container to stand upright.
(c) Anchor the tubing to the bottom of the container by forming a loose overhand knot in the tubing.
(d) Place the container upright in the sump.
(e) Extend the unanchored end of the tubing up, over, and beyond the lip of the hole.
(f) Place plastic sheeting over the hole and cover the edge with soil to hold it in place.
(g) Place a rock in the center of the plastic.
(h) Allow the plastic to lower into the hole until it is about 15 inches below ground level. The plastic now forms an inverted cone with the rock at its apex. Make sure that the apex of the cone is directly over your container. Also, make sure the plastic cone does not touch the sides of the hole, because the earth will absorb the condensed water.
(i) Put more soil on the edges of the plastic to hold it securely in place and to prevent loss of moisture.
(j) Plug the tube when not being used so that moisture will not evaporate.

> **NOTE:** You can drink water without disturbing the still by using the tube as a straw. You may want to use plants in the hole as a moisture source. If so, when you dig the hole you should dig out additional soil from the sides of the hole to form a slope on which to place the plants. Then proceed as above.

Figure 14-5. SURVIVAL WATER STILL

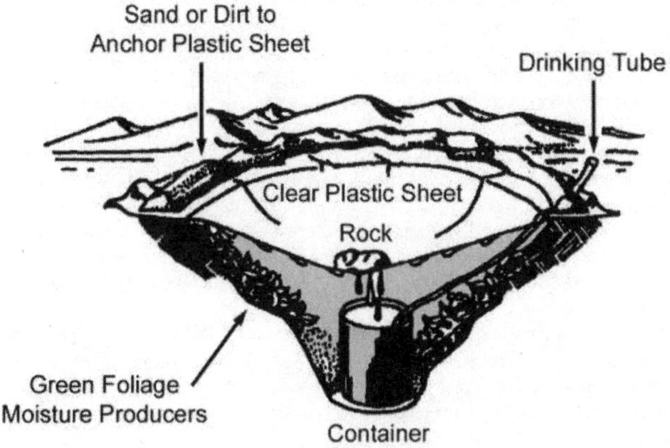

j. **Plant Food.** There are many plants throughout the world. Tasting or swallowing even a small portion of some can cause severe discomfort, extreme internal disorders, or death. Therefore, if you have the slightest doubt as to the edibility of a plant, apply the universal edibility test described below before eating any part of it.

(1) ***Universal Edibility Test.*** Before testing a plant for edibility, make sure there are a sufficient number of plants to make testing worth your time and effort. You need more than 24 hours to apply the edibility test outlined below:

- Test only one part of a potential food plant at a time.
- Break the plant into its basic components, leaves, stems, roots, buds, and flowers.
- Smell the food for strong or acid odors. Keep in mind that smell alone does not indicate a plant is edible.
- Do not eat for 8 hours before starting the test.
- During the 8 hours you are abstaining from eating, test for contact poisoning by placing a piece of the plant you are testing on the inside of your elbow or wrist. Usually 15 minutes is enough time to allow for reaction.
- During the test period, take nothing by mouth except purified water and the plant part being tested.
- Select a small portion and prepare it the way you plan to eat it.
- Before putting the prepared plant part in your mouth, touch a small portion (a pinch) to the outer surface of the lip to test for burning or itching.
- If after 3 minutes there is no reaction on your lip, place the plant part on your tongue, holding there for 15 minutes.
- If there is no reaction, thoroughly chew a pinch and hold it in your mouth for 15 minutes. Do not swallow.
- If no burning, itching, numbing, stinging, or other irritation occurs during the 15 minutes, swallow the food.
- Wait 8 hours. If any ill effects occur during this period, induce vomiting and drink a lot of water.
- If no ill effects occur, eat 1/ 2 cup of the same plant part prepared the same way. Wait another 8 hours. If no ill effects occur, the plant part as prepared is safe for eating.

(2) ***Poisonous Plants.*** Do not eat unknown plants that–

- Have a milky or discolored sap.
- Beans, bulbs, or seeds inside pods
- A bitter or soapy taste.
- Spines, fine hairs, or thorns.
- Foliage that resembles dill, carrot, parsnip, or parsley.
- An almond scent in woody parts and leaves.
- Grain heads with pink, purplish, or black spurs.
- A three leafed growth pattern.

k. **Insects.** Insects are the most abundant and easily caught life form on earth. Many insects provide 65 to 80 percent protein as compared to 20 percent beef. However you should avoid all insects that sting or bite, are hairy or bright colored, are common disease carriers (ticks, flies, and mosquitoes) and caterpillars and insects that have a pungent odor. Insects that have a hard outer shell such as beetles and grasshoppers should have their wings and barbed legs removed and must be cooked because they have parasites. Most soft shelled insects can be eaten raw. Insects can be ground into a paste and eaten or mixed with edible vegetation to improve or mask their taste.

l. **Animal Food.** Animal food contains the most food value per pound. Anything that creeps, crawls, swims, or flies is a possible source of food, however you must first catch, kill, and butcher it before this is possible. There are numerous methods for catching fish and animals in a survival situation. You can catch fish by using a net across a small stream, (**Figure 14-6**) or by making fish traps and baskets. Improvise fish hooks and spears as indicated in **Figure 14-7**, and use them for conventional fishing, spearing and digging.

Figure 14-6. SETTING A GILL NET IN THE STREAM

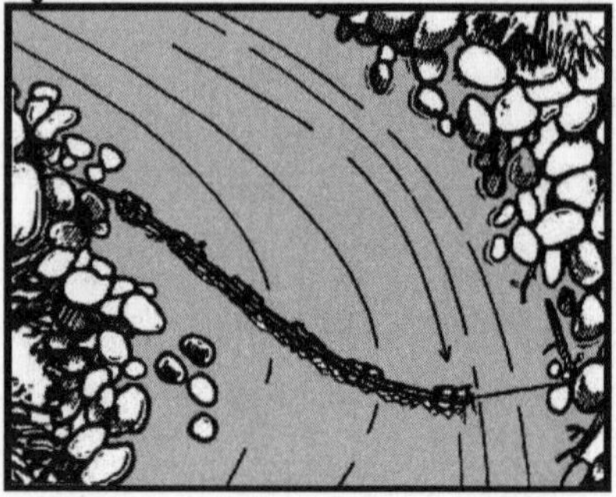

Figure 14-7. PEAR HOOKS AND FISH HOOKS

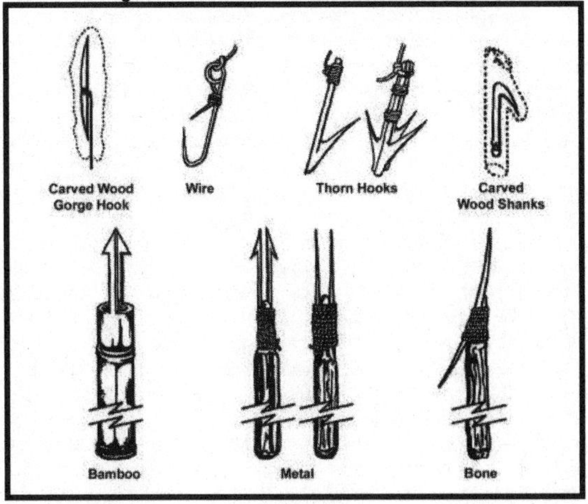

14-12. TRAPS AND SNARES. For an unarmed survivor or evader, or when the sound of a rifle shot could be a problem, trapping or snaring wild game is a good alternative. Several well placed traps have the potential to catch much more game than a Ranger with a rifle is likely to shoot.

a. To be effective with any type of trap or snare, you must

- Know the species of animal you intend to catch.
- Know how to construct a proper trap.
- Avoid alarming the prey with signs of your presence.

b. There are no catchall traps you can set for all animals. You must determine what species are in a given area and set your traps specifically with those animals in mind. Look for the following:

- Runs and trails.
- Tracks.
- Droppings.
- Chewed or rubbed vegetation.
- Nesting or roosting sites.
- Feeding and watering areas.

c. Position your traps and snares where there is proof that animals pass through. You must determine if it is a "run" or a "trail." A trail will show signs of use by several species and will be rather distinct. A run is usually smaller and less distinct and will only contain signs of one species. You may construct a perfect snare, but it will not catch anything if haphazardly placed in the woods. Animals have bedding areas, waterholes, and feeding areas with trails leading from one to another. You must place snares and traps around these areas to be effective.

d. An evader in a hostile environment must conceal traps and snares. It is equally important, however, to avoid making a disturbance that will alarm the animal and cause it to avoid the trap. Therefore, if you must dig, remove all fresh dirt from the area. Most animals will instinctively avoid a pitfall type trap. Prepare the various parts of a trap or snare away from the site, carry them in, and set them up. Such actions make it easier to avoid disturbing the local vegetation, thereby alerting the prey. Do not use freshly cut, live vegetation to construct a trap or snare. Freshly cut vegetation will "bleed" sap that has an odor the prey will be able to smell. It is an alarm signal to the animal.

e. You must remove or mask the human scent on and around the trap you set. Although birds do not have a developed sense of smell, nearly all mammals depend on smell even more than on sight. Even the slightest human scent on a trap will alarm the prey and cause it to avoid the area. Removing the scent from a trap is difficult but masking it is relatively easy. Use the fluid from the gall and urine bladders of previous kills. Do not use human urine. Mud, particularly from an area with plenty of rotting vegetation, is also good. Use it to coat your hands when handling the trap and to coat the trap when setting it. In nearly all parts of the world, animals know the smell of burned vegetation and smoke. It is only when a fire is actually burning that they become alarmed. Therefore, smoking the trap parts is an effective means to mask your scent. If one of the above techniques is not practical, and if time permits, allow a trap to weather for a few days and then set it. Do not handle a trap while it is weathering. When you position the trap, camouflage it as naturally as possible to prevent detection by the enemy and to avoid alarming the prey.

f. Canalize traps or snares you place on a trail or run. To build a channel, construct a funnel shaped barrier extending from the sides of the trail toward the trap, with the narrowest part nearest the trap. Canalization should be inconspicuous to avoid alerting the prey. As the animal gets to the trap, it cannot turn left or right and continues into the trap. Few wild animals will back up, preferring to face the direction of travel. Canalization does not have to be an impassable barrier. You only have to make it inconvenient for the animal to go over or through the barrier. For best effect, the canalization should reduce the trail's width to just slightly wider than the targeted animal's body. Maintain this constriction at least as far back from the trap as the animal's body length, and then begin the widening toward the mouth of the funnel.

(1) Use a treadle snare against small game on a trail (**Figure 14-8**). Dig a shallow hole in the trail. Then drive a forked stick (fork down) into the ground on each side of the hole on the same side of the trail. Select two fairly straight sticks that span the two forks. Position these two sticks so that their ends engage the forks. Place several sticks over the hole in the trail by positioning one end over the lower horizontal stick and the other on the ground on the other side of the hole. Cover the hole with enough sticks so that the prey must step on at least one of them to set off the snare. Tie one end of a piece of cordage to a twitch up or to a weight suspended over a tree limb. Bend the twitch up or raise the suspended weight to determine where you will tie a 5 centimeter or so long trigger. Form a noose with the other end of the cordage.

(2) Route and spread the noose over the top of the sticks over the hole. Place the trigger stick against the horizontal sticks and route the cordage behind the sticks so that the tension of the power source will hold it in place. Adjust the bottom horizontal stick so that it will barely hold against the trigger. As the animal places its foot on a stick across the hole, the bottom horizontal stick moves down, releasing the trigger and allowing the noose to catch the animal by the foot. Because of the disturbance on the trail, an animal will be wary. You must therefore use canalization.

Figure 14-8. TREADLE SNARE

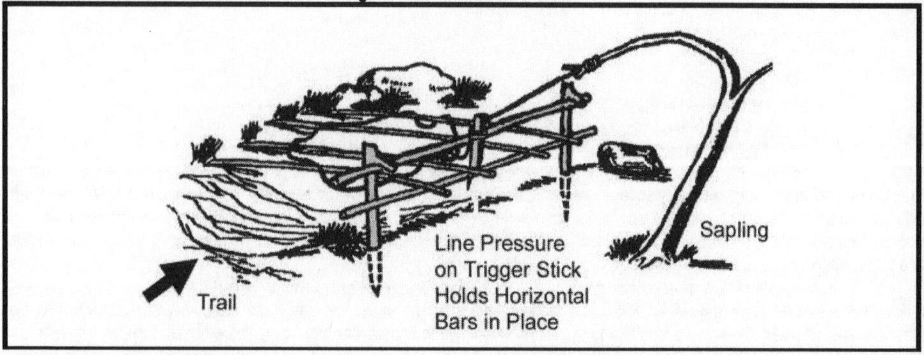

g. Trapping game can be accomplished through the use of snares, traps, or deadfalls. A snare is a noose that will slip and strangle or hold any animal caught in it. You can use inner core strands of parachute suspension lines, wire, bark of small hardwood saplings as well as hide strips from previously caught animals to make snares.

(1) ***Drag Noose Snare.*** The drag noose snare, **Figure 14-9**, is usually the most desirable in that it allows you to move away from the site, plus it is one of the easiest to make and fastest to set. It is especially suitable for catching rabbits. To make the drag noose snare–

(a) Make a loop in the string using a bow line or wireman's knot. When using wire, secure the loop by intertwining the end of the wire with the wire at the top of the loop.

(b) Pull the other end of the string (or wire) through the loop to form a noose that is large enough for the animal's head but too small for its body

(c) Tie the string (or attach the wire) to a sturdy branch. The branch should be large enough to span the trail and rest on the bush or support (two short forked sticks) you have selected. A snared animal will dislodge the drag stick, pulling it until it becomes entangled in the brush. Any attempt to escape tightens the noose, strangling or holding the animal.

Figure 14-9. DRAG NOOSE SNARE

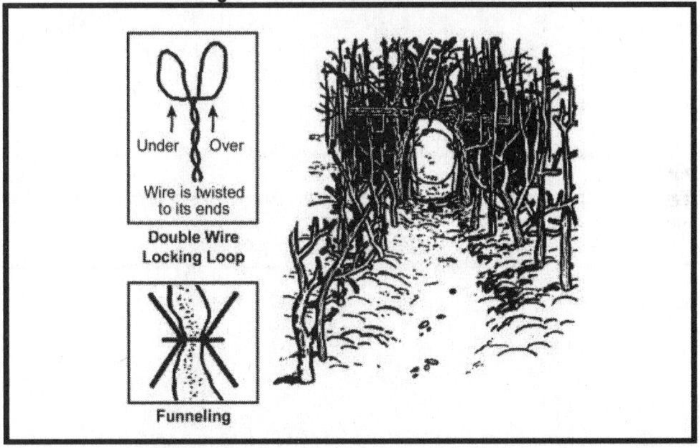

(2) ***Locking Snare Loop.*** Another type snare is the locking type snare loop (**Figure 14-10**) that will lock when pulled tight, ensuring the snared animal cannot escape.

(a) Use lightweight wire to make this snare such as trip wire, or the wire from a vehicle or aircraft electrical system. To construct this snare, cut a piece of wire twice the length of the desired snare wire.

(b) Double the wire and attach the running ends to a securely placed object, such as the branch of a tree. Place a stick about 1/ 2 inch in diameter through the loop end of the wire; holding the wire taut, turn the stick in a winding motion so that the wire is twisted together. You should have four to five twists per inch.

(c) Detach the wire from the branch and then remove the loop from the stick.

(d) Make a Figure 8 in the l/ 2 inch loop by twisting the loop over itself then fold the Figure 8 so the small loops are almost overlapping; run the loose wire ends through these loops. This forms a stiff noose that is strong. Tie the loose end to the stick (for a drag noose square) or branch you are using to complete the snare. This is an excellent snare for catching large animals.

Figure 14-10. LOCKING TYPE SNARE LOOP

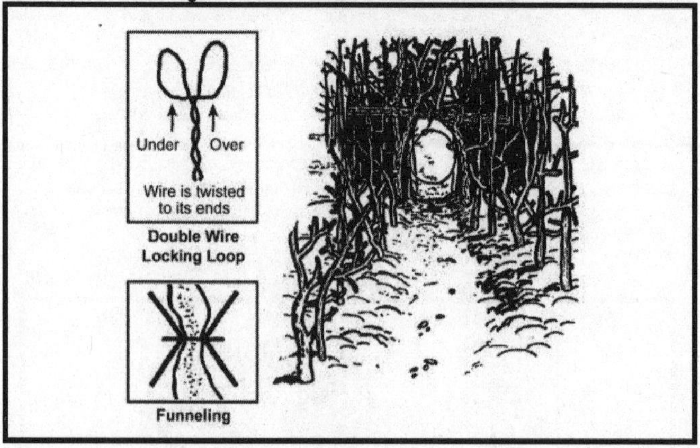

(3) ***Deadfall Trap***. Another means of obtaining game is the use of the deadfall trap (**Figure 14-11** and **Figure 14-12**).

Figure 14-11. TRIGGER WITH DEADFALL

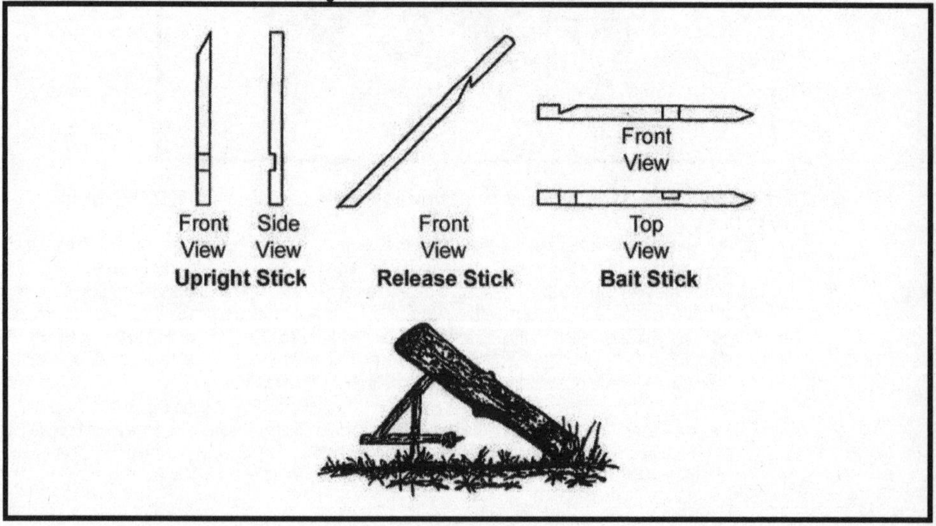

Figure 14-12. TRIGGER WITH DEADFALL

14-13 PROCESSING OF FISH OR GAME. Once you have obtained your fish or game, you must clean/ butcher and cook/ store it. Improper cleaning storing can result in inedible fish and game.

a. **Fish.** You must know how to tell if fish are free of bacterial decomposition that makes the fish dangerous to eat. Although cooking may destroy the toxin from bacterial decomposition, do not eat fish that appear spoiled.

(1) ***Spoilage.*** Eating spoiled or poisoned fish may cause diarrhea, nausea, cramps, vomiting, itching; paralysis, or a metallic taste in the mouth. These symptoms appear suddenly 1 to 6 hours after eating. If you are near the sea, drink sea water as soon as you notice this set of symptoms, or, force yourself to vomit. Signs of spoilage include–

- Peculiar odor.
- Suspicious color. Gills should be red or pink. Scales should be a pronounced–not faded–shade of gray.
- Dent that remains after pressing the thumb against the flesh then removing it.
- Slimy rather than moist or wet body.
- Sharp or peppery taste.

(2) ***Preparation.*** Fish spoil quickly after death, especially on a hot day, so prepare fish for eating as soon as possible after you catch them.

(a) Cut out the gills and large blood vessels that lie next to the backbone. (You can leave the head if you plan to cook the fish on a spit).

(b) Gut fish that are more than 4 inches long cut along the abdomen and scrape out the intestines.

(c) Scale or skin the fish.

(d) You can impale a whole fish on a stick and cook it over an "open fire." However, boiling the fish with the skin on is the best way to get the most food value. The fats and oil are under the skin, and by boiling the fish, you can save the juices for broth. Any of the methods used for cooking plant food can be used for cooking fish.

(e) Fish is done when the meat flakes off.

(f) To dry fish in the sun, hang them from branches or spread them on hot rocks. When the meat has dried, splash it with sea water, if available, to salt the outside. Keep seafood only if it is well dried or salted.

b. **Snakes.** All poisonous and nonpoisonous fresh water and land snakes are edible. To prepare snakes for eating use the following steps (**Figure 14-13**):

(1) Grip the snake firmly behind the head and cut off the head with a knife.

(2) Slit the belly and remove the innards. (You can use the innards for baiting traps and snares).

(3) Skin the snake. (You can use the skin for improvising, belts, straps, or similar items).

DANGER

VENOMOUS SNAKES

TAKE EXTREME CARE IN SECURING SNAKES--THE BITE OF SOME POISONOUS SNAKES CAN BE FATAL. EVEN AFTER A SNAKE'S HEAD IS CUT OFF, ITS REFLEX ACTION CAN CAUSE IT TO BITE, INJECTING POISON.

THE BEST TIME TO CAPTURE SNAKES IS IN THE EARLY MORNING OR LATE EVENING WHEN TEMPERATURES ARE LOW AND THEY MOVE SLOWLY.

KILL IT, OR USE A LONG STICK TO PIN DOWN ITS HEAD AND CAPTURE IT.

TO PICK UP A SNAKE, PLACE YOUR INDEX FINGER ON THE TOP REAR OF ITS HEAD WITH YOUR THUMB AND MIDDLE FINGER ON EITHER SIDE OF THE SNAKE'S HEAD BEHIND THE JAWS.

***KEEP YOUR INDEX FINGER ON TOP OF SNAKE'S HEAD* TO PREVENT IT FROM TURNING AND BITING YOU.**

Figure 14-13. CLEANING A SNAKE

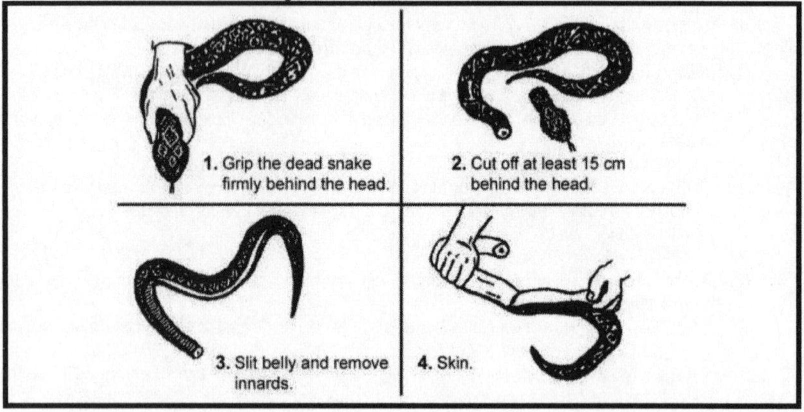

c. **Fowl.** Your first step after killing a fowl for eating or preserving is to pluck its feathers. If plucking is impractical, you can skin the fowl. Keep in mind, however, that a fowl cooked with the skin on retains more food value. Waterfowl are easier to pluck while dry, but other fowl are easier to pluck after scalding. After you pluck the fowl

(1) Cut off its neck close to the body.

(2) Cut an incision in the abdominal cavity and clean out the insides. Save the neck, liver, and heart for stew. Thoroughly clean and dry the entrails to use for cordage.

(3) Wash out the abdominal cavity with fresh clean water. You can boil fowl or cook it on a spit over a fire. You should boil scavenger birds such as vultures and buzzards for at least 20 minutes to kill any parasites. Use the feathers from fowl for insulating your shoes clothing, or bedding. You can also use feathers for fish lures.

d. Medium Sized Mammals. The game you trap or snare will generally be alive when you find it and is therefore dangerous. Be careful when you approach a trapped animal. Use a spear or club to kill it so you can keep a safe distance from it. After you kill an animal, immediately bleed it by cutting its throat. If you must drag the carcass any distance, do so before you cut off the hide so that the carcass is protected from dirt and debris that might contaminate it. Clean the animal near a stream if possible so that you can wash and cool the carcass and edible parts. Fleas and parasites will leave a cooled body so if the situation allows, wait until the animal cools before cleaning and dressing the carcass. To skin and dress the animal (**Figure 14-14** and **Figure 14-15**).

Figure 14-14. SKINNING AND BUTCHERING LARGE GAME

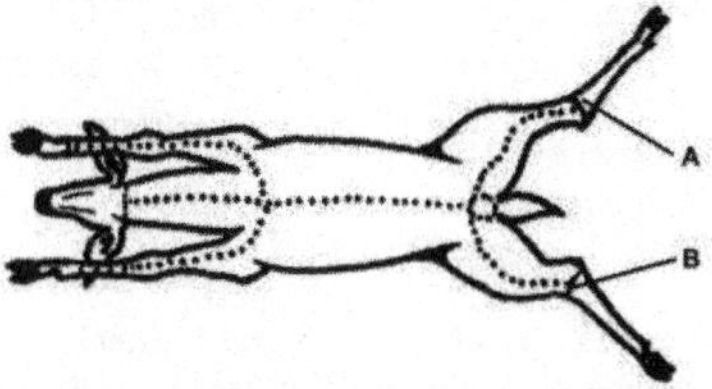

Figure 14-14. SKINNING SMALL GAME

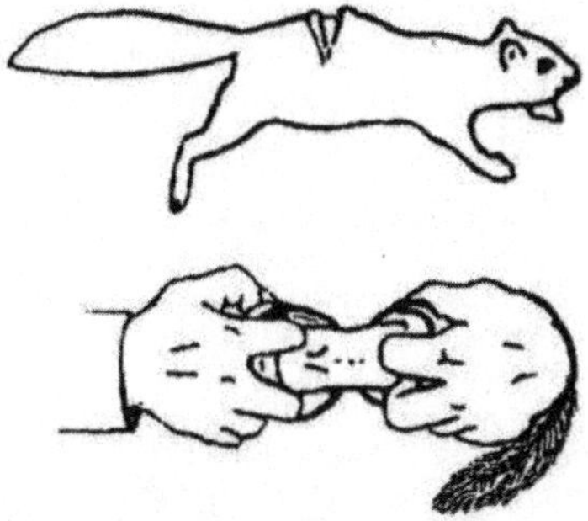

(1) ***Position.*** Place carcass, belly up, on a slope if available. You can use rocks or brush to support it.
(2) ***Genitals and or Udder.*** Remove genitals or udder.
(3) ***Musk Glands.*** Remove these to avoid tainting meat.
(4) ***Hide.*** Split hide from tail to throat. Make the cut shallow so that you do not pierce the stomach.
(5) ***Skin.*** Insert your knife under the skin, taking care not to cut into the body cavity. Peel the hide back several inches on each side to keep hair out of the meat.
(6) ***Chest Cavity.*** Open the chest cavity by splitting the sternum. You can do this by cutting to one side of the sternum where the ribs join.
(7) ***Windpipe and Gullet.*** Reach inside and cut the windpipe and gullet as close to the base of the skull as possible.
(8) Internal Organs. With the forward end of the intestinal tract free, work your way to the rear, lifting out internal organs and intestines. Cut only where necessary to free them.
(9) ***Bladder.*** Carefully cut the bladder away from the carcass so that you do not puncture the bladder (urine can contaminate meat). Pinch the urethra tightly and cut it beyond the point you are pinching. Remove the bladder.
(10) ***Anus.*** From the outside of the carcass, cut a circle around the anus. Pull the anus into the body cavity and out of the carcass.

(11) ***Blood.*** Lift or roll the carcass to drain all blood. Blood, which contains salts and nutrients, is a good base for soups (remember to boil the blood first).

> **NOTE:** Try to save as much blood as you can – it contains food and salt – and then boil the blood.

(12) ***Hide.*** Remove the hide, and make cuts along the inside of the legs to just above the hoof or paw. Peel the skin back, using your knife in a slicing motion to cut the membrane between the skin and meat. Continue this until the entire skin is removed.

(13) ***Entrails.*** Most of the entrails are usable. The heart, liver, and kidneys are edible. Cut open the heart and remove the blood from its chambers. Slice the kidneys and if enough water is available, soak or rinse them. In all animals except those of the deer family, the gall bladder (a small, dark colored, clear textured sac) is attached to the liver.

14) ***Sac.*** Sometimes, the sac looks like a blister on the liver. To remove the sac, hold the top portion of it and cut the liver around and behind the sac. If the gall bladder breaks and gall gets on the meat, wash the gall off the meat immediately so the gall does not taint the meat. Dispose of the gall.

(15) ***Preservation.*** Clean blood splattered on the meat will glaze over and help preserve the meat for a short time. However, if an animal is not bled properly, the blood will settle in the lowest part of its body and spoil the meat quickly. Cut out any meat contaminated this way. If the situation and time allow, you should preserve the extra meat for later use. If the air is cold enough, you can freeze the meat. In warmer climates, preserve by drying or smoking. One night of heavy smoking will make meat edible for about 1 week. Two nights will make it remain edible for 2 to 4 weeks. To prepare meat for drying or smoking, cut it with the grain in quarter inch strips. To air dry the meat, hang it in the wind and hot sun out the reach of animals; cover it so that blow flies cannot land on it.

(16) ***Temperature.*** When temperatures are below 40 degrees, you can leave meat hanging for several days without danger of spoilage. If maggots get on the meat, remove the maggots and cut out the discolored meat. The remaining meat is edible. Maggots, which are the larvae of insects, are also edible.

(17) ***Intestines.*** Thoroughly clean the intestines and use them for storing or smoking food or lashings for general use. Make sure they are completely dry to preclude rotting.

(18) ***Head.*** The head of most animals contains a lot of meat, which is fairly easy to get out. Skin the head and save the skin for leather. Clean the mouth thoroughly and cut out the tongue. After cooking the tongue, remove its outer skin. Cut or scrape the meat from the head. Or, you can roast the head over an open fire before cutting off the meat. Eyes are edible. Cook them but discard the retina (this is a plastic-like disc). The brain is also edible; in fact, some people consider it a delicacy. Each animal's brain matter is considered sufficient to tan the animal's hide.

(19) ***Tendons and Ligaments.*** Use the tendons and ligaments of the body of large animals for lashings.

(20) ***Bone Marrow.*** The marrow in bones is a rich food source. Crack the bones and scrape out the marrow, and use bones to make weapons or fish hooks.

(21) ***Smoke.*** To smoke meat, you will need an enclosed area – for instance, a teepee (**Figure 14-16**) or a pit. You will also need wood from deciduous trees, preferably green. Do not use conifer trees such as pines, firs, spruces, or cedars, as the smoke from these trees give the meat a disagreeable taste and the resin is inedible.

(a) *Para Teepee or Other Enclosed Area with a Vent at the Top.* When using this, set the fire in the center and let it burn down to coals, then stoke it with green wood. Place the strips of meat on a grate or hang them from the top of the enclosure so that they are about 2 feet above the smoking coals.

(b) *Pit Method.* To use the pit method of smoking meat, dig a hole about a yard/meter deep and 1/2 yard/meter in diameter. Make a fire at the bottom of the hole. After it starts burning well, add chipped green wood or small branches of green wood to make it smoke. Place a wooden grate about 1/ 2 yard/meter above the fire, and then lay the strips of meat on the grate. Cover the pit with poles, boughs, leaves, or other material. (A half a yard/meter is about 18 inches or 1 1/2 feet.)

Figure 14-16. SMOKING MEAT

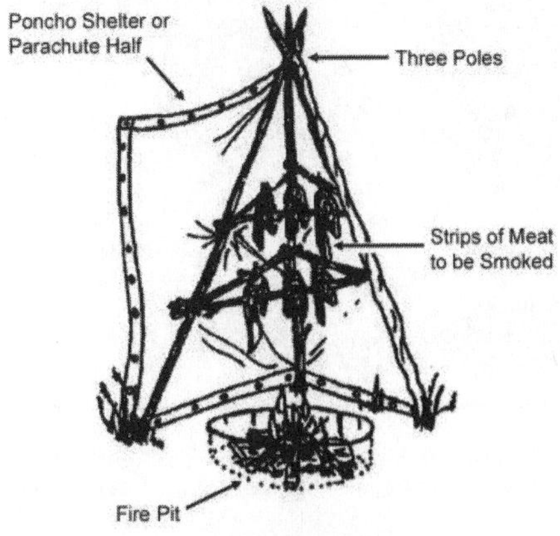

Tepee Smoker

14-14. SHELTERS. A shelter can protect you from the sun, insects, wind, rain, snow, hot or cold temperatures, and enemy observation. In some areas your need for shelter may take precedence over your need for food, possibly even your need for water. After determining your shelter site, you should keep in mind the type of shelter (protection) you need. You need to know how to make different types of shelters. Only two are described in this handbook. Additional information is available in FM 3 05.70, Survival (not releasable to foreigners).

a. **Planning Considerations.**

- How much time and effort are needed to build the shelter?
- Will the shelter adequately protect you from the elements (rain, snow, wind, sun, and so on)?
- Do you have tools to build it? If not, can you improvise tools from materials in the area?
- Do you have the type and amount of manmade materials needed to build it? If not, are there sufficient natural materials in the area?

b. **Poncho Lean To.** You need only a short time and minimal equipment to build this lean to (**Figure 14-17**). You need a poncho, 6 to 10 feet of rope, three stakes about 6 inches long, and two trees (or two poles) 7 to 9 feet apart. Before you select the trees you will use (or decide where to place the poles), check the wind direction. Make sure the back of your lean to will be into the wind. To make the lean to–

(1) Tie off the hood of the poncho. To do this, pull the draw cord tight; roll the hood long ways, fold it into thirds, and tie it with the draw cord.

(2) Cut the rope in half. On one long side of the poncho, tie half of the rope to one corner grommet, and the other half to the other corner grommet.

(3) Attach a drip stick (about a 4 inch stick) to each rope 1/ 4 to 3/ 4 inches away from the grommet. These drip sticks will keep rainwater from running down the ropes into the lean to. Using drip lines is another way to prevent dripping inside the shelter. Tie lines or string about 4 inches long to each grommet along the top edge of the shelter. This allows water to run to and down the line without dripping into the shelter.

(4) Tie the ropes about waist high on the trees (uprights). Use a round turn and two half hitches with quick release knot.

(5) Spread the poncho into the wind and anchor to the ground. To do this, put three sharpened sticks through the grommets and into the ground.

Figure 14-17. PONCHO LEAN TO

(6) If you plan to use the lean to for more than one night, or if you expect rain, make a center support to the lean to. You can do this by stretching a rope between two upright poles or trees that are in line with the center of the poncho.

(7) Tie another rope to the poncho hood; pull it upward so that it lifts the center of the poncho, and tie it firmly to the rope stretched between the two uprights.

(8) Another method is to cut a stick to place upright under the center of the lean to. This method, however, will restrict your space and movements in the shelter.

(9) To give additional protection from wind and rain, place boughs, brush, your rucksack, or other equipment at the sides of the lean to.

(10) To reduce heat loss to the ground, place some type of insulating material, such as leaves or pine needles, inside your lean to.

> **NOTE:** When at rest, as much as 80 percent of your body heat can be lost to the ground.

(11) To increase your security from enemy observation, lower the silhouette of the lean to by making two modifications.

(a) Secure the support lines to the trees at knee height rather than waist height.

(b) Use two knee high sticks in the two center grommets (sides of lean to), and angle the poncho to the ground, securing it with sharpened sticks as above.

c. Field Expedient Lean To. If you are in a wooded area and have sufficient natural materials, you can make an expedient lean to (**Figure 14-18**) without the aid of tools or with only a knife. You need more time to make it than the shelter previously mentioned, but it will protect you from most environmental elements. You will need two trees, (or two upright poles), about 6 feet apart; one pole about 7 feet long and 1 inch in diameter. Five to eight poles about 10 feet long and 1 inch in diameter for beams, cord or vines for securing, the horizontal support to the trees and other poles, saplings, or vines to crisscross the beams. To make this lean to:

Figure 14-18. FIELD EXPEDIENT LEAN TO

(1) Tie the 7 foot pole to the two trees at point about waist to chest high. This is your horizontal support. If there is a fork in the tree, you can rest the pole in it instead of tying the pole in place. If a standing tree is not available, construct a bipod using an Y shaped sticks or two tripods.
(2) Place one end of the beams (10 foot poles) one side of the horizontal support. As with all lean to type shelters, make sure the backside of the lean to is placed into the wind.
(3) Criss-cross sapling or vines on the beams.
(4) Cover the framework with brush, leaves, pine needles, or grass, starting at the bottom and working your way up like shingling.
(5) Place straw, leaves, pine needles, or grass inside the shelter for bedding.
(6) In cold weather, you can add to the comfort of your lean to (**Figure 14-18)** by building a fire reflective wall. Drive four stakes about 4 feet long into the ground to support the wall. Stack green logs on top of one another between the support stales. Bind the top of the support stakes so the green logs will stay in place. Fill in the spaces between the logs with twigs or small branches. With just a little more effort, you can have a drying rack. Cut a few 3/ 4 inch diameter poles. The length depends on distance between the lean to support and the top of the fire reflective wall. Lay one end of the poles on the lean to horizontal support and the other ends on top of the reflector wall. Place and tie into place smaller sticks across these poles. You now have a place to dry clothes, meat, or fish.

14-15 FIRES. A fire can full fill several needs. It can keep you warm, it can keep you dry: you can use it to cook food, to purify water, and to signal. It can also cause you problems when you are in enemy territory: it creates smoke, which can be smelled and seen from a long distance: It causes light which can be seen day or night and it leaves signs of your presence. Remember you should always weigh your need for a fire against your need to avoid enemy protection. When operating in remote areas you should always take a supply of matches in a waterproof case and keep them on your person.

a. **Selection**. When selecting a site to build a fire, you should consider the following:

- Where (terrain and climate) you are operating.
- What materials and tools are available.
- How much time you have.
- Why you need a fire.
- Where is the enemy–how near is he?

b. **Preparation.** If you are in a wooded or brush covered area, clear brush away and scrape the surface soil from the spot you selected. The cleared circle should be at least 3 feet (1 meter) in diameter so that there is little chance of the fire spreading. To prepare the site for a fire, ensure that it is dry and that it look for a dry spot that

- Offers protection from the wind.
- Is suitably placed in relation to your shelter (if any).
- Concentrates the heat in the direction you desire.
- Has a supply of wood or other fire burning material.

c. **Dakota Fire Hole.** In some situations, an underground fireplace will best meet your needs. It conceals the fire to some extent and serves well for cooking food. To make an underground fireplace or Dakota fire hole (**Figure 14-19**):

(1) Dig a hole in the ground.

(2) On the upwind side of this hole, poke one large connecting hole for ventilation.

Figure 14-19. DAKOTA FIRE HOLE

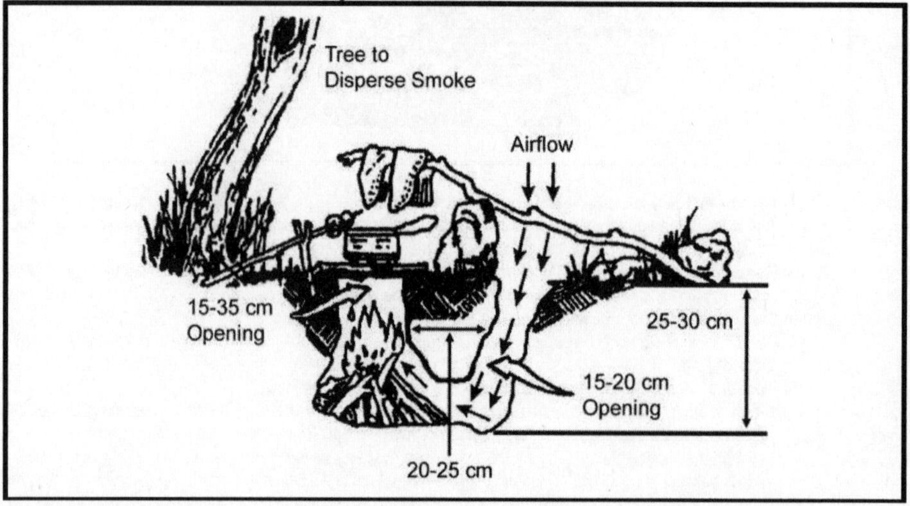

d. **Aboveground Fire**. If you are in a snow covered or wet area, you can use green logs to make a dry base for your fire (**Figure 14-20**). Trees with wrist-sized trunks are easily broken in extreme cold. Cut or break several green logs and lay them side by side on top of the snow. Add one or two more layers, laying the top layer logs in a direction opposite those of the layer below it.

Figure 14-20. BASE FOR FIRE IN SNOW COVERED AREA

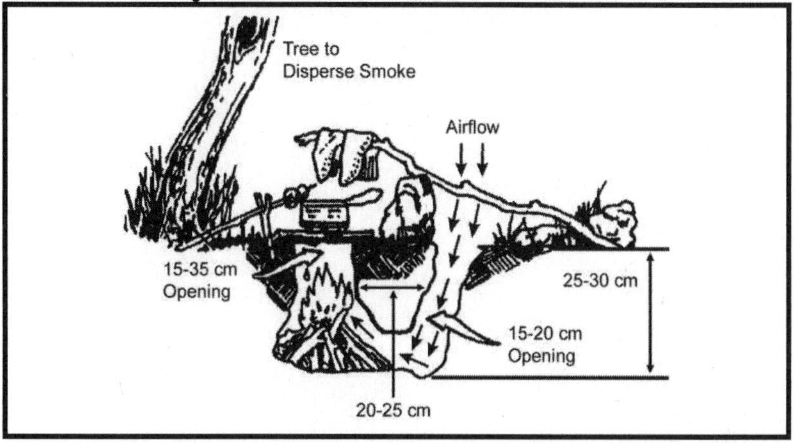

14-16. METHODS. There are several methods for laying a fire for quick fire making. Three easy methods follow (**Figure 14-21**):

a. Tepee. Arrange tinder and a few sticks of kindling in the shape of a cone. Fire the center. As the cone burns away, the outside logs will fall inward, feeding the heart of the fire. This type of fire burns well even with wet wood.

b. Lean To. Push a green stick into the ground at a 30-degree angle. Point the end of the stick in the direction of the wind. Place some tinder (at least a handful) deep inside this lean to stick. Light the tinder. As the kindling catches fire from the tinder, add more kindling.

c. Cross Ditch. Scratch a cross about 1 foot in size in the ground. Dig the cross 3 inches deep. Put a large wad of tinder in the middle of the cross. Build a kindling pyramid above the tinder. The shallow ditch allows air to sweep under the fire to provide a draft.

Figure 14-21. METHODS FOR LAYING A FIRE

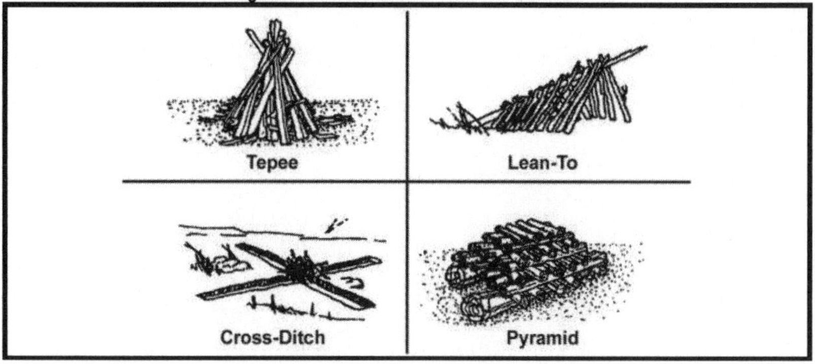

Chapter 15
AVIATION

Army aviation and Infantry units can be fully integrated with other members of the combined arms team to form powerful and flexible air assault task forces. These forces can project combat power throughout the depth and width of the modern battlefield, with little regard for terrain barriers. These combat operations are deliberate, precisely planned, and vigorously executed. They strike the enemy when and where he is most vulnerable. See **Figure 2-11**, *Air Movement Annex,* and **Figure 2-12**, *Coordination Checklists*, which include the *Army Aviation Coordination Checklist*.

15-1. REVERSE PLANNING SEQUENCE. Successful air assault execution is based on a careful analysis of METT-TC and detailed, precise reverse planning. Five basic plans that comprise the reverse planning sequence are developed for each air assault operation. The battalion is the lowest level that has sufficient personnel to plan, coordinate, and control air assault operations. When company size or lower operations are conducted, most of the planning occurs at battalion or higher headquarters. The five plans are–

a. **Ground Tactical Plan**. The commander's ground tactical plan forms the foundation of a successful air assault operation. All additional plans must support this plan. It specifies actions in the objective area to ultimately accomplish the mission and address subsequent operations.

b. **Landing Plan.** The landing plan must support the ground tactical plan. This plan outlines a sequence of events that allows elements to move into the area of operations, and ensures that units arrive at designated locations at prescribed times, and that as soon as they arrive, they are prepared to execute the ground tactical plan.

c. **Air Movement Plan**. The air movement plan is based on the ground tactical and landing plans. It specifies the schedule and provides instructions for air movement of troops, equipment, and supplies from PZs to LZs.

d. **Loading Plan**. The loading plan is based on the air movement plan. It ensures that troops, equipment, and supplies are loaded on the correct aircraft. Unit integrity is maintained when aircraft loads are planned. Cross loading may be necessary to ensure survivability of command and control assets, and that the mix of weapons arriving at the LZ is ready to fight.

e. **Staging Plan**. The staging plan is based on the loading plan and prescribes the arrival time of ground units (troops, equipment and supplies) at the PZ in the order of movement

15-2. SELECTION AND MARKING OF PICKUP AND LANDING ZONES

a. Considerations. Small unit leaders should consider the following when selecting a PZ/ LZ:

(1) ***Size***. Minimal circular landing point separation from other aircraft and obstacles is needed:
OH 58D – 25 meters.
UH 1, AH 1 – 35 meters.
UH 60, AH 64 – 50 meters.
Cargo helicopters – 80 meters.

(2) ***Surface Conditions***. Avoid potential hazards such as sand, blowing dust, snow, tree stumps, or large rocks.

(3) ***Ground Slope***.
0 to 6 percent—land upslope.
7 to 15 percent—land sideslope.
Over 15 percent—no touchdown (aircraft may hover).

(4) ***Obstacles***. An obstacle clearance ratio of 10 to 1 is used in planning approach and departure of the PZ and LZ. For example, a tree that is 10 feet tall requires 100 feet of horizontal distance for approach or departure. Mark obstacles with a red chemlight at night or red panels in daytime. *Avoid using markings if the enemy would see them.*

(5) ***Approach/ Departure.*** Approach and depart into the wind and along the long axis of the PZ/ LZ.

(6) ***Loads.*** The greater the load, the larger the PZ/ LZ must be to accommodate the insertion or extraction.

b. **Marking of PZs and LZs**.

(1) ***Day***. A ground guide will mark the PZ or LZ for the lead aircraft by holding an M16/ M4 rifle over his head, by displaying a folded VS 17 panel chest high, or by other coordinated and identifiable means.

(2) ***Night***. The code letter "Y" (inverted "Y") is used to mark the landing point of the lead aircraft at night (**Figure 15-1**). Chemical lights or "beanbag" lights are used to maintain light discipline. A swinging chemlight may also be used to mark the landing point.

Figure 15-1. INVERTED "Y"

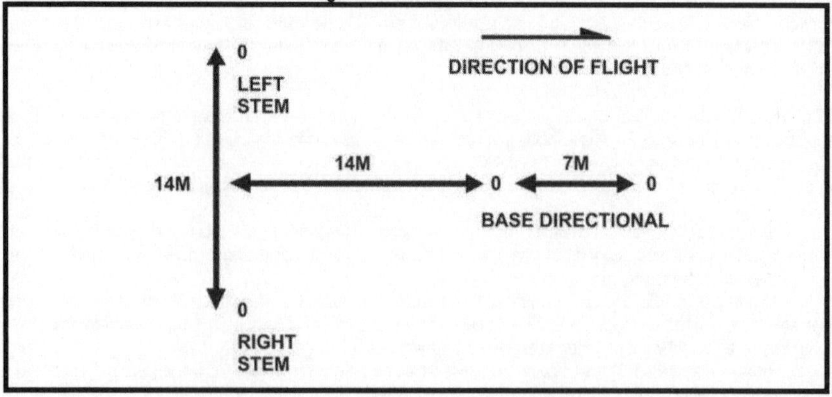

15-3. AIR ASSAULT FORMATIONS. Aircraft supporting an operation may use any of the following PZ/ LZ configurations (**Table 15-1**), which are prescribed by the air assault task force (AATF) commander working with the air mission commander (AMC).

Table 15-1. AIR ASSAULT FORMATIONS

FORMATION	*PROS*	*CONS*
Heavy Left or Right	Provides firepower to front and flank	Requires a relatively long, wide landing area Presents difficulty in pre-positioning loads Restricts suppressive fire by inboard gunners
Diamond	Allows rapid deployment for all round security Requires only a small landing area	Presents some difficulty in pre-positioning loads Restricts suppressive fire of inboard gunners
Vee	Requires a relatively small landing area Allows rapid deployment of forces to the front	Presents some difficulty in pre-positioning loads
Echelon Left or Right	Allows rapid deployment of forces to the flank Allows unrestricted suppressive fire by gunners	Presents some difficulty in pre-positioning loads Requires a relatively long, wide landing area
Trail	Requires a relatively small landing area Allows rapid deployment of forces to the flank Simplifies pre-positioning of loads Allows unrestricted suppressive fire by gunners.	None
Staggered Trail Left or Right	Simplifies pre-positioning of loads Allows rapid deployment for all round security	Requires a relatively long, wide landing area Somewhat restricts gunners' suppressive fire

a. **Heavy Left or Right Formation (Figure 15-2).**

PROS: Provides firepower to front and flank.

CONS: Requires a relatively long, wide landing area; presents difficulty in pre positioning loads; restricts suppressive fire by inboard gunners.

Figure 15-2. HEAVY LEFT/ HEAVY RIGHT FORMATION

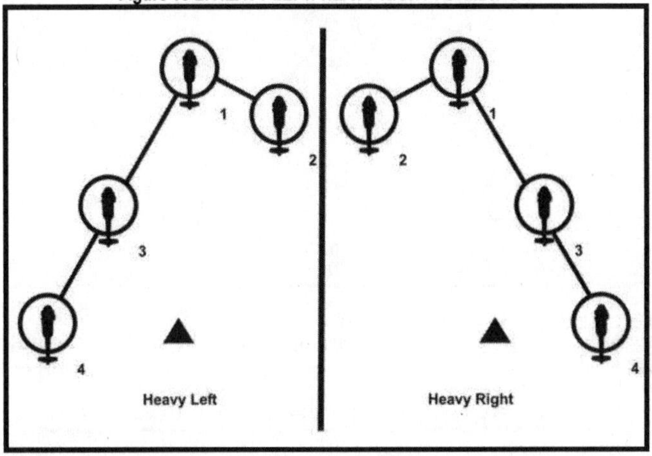

b. **Diamond Formation (Figure 15-3).**

PROS: Allows rapid deployment for all round security; requires a small landing area.

CONS: Presents some difficulty in pre positioning loads; restricts suppressive fire by inboard gunners.

Figure 15-3. DIAMOND FORMATION

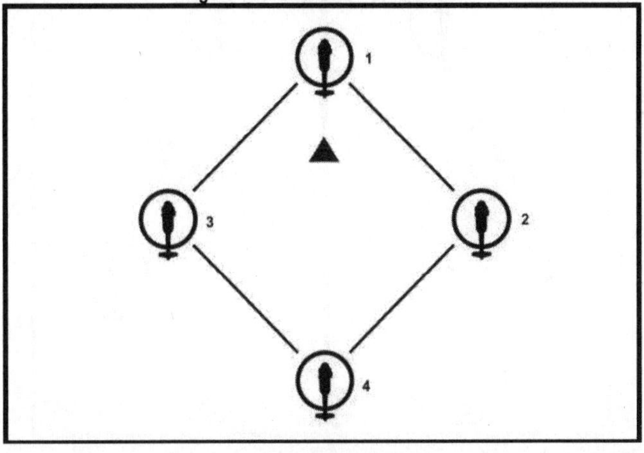

c. **Vee Formation (Figure 15-4).**

PROS: Requires a relatively small landing area; allows rapid deployment of forces to the front; restricts suppressive fire of inboard gunners.

CONS: Presents some difficulty in pre positioning loads.

Figure 15-4. VEE FORMATION

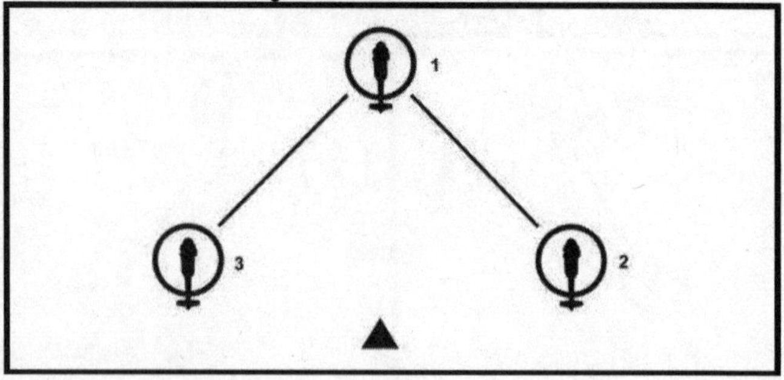

d. **Echelon Left or Right Formation (Figure 15-5).**

PROS: Allows rapid deployment of forces to the flank; allows unrestricted suppressive fire by gunners.

CONS: Requires a relatively long, wide landing area; presents some difficulty in pre positioning loads.

Figure 15-5. ECHELON LEFT/ ECHELON RIGHT FORMATION

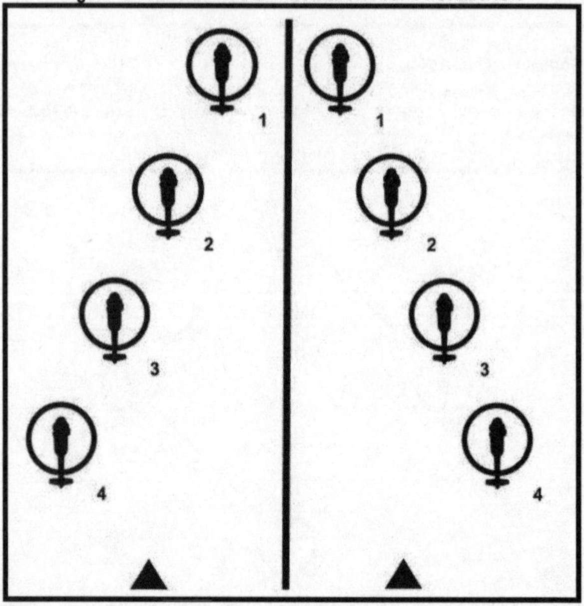

e. **Trail Formation (Figure 15-6).**

PROS: Requires a relatively small landing area; allows rapid deployment of forces to the flank; simplifies pre positioning loads; allows unrestricted suppressive fire by gunners.

CONS: None.

Figure 15-6. TRAIL FORMATION

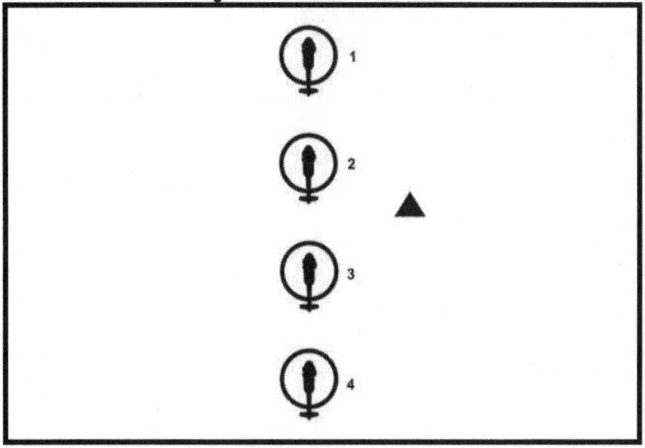

f. **Staggered Trail Left or Right Formation (Figure 15-7).**

PROS: Requires a relatively long, wide landing area; gunners' suppressive fire restricted somewhat.

CONS: Simplifies pre positioning loads; allows rapid deployment for all round security.

Figure 15-7. STAGGERED TRAIL LEFT/ STAGGERED TRAIL RIGHT FORMATION

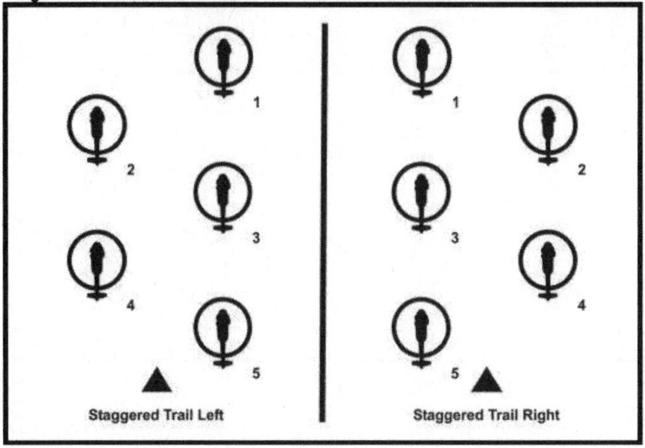

15-4. PICKUP ZONE OPERATIONS

Prior to arrival of aircraft, the PZ is secured, PZ control party is positioned, and the troops and equipment are positioned in platoon/ squad assembly areas. In occupying a patrol/ squad assembly area, the patrol/ squad leader does the following steps. **Figure 15-8** shows an example of a large, one sided PZ. Figures 15-9 through 15-12 show loading/ unloading procedures and techniques:

- Maintain all round security of the assembly area.
- Maintain communications.
- Organize personnel and equipment into chalks and loads.
- Conduct safety briefing and equipment check of troops.

Figure 15-8. LARGE, ONE-SIDED PICKUP ZONE

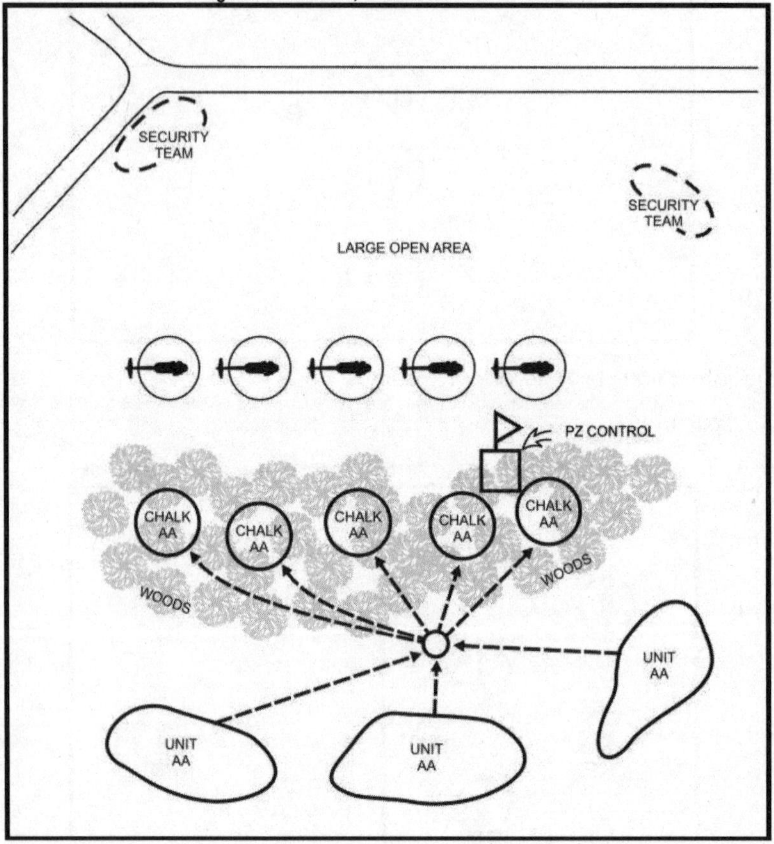

Figure 15-9. UH 60 LOADING SEQUENCE

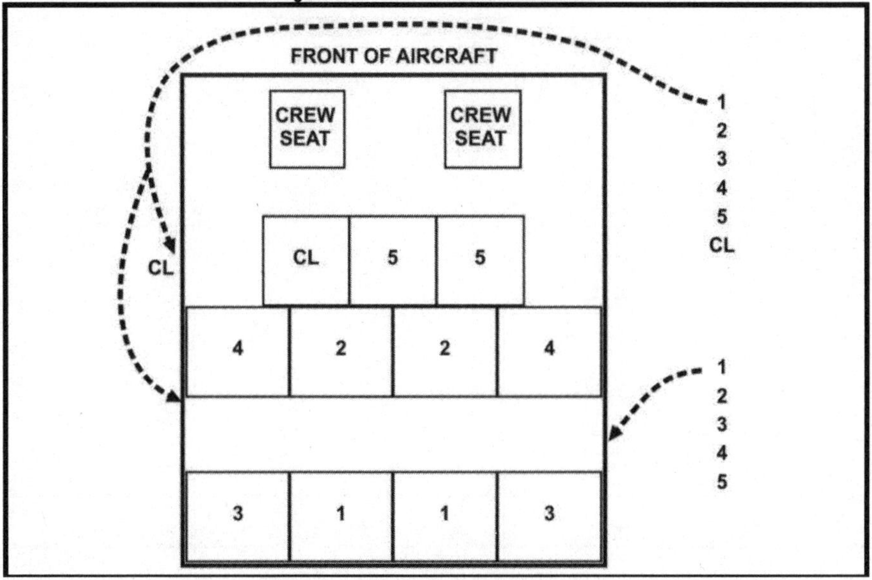

Figure 15-10. UNLOADING SEQUENCE

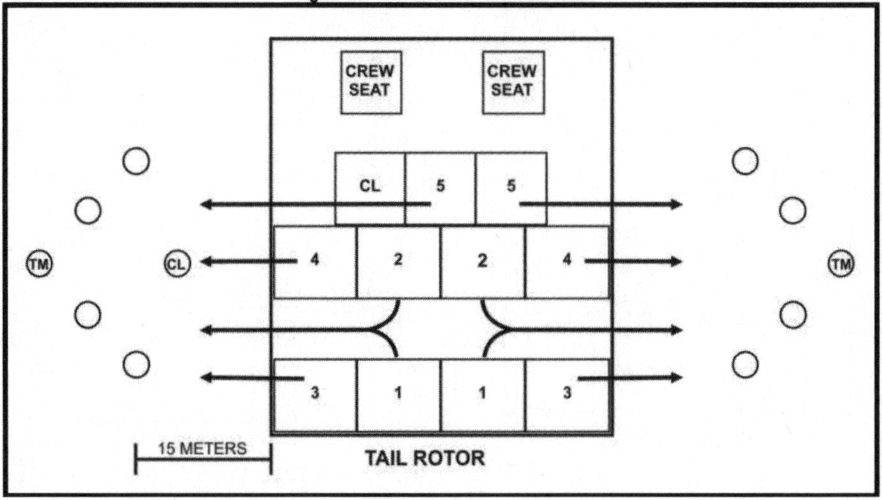

Figure 15-11. TACTICAL LOADING SEQUENCE

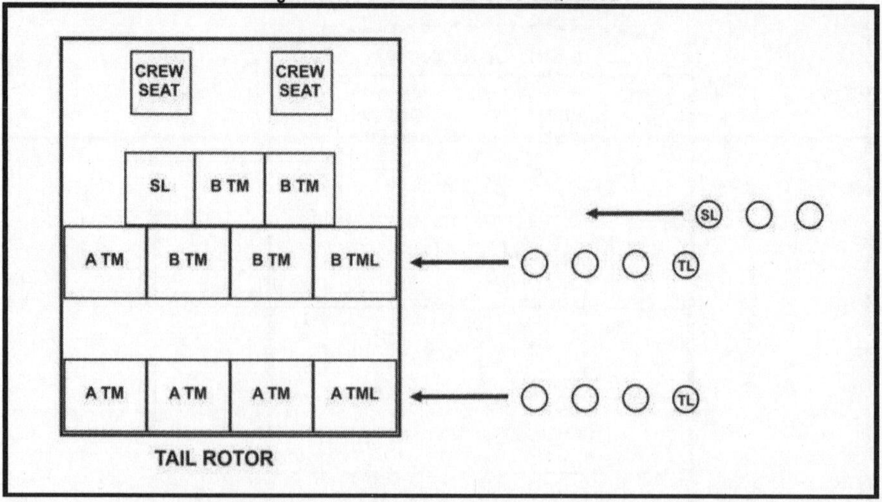

Figure 15-12. TACTICAL UNLOADING USING DOOR NEAREST COVER, CONCEALMENT

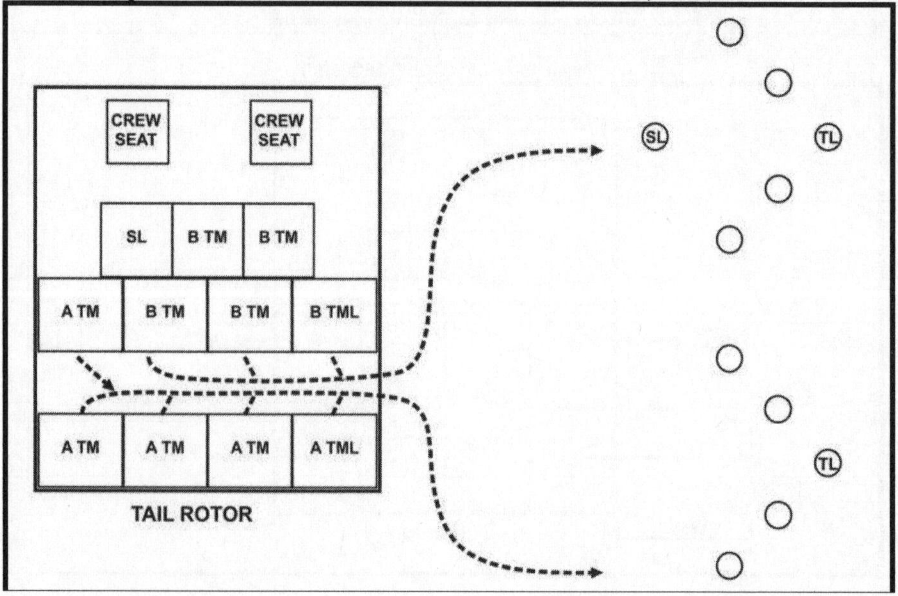

15-5. SAFETY. Safety is the primary concern of all leaders when operating in/around aircraft. The inclusion of aircraft into Ranger operations brings high risks. Consider the following:

a. Approach the aircraft from 45 to 90 degrees off the nose.
b. Point upward the muzzles of weapons with blank firing adapters.
c. Point downward the muzzles of weapons loaded with live ammunition.
d. Wear the ballistic helmet.
e. When possible, conduct an air crew safety brief with all personnel.
f. At a minimum, cover loading/ off loading, emergency, and egress procedures.
g. [Leaders] Carry a manifest and turn in a copy to higher.

15-6. REQUIREMENTS. Minimum landing space requirements and minimum distance between helicopters on the ground depend on many factors. If the aviation unit SOP does not spell out these requirements, the aviation unit commander works with the Pathfinder leader. The final decision about minimum landing requirements rests with the aviation unit commander. In selecting helicopter landing sites from maps, aerial photographs, and actual ground or aerial reconnaissance, he considers the following factors:

a. **Number of Helicopters.** To land a large number of helicopters at the same time, the commander might have to provide another landing site(s) nearby. Or, he can land the helicopters at the same site, but in successive lifts.

b. **Landing Formations.** Helicopter pilots should try to match the landing formation to the flight formation. Pilots should have to modify their formations no more than necessary to accommodate the restrictions of a landing site. However, in order to land in a restrictive area, they might have to modify their formation somewhat.

c. **Surface Conditions.** Rangers choose landing sites that have firm surfaces. This prevents helicopters from bogging down, creating excessive dust, or blowing snow. Rotor wash stirs up any loose dirt, sand, or snow. This can obscure the ground, especially at night. Rangers remove these and any other debris from landing points, since airborne debris could damage the rotor blades or turbine engine(s).

d. **Ground Slope.** Rangers choose landing sites with relatively level ground. For the helicopter to land safely, the slope should not exceed 7 degrees. Whenever possible, pilots should land upslope rather than downslope. All helicopters can land where ground slope measures 7 degrees or less (**Figure 15-13**).

(1) ***Day Operation Signals.*** For daylight operations, you can use different smoke colors for each landing site. You can use the same color more than once, just spread them out. Use smoke only if you have to, because the enemy can see it, too. Try to use it only when the pilot asks for help locating his helicopter site.

(2) ***Night Operation Signals.*** For night operations, use pyrotechnics or other visual signals in lieu of smoke. As in daylight, red signals mean "Do not land," but you can also use them to indicate other emergency conditions. All concerned must plan and know emergency codes. Each flight lands at the assigned site according to CC messages and the visual aids displayed. You can use arm and hand signals to help control the landing, hovering, and parking of helicopters.

Figure 15-13. GROUND SLOPE

GROUND SLOPE EXPRESSED IN DEGREES

The approximate slope angle may be calculated by multiplying the gradient by 57.3. This method is reasonably accurate for slope angles under 20 degrees.

VD = B-A = 150
HD = 3,000

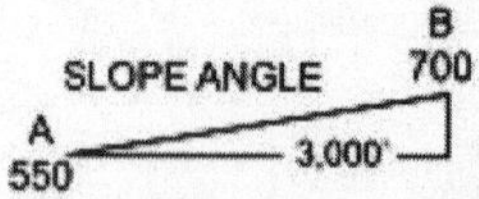

DEGREE OF SLOPE $\frac{150 \times 57.3}{3,000}$

$\frac{8,595}{3,000}$ = APPROXIMATELY 3° OF SLOPE

GROUND SLOPE EXPRESSED AS PERCENTAGE

To determine the percent of ground slope, divide the vertical distance (VD) by the horizontal distance (HD) and multiply by 100.

$$\text{PERCENT of SLOPE} = \frac{VD}{HD} \times 100$$

Verticle distance is the difference in field elevation between the two ends of the landing site. Always round number up to the next whole number.

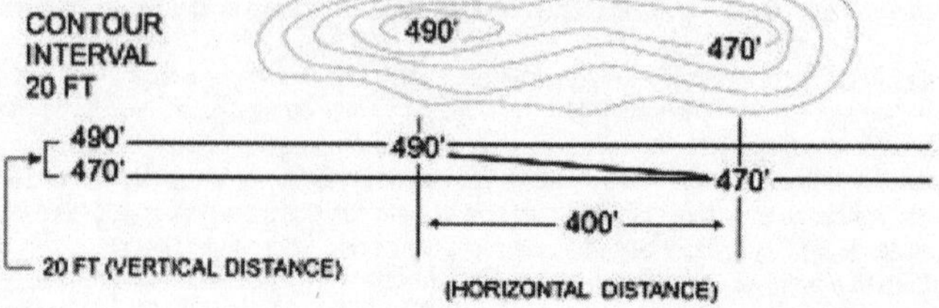

$400\overline{)20.00}$ = .05

.05 x100 = 05.00 = 5 PERCENT

PATHFINDER SLOPE LANDING RULES

Do not land small utility and observation aircraft on slopes exceeding 7 degrees.

Give large utility and cargo aircraft an advisory if ground-slope is between 7 and 15 degrees.

Always advise pilot when landing wheeled aircraft on a sideslope.

15-7. DESERT. The typical desert is a dry, barren region, generally treeless and sandy. It suffers environmental extremes, with violent and unpredictable weather changes. Its terrain conforms to no particular model. Frequent clear days offer unequaled visibility and flight conditions, but a sudden sandstorm immediately halts all operations. Successful desert operations require special training, acclimatization, and great self discipline.

a. **Communications.** In desert operations, the radio offers the best way to communicate. The low, rolling terrain allows good radio range. Due to the increased distances involved in military desert operations, FM radio communications may prove inadequate, especially in the higher FM frequencies. Rangers, aircraft, and ground crew must all have high frequency radio equipment. Sand or dust in equipment or a poor electrical ground cause most communication problems. Due to the increased distances between land force units engaged in desert operations, helicopters may provide air or ground relay or help deploy ground radio rebroadcast facilities. **Table 15-2** shows an example ground-to-air radio transmission.

Table 15-2. GROUND TO AIR TRANSMISSIONS

Pilot:	***ALPHA ONE LIMA ONE SIX [A1L16], THIS IS ROMEO TWO BRAVO TWO SEVEN [R2B27], OVER.***
Ranger:	ROMEO TWO BRAVO TWO SEVEN, THIS IS ALPHA ONE LIMA ONE SIX, OVER.
Pilot:	***THIS IS BRAVO TWO SEVEN, CCP INBOUND, OVER.***
Ranger:	THIS IS LIMA ONE SIX, STATE TYPE, NUMBER, AND INTENTIONS, OVER.
Pilot:	***THIS IS BRAVO TWO SEVEN, FOUR UNIFORM HOTEL SIXTIES [UH-60s], TROOP DROP-OFF AND SLING LOAD, FOR YOUR SITE, OVER.***
Ranger:	THIS IS LIMA ONE SIX, ROGER, HEADING THREE TWO FIVE [325], THREE THOUSAND [3,000] METERS. LAND THREE TWO FIVE, SIGNAL ON CALL, LAND ECHELON RIGHT, SLING-LOAD AIRCRAFT USE NUMBER FOUR LANDING POINT, CONTINUE APPROACH FOR VISUAL CONTACT, OVER.

b. **Navigation.** Many of the conditions experienced in cold weather operations resemble those in desert operations. Rangers and pilots find distances and altitudes hard to judge in the desert. The lack of definable terrain features makes navigation difficult, especially at night and over long distances. Also, the sameness of the terrain can influence a pilot to pay less attention to his surroundings. Rangers may have to mark and man release points.

c. **Landing Sites.** The climatic conditions in the desert profoundly affect the setup and operation of landing sites. Most importantly, the Ranger must consider density altitude, wind, and sand (dust). Sand on a landing site can produce brownout conditions similar to those in snowy areas, so the same precautions apply. This makes a rocky area a better landing site than a sandy hollow, depression, or valley.

d. **Wind.** Desert winds generally calm down for an hour or two around sundown. Another calm occurs before sunrise. Other than those times, desert winds can drive dense clouds of dust and sand with hurricane force, and rapid temperature changes often follow strong winds. The Pathfinder leader must consider what times of day the wind will allow him to operate the landing site.

(1) The extreme heat often experienced in the desert also affects the aircraft's ACL. When supporting a ground unit, the Ranger leader coordinates with the aviation element to determine the ACL for each type of aircraft. Both the minimum distance between aircraft and the size of the landing point increase in desert operations: 100 meters between aircraft, 100 meter diameter landing points. In daylight hours, ground crew members mark the touchdown points. They paint sandbags a bright color or mark them using some other quick method. Ideally, they use signalmen.

2) When establishing a landing site, the Ranger leader considers taxi procedures. When an aircraft must taxi, the pilot moves it into a vertical position as quickly as possible to reduce the amount of sand (dust) the engine sucks in as well as to avoid a brownout. Pilots should avoid taxiing over the same area repeatedly.

e. **Liftoffs.** Pilots will not try a normal liftoff in a sandstorm. Helicopters with wheels and airplanes should make a running type takeoff. Helicopters with skids should make a maximum performance liftoff.

f. **Landings.** When they can, pilots should use a running type landing to reduce sand intake. If a pilot can make a running landing, he keeps the touchdown roll to a minimum to keep from overloading the landing gear. If the terrain does not permit a running landing, the pilot lands at a greater than normal angle. He should never land from a hover.

g. **Safety.** Ground crew personnel should wear clothing that will protect them against the sand blown around by the rotor wash. Each person on the ground should take special care to keep the sand out of his eyes, ears, nose, and mouth. Goggles, earplugs, and cloth masks provide adequate protection for facial areas. Other ground crew procedures resemble those for cold weather operations.

15-8. MOUNTAINS. Mountains have rugged, divided terrain with steep slopes and few natural or manmade lines of communication. Weather fluctuates seasonally from extreme cold, with ice and snow, to extreme heat. Also, it can switch between the two extremes very quickly. This unpredictability greatly affects operations.

a. **Communications**. Mountainous terrain often limits or restricts communications. To maintain communications within the AO, aircraft may have to limit operations to the vicinity of the unit. Other aircraft can serve as radio relay stations. Ranger units may also have to set up radio relays at the RP, CCP, or both.

(1) Mountain conditions challenge aviators in Ranger operations more than any other conditions. For precise flying in mountainous areas, pilots need large scale terrain maps.

(2) Since intervening terrain degrades GTA communications, providing navigational aid and control over extended ranges might prove difficult.

b. **Wind.** The main weather hazard in the mountains is wind. Even moderate winds (11 to 20 knots) can produce significant turbulence over mountain ridges. Predicting wind conditions is difficult. The windward side of a mountain maintains a steady direction of airflow, though the strength of the wind may vary. The leeward side has turbulent winds with strong vertical currents. This turbulence might prevent assault landings and require pilots to fly at higher altitudes. This naturally increases the risk of detection and destruction.

c. **Density Altitude.** In the mountains, density altitude can vary a lot between PZs and LZs. It can also vary greatly from one time of day to another. It normally peaks in the late afternoon, and drops to its lowest point at dawn.

d. **Navigation.** In the mountains, the helicopter offers the best way to rapidly move forces. In the offense, air assault operations can insert forces into the enemy's rear area and bypass or envelop his defenses. In the defense, helicopters can move reinforcements and reserves rapidly.

e. **Landing Sites.** Mountainous regions offer few, if any airfields for fixed wing aircraft, and few LZs suitable for multiple helicopters.

(1) If the enemy situation allows, Rangers to set up LZs on the windward side of the mountain, since that side offers more stable winds.

(2) When they can only find LZs designed for single aircraft, planners increase in flight spacing. This places an extra load on each crew. When conducting multiship operations into a small LZ, the Ranger controller should allow sufficient time between liftoff and landing for the turbulent air generated during the departure of the previous helicopter to stabilize. Otherwise, the pilot of the incoming craft will experience that turbulence and lose lift.

(3) A pilot must touchdown very carefully on the typical small, rough, sloped mountain LZ. Depending on the angle of the slope and on the aircraft's available torque, the pilot might be able to make a normal slope landing. Pilots of larger craft, such as cargo helicopters, may have trouble positioning the entire fuselage in the available area. Once the cockpit extends over the landing area, the pilot cannot see the ground. He must rely on the crew chief and signalman to direct him.

(4) During a mountain approach to an LZ surrounded by uneven terrain, the pilot has a hard time determining the actual aircraft altitude and rate of closure. Where the terrain slopes up to the LZ, a visual illusion occurs. The pilot may think he is flying too high and closing too slowly. If the terrain slopes down to the LZ, he may feel he is flying too low and closing too fast. Employing a signalman on the ground gives the pilot a visual reference to adjust his controls. He may need more than one signalman.

f. **Site Assessment**. Rangers should determine the following information while reconnoitering and selecting a mountain site:

(1) The size, slope, amount of surface debris, and the area covered by shadows and obstacles in and around the site.
(2) The approximate direction, speed, and characteristics of the wind.
(3) The inbound route, if necessary. When the pilot cannot land due to a steep slope, the aircraft may terminate at a hover to off load troops and supplies.
(4) The departure route, which should orient into the wind and over the lowest obstacles.

15-9. OBSERVATION HELICOPTERS. This category includes the OH 58D Kiowa and the OH 6A Cayuse.

a. **OH 58D Kiowa. Table 15-3** shows specifications for the Kiowa; **Figure 15-14** shows the aircraft from three angles.

Table 15-3. SPECIFICATIONS FOR THE OH 58D (KIOWA)

Rotor Diameter	35 feet
Length:	
Rotor Operating	42 feet 2 inches
Blades Removed	33 feet 10 inches
Height to Top of Turret	12 feet 9-1/2 inches
Tread (Skids)	6 feet 2 inches
Main Rotor	
Disk Area	0.962 square feet
Blade Area	38.26 square feet
Clear Area Needed for Rotor	12.5 meters
TDP # 1	25 meters diameter

Figure 15-14. OH 58D (KIOWA)

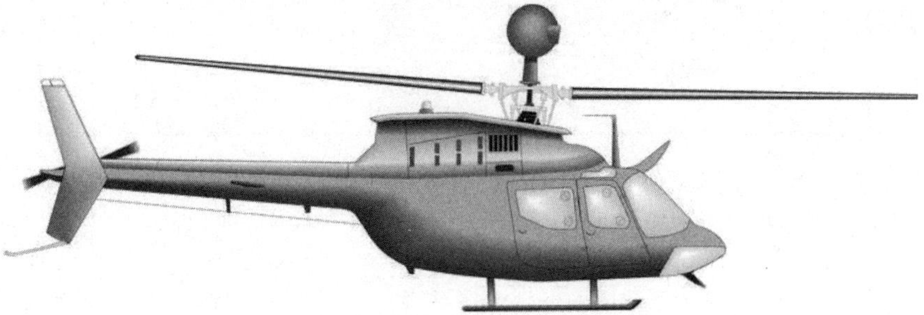

b. **OH 6A Cayuse. Table 15-4** shows specifications for the Cayuse; **Figure 15-15** shows the aircraft from three angles.

Table 15-4. SPECIFICATIONS FOR THE OH 6A CAYUSE

Rotor Diameter	26 feet 4 inches
Length: Rotor Operating Rotor Folded	 39 feet 3-3/4 inches 22 feet 9 1/2 inches
Span, Maximum Lateral	26 feet
Height	8 feet 9 inches
Tread (Skids)	6 feet 3 inches
Clear Area Needed for Rotors	9.3 meters
TDP # 1	25 meters diameter

Figure 15-15. OH 6A (CAYUSE)

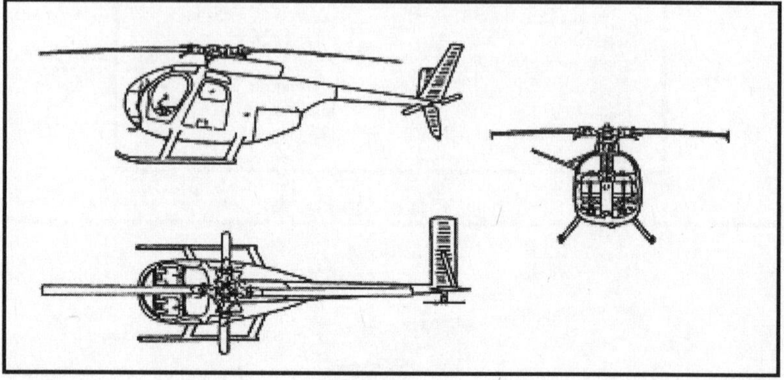

15-10. ATTACK HELICOPTERS

This category of helicopters includes the AH 1S Cobra and the AH 64A Apache.

a. **AH 1S Cobra. Table 15-5** shows specifications for the Cobra; **Figure 15-16** shows the aircraft from three angles.

Table 15-5. SPECIFICATIONS FOR THE AH 1S (COBRA)

Rotor Diameter	44 feet
Length:	
Rotor Operating	53 feet 1 inch
Fuselage	44 feet 9 inches
Span, Maximum Lateral	11 feet 8 inches
Height	11 feet 7 inches
Tread (Skids)	7 feet
Rotor Ground Clearance (Static)	7 feet 10 inches
Clear area needed for Rotors	16.18 meters
TDP # 2	35 meters diameter

Figure 15-16. AH 1S (COBRA)

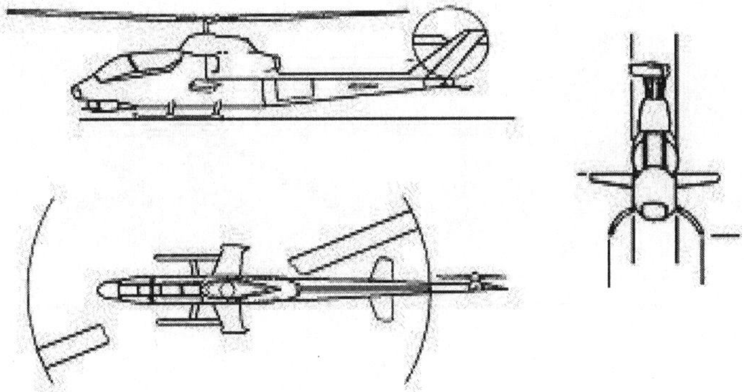

a. **AH 64A (Apache). Table 15-6** shows three views and final specifications for the AH 64A Apache; **Figure 15-17** shows the aircraft from three angles.

Table 15-6. SPECIFICATIONS FOR THE AH 64A (APACHE)

Rotor Diameter	48 feet
Length:	
Rotors Operating	58 feet 3-1/8 inches
Rotors Static	57 feet 4 inches
Fuselage	48 feet
Height	15 feet 3-1/2 inches
Clear area needed for Rotors	17.9 meters
Minimum TDP without commander's approval is #3	50 meters

Figure 15-17. AH 64A (APACHE)

15-11. UTILITY HELICOPTERS

This category of helicopters includes the UH 1H Iroquois and the UH 60A Blackhawk.

a. **UH 1H Iroquois. Table 15-7** shows specifications for the Iroquois; **Figure 15-18** shows the aircraft from three angles.

Table 15-7. SPECIFICATIONS FOR THE UH 1H (IROQUOIS)

Rotor Diameter	48 feet
Length:	
Rotors Operating or Static	57 feet 1 inch
Fuselage	41 feet 10-3/4 inches
Span, Maximum Lateral	9 feet 4 inches
Height	14 feet 6 inches
Tread	8 feet 6-1/2 inches
Ground Clearance (Static)	
Against Stops	6 feet 6 inches
Clear area needed for Rotors	17.4 meters
TDP # 2	35 meters diameter
Allowable Cargo Load	4,000 pounds

Figure 15-18. UH 1H (IROQUOIS)

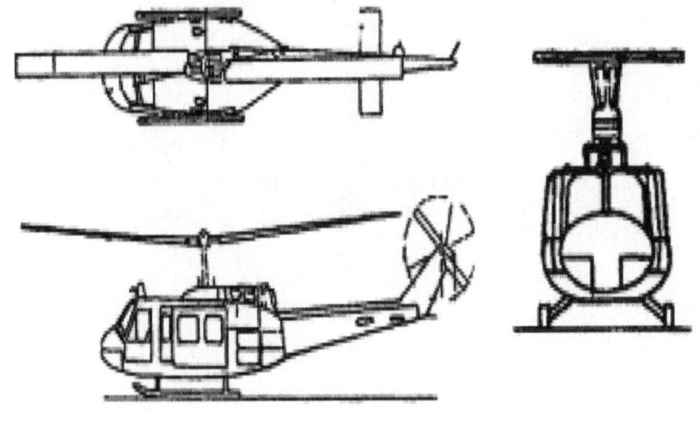

b. **UH 60A Blackhawk. Table 15-8** shows specifications for the UH 60A Blackhawk; **Figure 15-19** shows the aircraft from above and from the left side.

Table 15-8. SPECIFICATIONS FOR THE UH 60A (BLACKHAWK)

Rotor Diameter	53 feet 8 inches
Length:	
Rotors Operating or Folded	64 feet 10 inches
Fuselage	50 feet 7–1/2 inches
Span, Maximum Lateral	9 feet 8–1/2 inches
Height	16 feet 5 inches
Tread	8 feet 10–1/2 inches
Ground Clearance (Static), Against Stops	8 feet 9 inches
Clear area needed for Rotors	19.5 meters
TDP # 3	50 meters diameter
Allowable Cargo Load	8,000 pounds

Figure 15-19. UH 60A (BLACKHAWK)

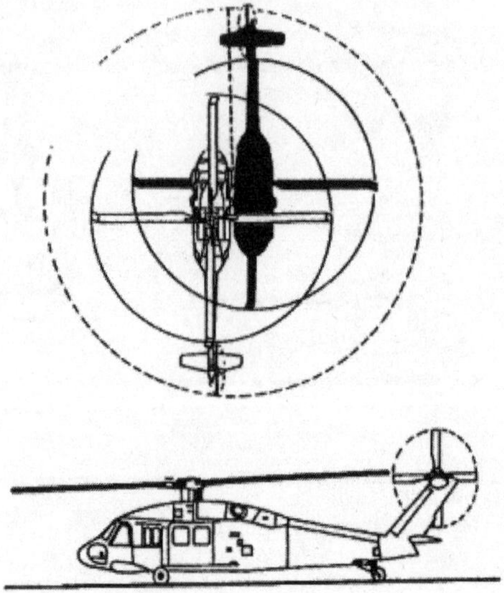

15-12. CARGO HELICOPTERS

This category of helicopters includes the CH 47B/C and the CH 47 B/D Chinooks. With slingload, cargo helicopter TDP is #5 (100 meter diameter); without slingload, it is #4 (80 meter diameter).

a. **CH 47B/ C Chinook. Table 15-9** shows specifications for the CH 47B/C Chinook; **Figure 15-20** shows the aircraft from three angles.

Table 15-9. SPECIFICATIONS FOR THE CH 47 B/ C (CHINOOK)

Rotor Diameter	60 feet
Length:	
Rotors Operating	98 feet 10 3/4 inches
Rotors Folded	50 feet 9 inches
Height (Overall)	18 feet 11 1/2 inches
Tread	11 feet 11 inches
Rotor Ground Clearance	
Static Forward	7 feet 4 3/4 inches
Idling Forward	10 feet 11 inches
Clear Area Needed for Rotors	30.4 meters
TDP # 4	80 meters diameter
Allowable Cargo Load	21,000 pounds

Figure 15-20. CH-47B (CHINOOK)

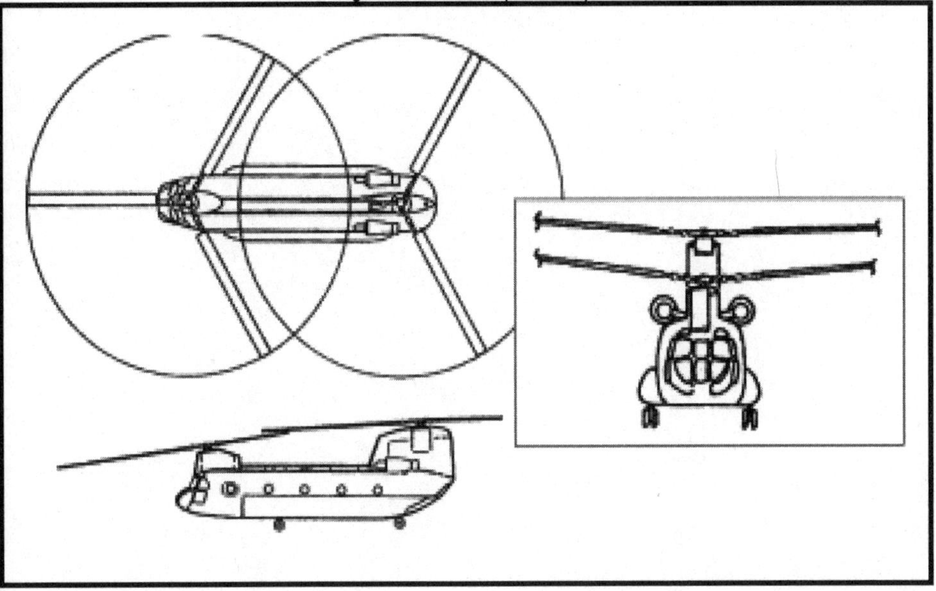

Chapter 16
FIRST AID

Patrolling, more than some other types of missions, puts Rangers in harm's way. CASEVAC planning is vital. Also, because trained medical personnel might be unavailable at the initial point of injury, everyone must know how to diagnose and treat injuries, wounds, and common illnesses. The unit should also have a plan for handling KIAs.

16-1. LIFESAVING STEPS. Whatever the injury, (1) stop life-threatening bleeding; (2) open the airway and restore breathing; (3) stop the bleeding and protect the wound; (4) check, treat, and monitor for shock; and (5) MEDEVAC the casualty.

16-2. CARE UNDER FIRE. When still under fire, (1) maintain situational awareness; (2) return fire; (3) protect the casualty; (4) move the casualty to cover; and (5) identify and control severe bleeding with bandage or tourniquet.

16-3. PRIMARY SURVEY. Use the alphabet to remember how to deal with life threatening injures such as blocked airway, not breathing, or uncontrollable bleeding (hemorrhaging) **(Table 16-1).**

Table 16-1. The ABC's.

A	**AIRWAY** Open airway by patient position or with airway adjuncts.
B	**BREATHING** Seal open chest wounds with occlusive dressing.
C	**CIRCULATION**..Identify uncontrolled bleeding and control with pressure or tourniquet. Start IV if needed.
D	**DISABILITY** Determine Level of consciousness.
E	**EXPOSURE** Fully expose patient. (Environment dependent)

16-4. AIRWAY MANAGEMENT. The airway is usually obstructed (blocked) at the base of the tongue.

a. If this happens, open the airway using the chin lift (for nontraumatic injuries, shown in **Figure 16-1**) or the jaw thrust (for trauma, **Figure 16-2**).

b.

Figure 16-1. CHIN LIFT

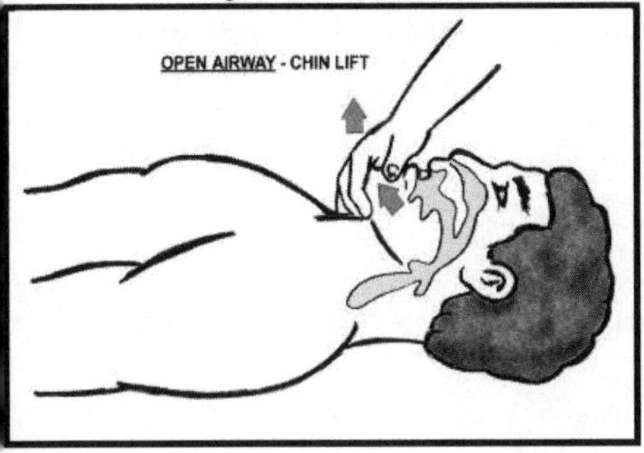

Figure 16-2. JAW THRUST

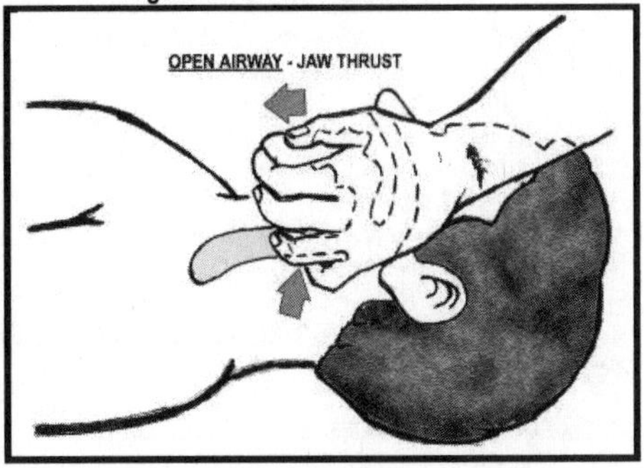

b. Remove debris (teeth, blood clots, bone) from the oral cavity; use suction if you have it; and place airway adjuncts to allow the victim to breathe through their nose (**Figure 16-3**) or mouth (**Figure 16-4**).

Figure 16-3. NASAL AIRWAY

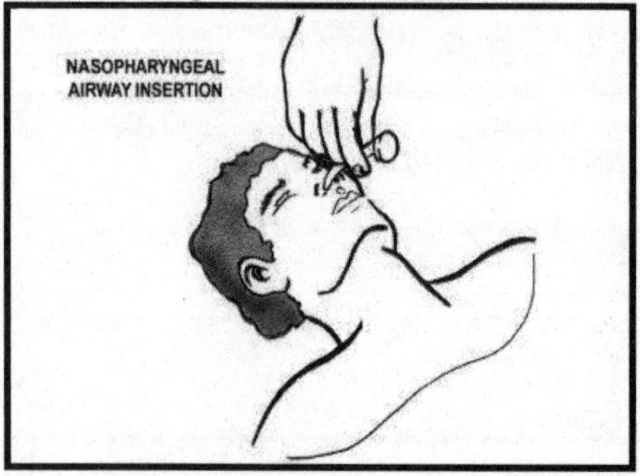

Figure 16-4. MOUTH AIRWAY

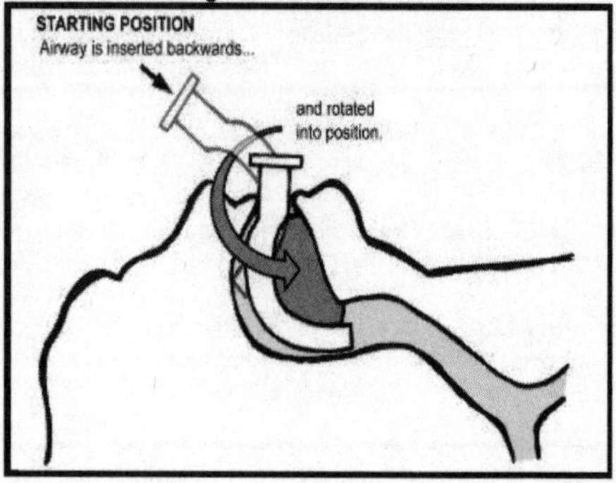

16-5 BREATHING. If the patient is having trouble breathing–

a. Expose the chest and identify open chest injuries

b. Apply a dressing to seal open entry and exit chest wounds

c. Place the patient on the injured side, or position him where he can breathe most comfortably.

16-6. BLEEDING. Quickly identify and control bleeding.

a. Apply a tourniquet to arterial bleeding of the extremities

b. If this does not control the bleeding, apply a second tourniquet above the first and apply a pressure dressing.

c. Control all other bleeding with either a standard or pressure dressing.

d. Check dressings often to ensure bleeding is under control.

16-7. SHOCK. Shock is caused by an inadequate flow of oxygen to body tissues.

a. The most common form of shock is hemorrhagic (due to uncontrolled bleeding).

b. Signs and symptoms of shock include altered mental state, increased pulse and respiration, reduced or no pulse, and profuse sweating.

c. Basic treatment

(1) Control bleeding

(2) Open airway

(3) Restore breathing

(4) Initiate IV or saline lock

(5) Monitor condition.

16-8. EXTREMITY INJURIES. Identify and control bleeding. If you suspect a fracture, splint it as it lies. Do not reposition the injured extremity.

1

16-9. ABDOMINAL INJURIES. Identify and control bleeding, and then–

a. Treat for shock.

b. If internal organs are exposed, cover them with dry, sterile dressing. Do not place them back in the abdominal cavity.

c. Place patient in comfortable position. Flex knees to relax abdomen.

d. Do not give anything by mouth to the patient.

16-10. **BURNS**

a. Remove patient from burn source.

b. Remove all clothing and jewelry from the areas of the body with burns.

c. Cover burns with dry, sterile dressings. Ensure fingers and toes have dressings between them before covering entire area.

d. Evacuate immediately any casualties with burns of the face, neck, hands, genitalia, or over 20 percent (one fifth) of his body surface (**Figure 16-5**).

Figure 16-5. Percentages of body area.

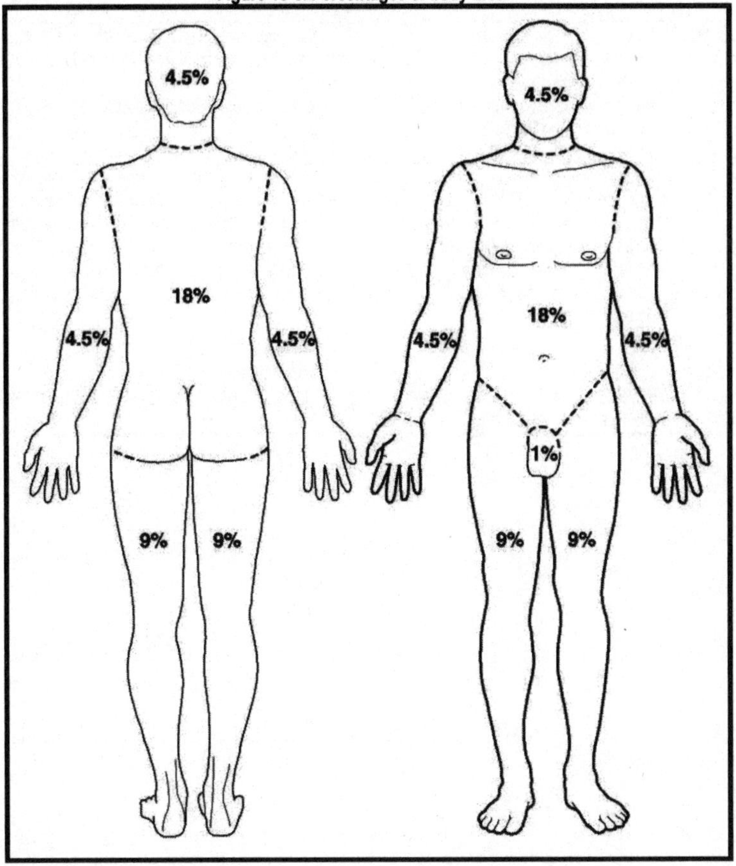

16-11. HOT WEATHER (HEAT) INJURIES. Table 16-2, Table 16-3, and **Table 16-4** show first aid for heat injuries, cold injuries, and environmental injuries.

Table 16-2. HEAT INJURIES

INJURY	SIGNS/ SYMPTOMS	FIRST AID
Heat Cramps	Casualty experiences muscle cramps in arms, legs and/ or stomach, may also have wet skin and extreme thirst.	1. Move the casualty to a shaded area and loosen clothing. 2. Allow casualty to drink 1 quart of cool water slowly per hour. 3. Monitor casualty and provide water as needed. 4. Seek medical attention if cramps persist.
Heat Exhaustion	Casualty experiences loss of appetite, headache, excessive sweating, weakness or faintness, dizziness, nausea, muscle cramps. The skin is moist, pale, and clammy.	1. Move the casualty to a cool, shaded area and loosen clothing. 2. Pour water on casualty and fan to increase cooling effect of evaporation. 3. Provide at least one quart of water to replace lost fluids. 4. Elevate legs. 5. Seek medical aid if symptoms continue.
Heat Stroke (**Sunstroke**)	Casualty stops sweating (hot, dry skin), may experience headache, dizziness, nausea, vomiting, rapid pulse and respiration, seizures, mental confusion. Casualty may suddenly collapse and lose consciousness.	1. Move casualty to a cool, shaded area, loosen clothing, and remove outer clothing if the situation permits. 2. Immerse in cool water. If cool bath is not available, massage arms and legs with cool water. Fan casualty to increase the cooling effect of evaporation. 3. If conscious, slowly consume one quart of water.
	DANGER **SUNSTROKE** ***THIS IS A MEDICAL EMERGENCY! SEEK MEDICAL AID AND EVACUATE ASAP. PERFORM ANY LIFESAVING MEASURES.***	

Table 16-3. COLD INJURIES

INJURY	SIGNS/ SYMPTOMS	FIRST AID
Chilblain	Red, swollen, hot, tender, itchy skin. Continued exposure may lead to infected (bleeding, ulcerated) skin lesions.	1. Area usually responds to locally applied warming (body heat). 2. Do Not rub or massage area. 3. Seek medical treatment.
Immersion (trench) foot	Affected parts are cold and numb. As body parts warm, they may become hot, with burning and shooting pains. ***Advanced stage:*** Skin is pale with bluish cast; pulse decreases; blistering and swelling occur, swelling, heat hemorrhages, and gangrene may follow.	1. Gradual warming by exposure to warm air. 2. DO NOT massage or moisten skin. 3. Protect affected parts from trauma. 4. Dry feet thoroughly: avoid walking. 5. Seek medical treatment.
Frostbite	***Superficial:*** Redness, blisters in 24 to 36 hours followed by peeling skin ***Deep:*** Preceded by superficial frostbite; skin is painless, pale-yellowish, waxy, "wooden" or solid to touch, blisters form in 12-36 hours	***SUPERFICIAL*** 1. Keep casualty warm; gently warm affected parts. 2. Decrease constricting clothing, increase exercise and insulation. ***DEEP*** 1. Protect the part from additional injury. 2. Seek medical treatment as fast as possible.
Snow Blindness	Red, scratchy, or watery eyes; headache; increased pain in eyes with exposure to light.	1. Cover the eyes with a dark cloth. 2. Seek medical treatment.
Dehydration	Similar to heat exhaustion.	1. Keep warm, loosen clothes. 2. Replace lost fluids, rest, and additional medical treatment.
Hypothermia	Casualty is cold, shivers uncontrollably until shivering stops. A core (rectal) temp below 95° F can affect consciousness. Uncoordinated movements, shock, and coma may occur as body temperature drops.	***MILD HYPOTHERMIA*** 1. Warm body evenly and without delay. (Heat source must be provided.) 2. Keep dry, protect from elements. 3. Warm liquids may be given to conscious casualty only. 4. Be prepared to start CPR. 5. Seek medical treatment immediately. ***SEVERE HYPOTHERMIA*** 1. Quickly stabilize body temperature. 2. Attempt to prevent further heat loss. 3. Handle the casualty gently. 4. Evacuate to nearest medical treatment facility as soon as possible.

Table 16-4. ENVIRONMENTAL INJURIES

TYPE	FIRST AID
Snake bite	1. Get the casualty away from the snake. 2. Remove all rings and bracelets from the affected extremity. 3. Reassure the casualty and keep him quiet. 4. Apply constricting band(s) 1 to 2 finger widths close to the bite. You should be able to slip 1 finger between the band and skin. *ARM OR LEG BITE* - Place one band above and one band below the bite site. *HAND OR FOOT BITE* - Place one band above the wrist or ankle. 5. Immobilize the affected limb below the level of the heart. 6. **Kill the snake, if possible, (without damaging its head or endangering yourself) and send it with the casualty.** 7. Seek medical treatment immediately.
Brown recluse or black widow, spider bite	1. Keep the casualty calm. 2. Wash the area. 3. Apply ice or a freeze pack, if available. 4. Seek medical treatment.
Tarantula bite, scorpion sting, ant bite	1. Wash the area. 2. Apply ice or a freeze pack, if available. 3. Apply baking soda, calamine lotion, or meat tenderizer to the bite site to relieve pain and itching. 4. If site of bite(s) or sting(s) is on the face, neck (possible airway blockage), or genital area, or if reaction is severe, or if the sting is by the dangerous Southwestern scorpion, keep the casualty as quiet as possible, administer an antidote if needed and seek immediate medical aid.
Wasp or bee sting	1. If the stinger is present, remove by scraping with a knife or finger nail. DO NOT squeeze venom sack on stinger, more venom may be injected. 2. Wash the area. 3. Apply ice or freeze pack, if available. 4. If allergic signs or symptoms appear, be prepared to administer an antidote and seek medical assistance.
Human and Other animal Bites	1. Cleanse the wound thoroughly with soap or detergent solution. 2. Flush bite well with water. 3. Cover bite with a sterile dressing. 4. Immobilize injured extremity. 5. Transport casualty to a medical treatment facility. 6. For animal bites, without endangering yourself or damaging the animal's head, kill the animal and send its head with the casualty. 7. For human bites, try to extract some of the attacker's saliva from the wound and send that in a sealed, identified container with the casualty.
Poison Ivy, Oak, Sumac	1. Gently clean affected area two to three times daily. Wash clothing. 2. Apply topical anti-itch lotion or ointment as needed, and cover. 3. Avoid scratching the area. 4. Observe for signs of infection (increasing redness, tenderness, warmth to touch). 5. Seek medical attention if rash persists or signs of infection develop.

16-12. POISONOUS PLANT IDENTIFICATION. Poison plants include, among others, poison ivy, oak, and sumac, as well as a few more such as stinging nettles, which we will not discuss here (Figure 16 6).

a. Poison Ivy. Poison ivy is grows as a vine or shrub. The compound leaves of poison ivy have three pointed leaflets. The middle one has a much longer mini-stalk than the two side ones. The leaflet edges can be smooth or toothed but are rarely lobed (lobed leaves look something like a hand with fingers). The leaves vary greatly in size, from 1/3 inch to just over 2 inches long. In spring, the leaves appear reddish. They turn green in the summer, and turn red, orange, and yellow in fall. Small greenish flowers grow in bunches right where the leaf joins the main stem. The flowers are later replaced by clusters of poisonous white, waxy, plump, droopy fruit.

b. Poison Oak. Poison oak is a widespread deciduous shrub throughout mountains and valleys of North America, generally below 5,000 feet elevation. It commonly grows as a climbing vine with airy roots that cling to the trunks of oaks and sycamores. Poison oak can also form dense thickets. Leaves typically have three leaflets (sometimes five), with the terminal one on a slender mini-stalk, as opposed to Eastern poison ivy, whose terminal leaf is often on a longer mini-stalk, and whose leaves tend to be less ragged and serrated (less "oak like"). Like many members of the sumac family (Anacardiaceae) new foliage and autumn leaves often turn brilliant shades of pink and red.

c. Poison Sumac. Poison sumac is a woody perennial shrub or small tree. It grows from 5 to 25 feet tall, and favors swampy areas. To identify it, look for the fruit that grows between the leaf and the branch. Look for red stems that stay red all year. Leaves grow adjacent to each other and grow in odd numbers totaling 5- 13 per stem. They have a glossy, waxy look and they turn bright red and orange during the fall.

Figure 16-6. POISONOUS PLANTS

16-13. FOOT CARE. Use moleskin to prevent blisters prior to movement or foot march. Drain large blisters. Clean area, puncture with needle, drain blister. Place moleskin over area. Observe for signs of infection. Keep feet as clean and dry as possible. Use foot powder and change socks. Let feet air dry as mission permits.

16-14. LITTER. The proper procedures for employing a litter follow:

a. **Unroll Stretcher**

(1) Remove the stretcher from the pack and place on the ground.

(2) Unfasten retainer strap, step on foot end of the stretcher, and unroll the stretcher completely, to the opposite end.

(3) Bend the stretcher in half and back roll. Repeat with opposite end. The stretcher will lay flat and is ready to load patient.

b. **Place Patient on the Stretcher**

(1) ***Log Roll Method.***

(a) Place stretcher next to patient. Ensure head end of stretcher is adjacent to head of patient. Place all straps under stretcher.

(b) Log roll patient and slide stretcher as far under patient as possible. Gently roll patient down onto stretcher.

(c) Slide patient to center of stretcher. Be sure to keep patient's spinal column as straight as possible.

(d) Pull straps out from under stretcher and secure patient.

(2) ***Slide Method.***

(a) Position foot end of stretcher at head of patient.

(b) Have one rescuer straddle stretcher and support patient's head, neck, and shoulders.

(c) Grasp foot straps of stretcher and slide stretcher under patient.

(d) Center patient on stretcher and secure patient.

c. **Secure Patient**

(1) Lift sides of stretcher and fasten the four cross straps to the buckles directly opposite the straps.

(2) Feed foot straps thru unused grommets at foot end of stretcher, and then fasten the straps to the buckles.

(3) For horizontal prepare stretcher for horizontal lift/ descent, use two nylon webbing straps rated at 9,000 pounds each. The head strap is 6" shorter than foot strap and is used at head end of stretcher only.

(4) Insert one end of head strap thru lift slot at head end of stretcher.

(5) Bring strap under stretcher and thru slot on opposite side of stretcher.

(6) Equalize length of strap. Repeat procedure with other strap at foot end of stretcher.

(7) Equalize all four straps and secure to large steel locking carabiner.

NOTE: Use this procedure only after loading the patient and properly securing them in the stretcher.

d. **Prepare Stretcher for Vertical Lift/ Descent.**

NOTE: A 30-foot length of 3/ 8" static kernmantle rope with a figure 8 knot tied in the center is used to configure stretcher for vertical left/ descent.

(1) Pass each end of the rope thru grommets at the head end of the stretcher. Pull the knot up against the stretcher.

(2) Continue feeding rope thru unused grommets and carrying handles all the way to the foot end of the stretcher. Ensure both ends of rope are even.

(3) Pass the rope ends thru grommets at the foot end of stretcher from the inside outward. Tie the ends of the rope together with a square knot.

(4) Bring ends of rope up over end of stretcher. Pass thru carrying handles and secure with a square knot. Safety each side with an overhand knot.

e. **Use Carabiner to Complete Lift.** Fasten a large locking carabiner on the Figure 8 located on the head end of the stretcher to accomplish the lift

16-15. HYDRATION AND ACCLIMATIZATION. Table 16-5 shows strategies for minimizing dehydration and increasing acclimatization and good hydration practices.

Table 16-5. HYDRATION MANAGEMENT AND ACCLIMATIZATION

STRATEGY	SUGGESTIONS FOR IMPLEMENTATION
Start early	1. Start at least 1 month prior to school. 2. Be flexible and patient: performance benefits take longer than physiological benefits.
Mimic the training environment climate	1. In warm climates, acclimatize in the heat of day. 2. In temperate climates, work out in a warm room wearing sweats.
Ensure adequate heat stress	1. Induce sweating. 2. Work up to 100 minutes of continuous physical exercise in the heat. Be patient. The first few days, you may not be able to go the full 100 minutes without resting. 3. Once you can comfortably exercise for 100 minutes in the heat, then continue doing so for seven days. Work up to at least fourteen days, and increase your exercise intensity each day (loads, or training runs).
Teach yourself to drink and eat	1. Your thirst mechanism will improve as you acclimatize to the heat, but you will still under drink if you wait until you feel thirst. 2. Acclimatizing to heat increases your water requirements! 3. Dehydration offsets most benefits of physical fitness and heat acclimatization. 4. You will sweat out more electrolytes during the first week of heat acclimatization, so add salt to your food or drink electrolyte solutions. 5. A convenient way to learn how much water your body needs to replace is to weigh yourself before and after the 100 minutes of exercise in the heat. For each pound lost, you should drink about one-half quart of fluid so, for example, if you lose 8 pounds, 8 times 1/2 quart = 4 quarts or one gallon of fluid. 6. Do not skip meals, as this is when your body replaces most of its water and salt losses.

16-16. WORK, REST, AND WATER CONSUMPTION. Table 16-6 shows a work, rest, and water consumption table. The guidance applies to the average size, heat-acclimated Ranger wearing ACU (not hot weather gear, except as noted). The work and rest times and fluid replacement volumes shown will help the Ranger sustain his performance and hydration for at least 4 hours of work in the specified heat category. Fluid needs can vary based on individual differences (give or take one quart per hour).

a. ***"NL"*** means that there is no limit to work time per hour. ***"Rest"*** means minimal physical activity such as sitting or standing, preferably in the shade.

b. Consume no more than 1.5 quarts of fluid per hour, and no more than 12 quarts per day.

c. If you are wearing body armor in a humid climate, then add 5° F to the WBGT. If wearing MOPP 4 clothing, add 10° F to the WBGT.

d. Work categories include easy, moderate, and hard.

1) ***Easy Work.*** This includes, for example, maintaining weapons; walking on hard surfaces at 2.5 mph with a load of no more than 30 pounds; participating in marksmanship training; and participating in drills or ceremonies.

2) ***Moderate Work.*** This includes, for example, walking in loose sand at 2.5 mph (maximum) or with no load; walking on a hard surface at 3.5 mph (maximum) with a load weighing no more than 40 pounds; performing calisthenics; patrolling; or conducting individual movement techniques such as the low or high crawl.

3) ***Hard Work.*** This includes, for example, walking on a hard surface at 3.5 mph with a load weighing 40 or more pounds; walking in loose sand at 2.5 mph while carrying a load; and conducting field assaults.

Table 16-6. WORK, REST, AND WATER CONSUMPTION TABLE

HEAT CATEGORY	WBGT INDEX, IN DEGREES FAHRENHEIT	EASY WORK		MODERATE WORK		HARD WORK	
		WORK/ REST	WATER INTAKE (QT/ H)	WORK/ REST	WATER INTAKE (QT/ H)	WORK/ REST	WATER INTAKE (QT/ H)
1	78 to 81.9	NL	0.50	NL	0.75	40/ 20	0.75
2 (GREEN)	82 to 84.9	NL	0.50	50/ 10	0.75	30/ 30	1.00
3 (YELLOW)	95 to 87.9	NL	0.75	40/ 20	0.75	30/ 30	1.00
4 (RED)	88 to 89.9	NL	0.75	30/ 30	0.75	20/ 40	1.00
5 (BLACK)	90 or more	50/ 10 min	1.00	20/ 40	1.00	10/ 50	1.00

Appendix A
RESOURCES

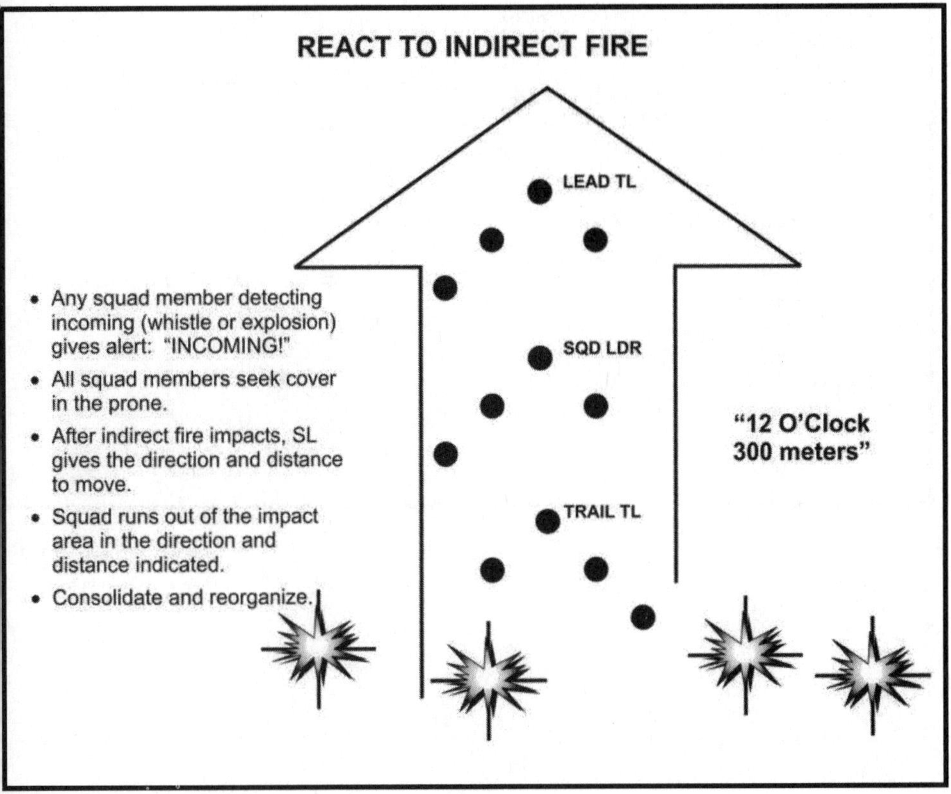

REACT TO CONTACT

- Seek nearest cover.
- Return fire (known or suspected enemy position).
- TM LDRs control fire by using fire commands.
- Report enemy situation (number/position).
- Maintain contact (visual/oral) with team members.
- SQD LDR moves to team in contact and makes an assessment of the situation.
- Factors of his assessment:
  - Can squad move out to engagement area.
  - Can squad gain and maintain suppressive fire.
  - Location of enemy.
  - Size of enemy.
  - Vulnerable flanks.
  - Covered/concealed flanking routes.
- SQD LDR determines COA (break contact, squad attack, etc.)
- Report situation to PL

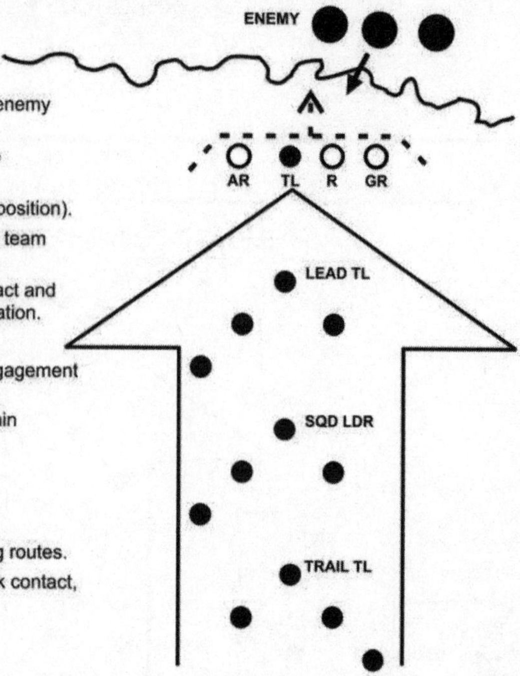

REACT TO A NEAR AMBUSH

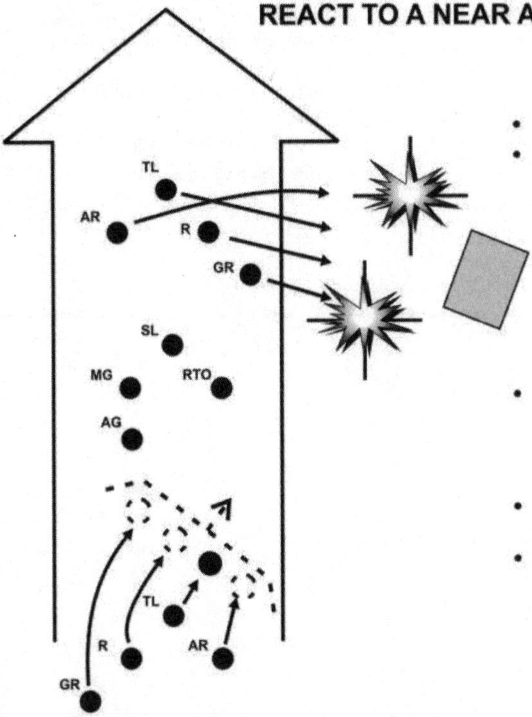

- Within hand grenade range - 35 meters.
- Soldiers in the kill zone: (without orders)
  - Return fire immediately.
  - Seek nearest available cover.
  - Assume prone position.
  - Throw concussion, frag, or smoke grenades.
  - After explosion of grenades, assault through ambush using fire and movement..
- Soldiers not in kill zone:
  - Identify enemy location.
  - Place accurate suppressive fire.
  - Shift fires as assault begins.
- Soldiers in kill zone continue to assault to eliminate ambush or until contact is broken.
- Consolidate and reorganize.

REACT TO A FAR AMBUSH

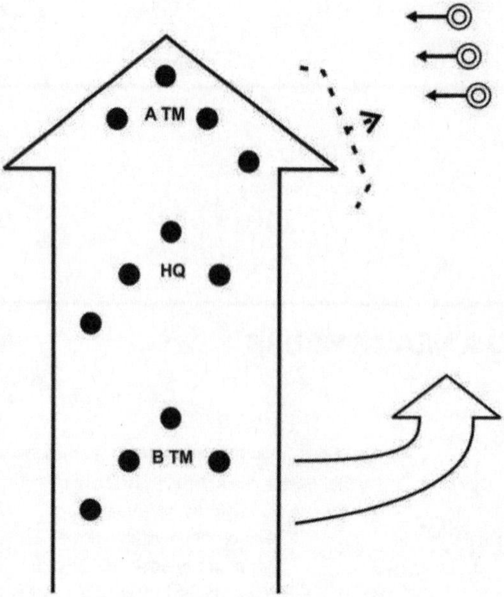

- More than 35 meters.
- TM in the kill zone: (without orders)
  - Return fire immediately.
  - Seek cover and concealment.
  - Suppress enemy (overwatch).
- SL assess situation.
  - Determine COA (flank).
- TM not in contact:
  - Move along covered and concealed route.
  - Assault enemy on weak flank.
  - Suppress enemy (overwatch).
- Overwatch TM continues to suppress, shifts/cease fire as bounding team enters sector.
- Bounding team continues to assault through enemy.
- SL may request indirect fire.
- Consolidate and reorganize.

BREAK CONTACT

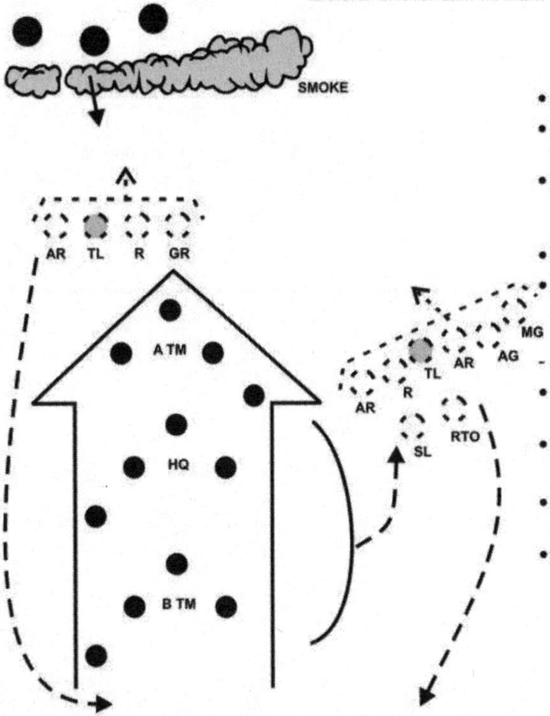

- Squad Leader orders: "Break Contact."
- Squad Leader designates SPT element and maneuver element.
- SL issues distance and direction or a terrain feature for the maneuver element.
- SBF suppresses enemy position.
- Maneuver uses smoke to mask movement.
  - Takes up overwatch position.
  - Begins to suppress enemy
- Squad Leader directs SBF to break contact.
- SBF uses smoke to screen movement.
  - Takes up overwatch position.
- Squad continues to bound away until contact is broken.
- Consolidate / reorganize.

FORMATIONS AND ORDER OF MOVEMENT

I. Movement Formation: Fire Team Wedge; MG Team attached.

II. Three Movement Techniques used:

A. Traveling technique used behind FFL when contact is not likely.

B. Traveling Overwatch forward of the FFL when enemy contact is possible.

C. Bounding Overwatch used forward of the FFL when enemy contact is expected.

III. Distances are based on but not dictated by visibility, terrain, and vegetation.

IV. Actions at Night: Modified Wedge

V. Actions at the Halt: Short and Long Halt (GV/LV)

VI. Leader Location: Fixed/Unfixed

FIRE TEAM WEDGE

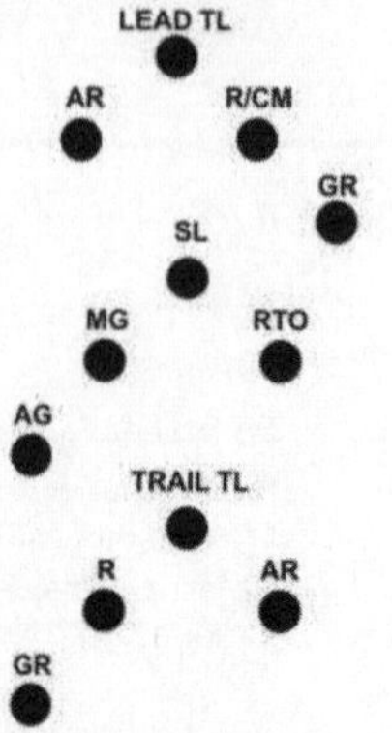

MODIFIED WEDGE

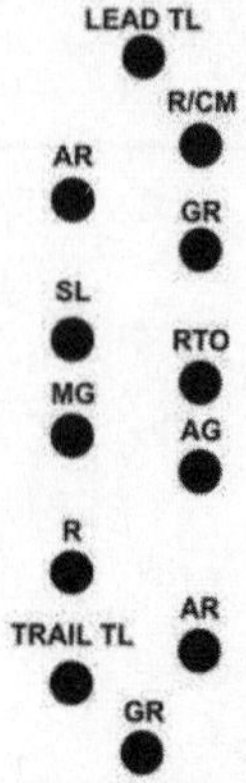

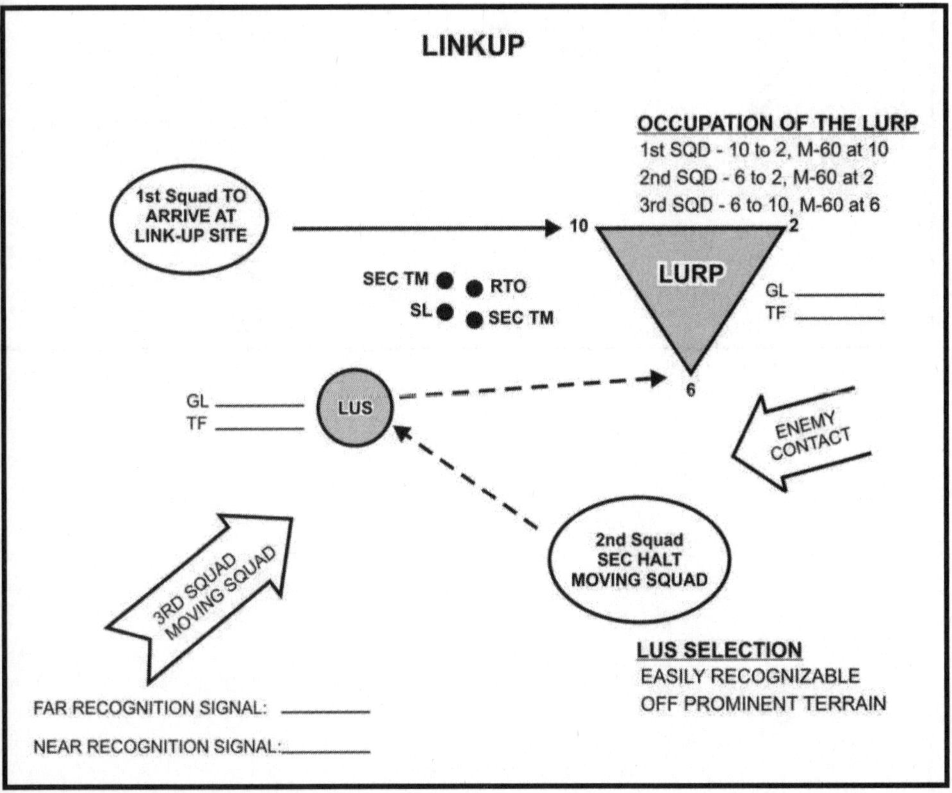
LINKUP
OCCUPATION OF THE LURP
1st SQD - 10 to 2, M-60 at 10
2nd SQD - 6 to 2, M-60 at 2
3rd SQD - 6 to 10, M-60 at 6
1st Squad TO ARRIVE AT LINK-UP SITE
10
2
LURP
6
GL
TF
SEC TM
RTO
SL
SEC TM
LUS
GL
TF
ENEMY CONTACT
3RD SQUAD MOVING SQUAD
2nd Squad SEC HALT MOVING SQUAD
LUS SELECTION
EASILY RECOGNIZABLE
OFF PROMINENT TERRAIN
FAR RECOGNITION SIGNAL:
NEAR RECOGNITION SIGNAL:

LINEAR DANGER AREA

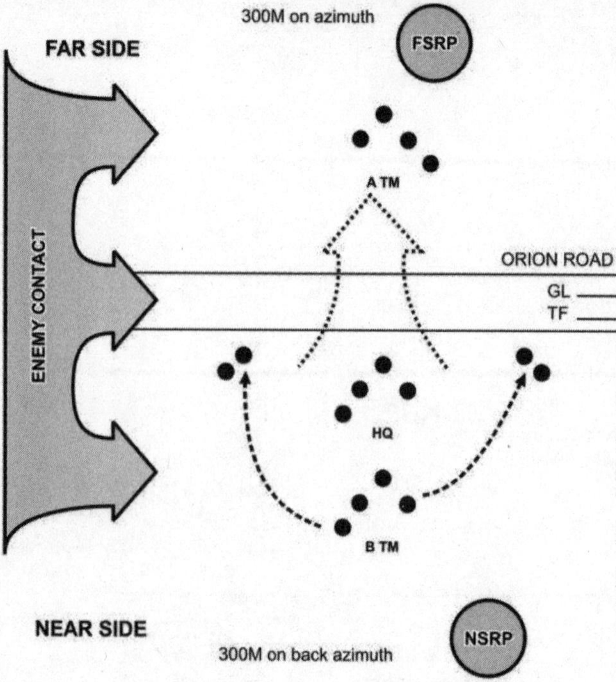

Far side Rally Point info for a known danger area.

GL ____________
TF ____________
DIR ____________
DIS ____________

1. Designate near and far side rally points.
2. Secure near side and emplace flank security.
3. Clear far side.
4. Continue unit crossing.
5. Retrieve near side security and complete unit crossing.
6. Accountability/head count.

Near side Rally Point info for a known danger area.

GL ____________
TF ____________
DIR ____________
DIS ____________

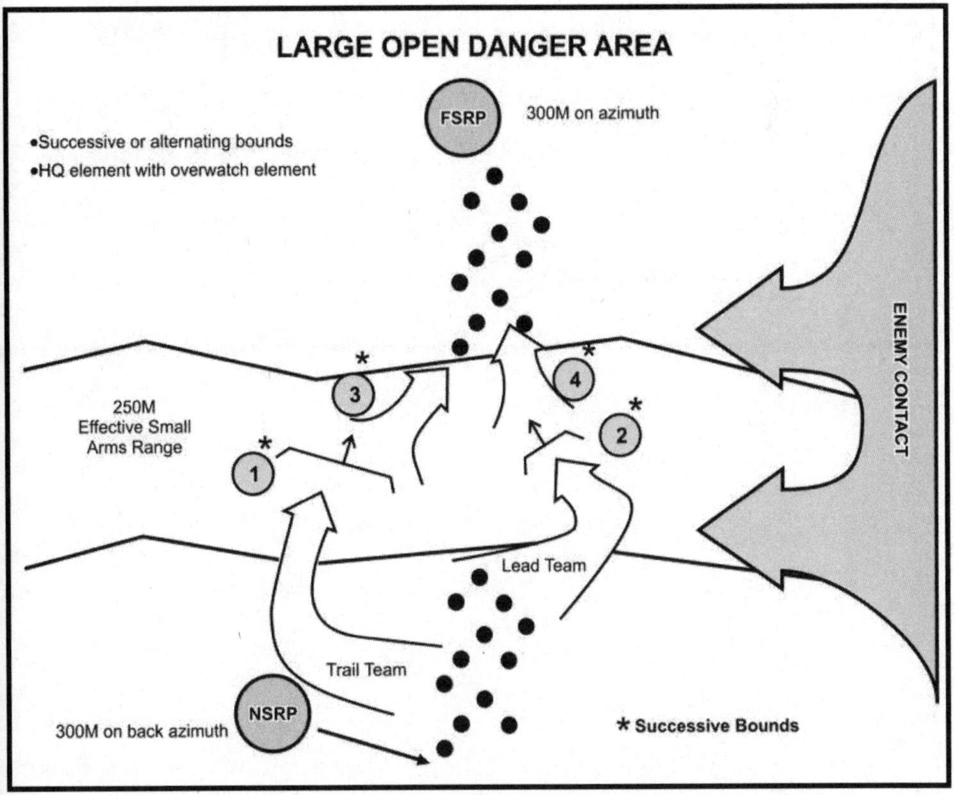
LARGE OPEN DANGER AREA
FSRP
300M on azimuth
•Successive or alternating bounds
•HQ element with overwatch element
ENEMY CONTACT
3
4
2
1
250M
Effective Small
Arms Range
Lead Team
Trail Team
NSRP
300M on back azimuth
* Successive Bounds

CROSSING A SMALL OPEN AREA

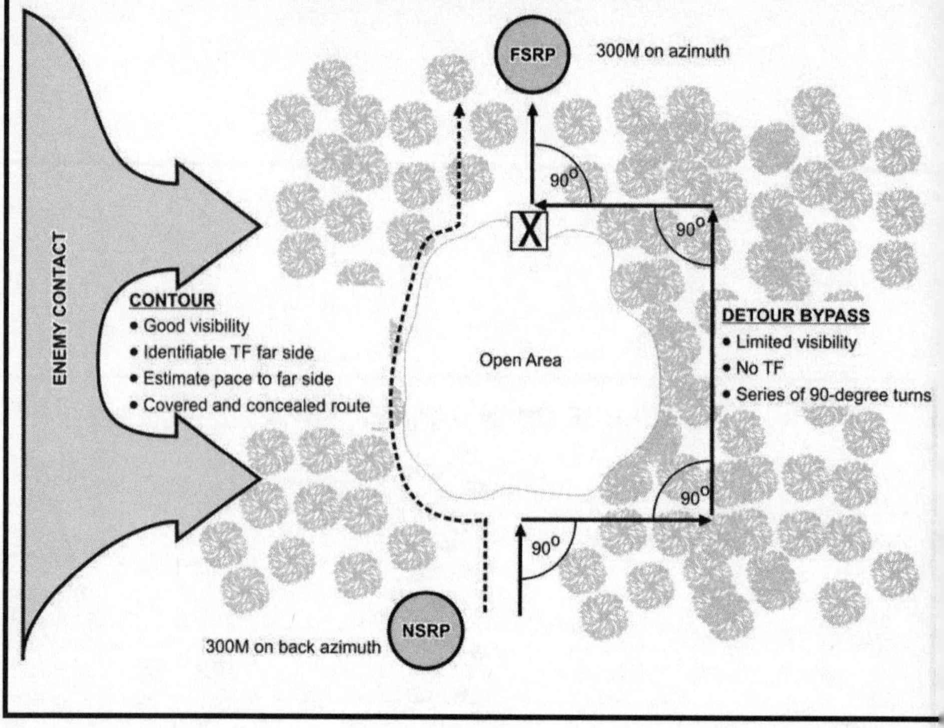

SQUAD ATTACK

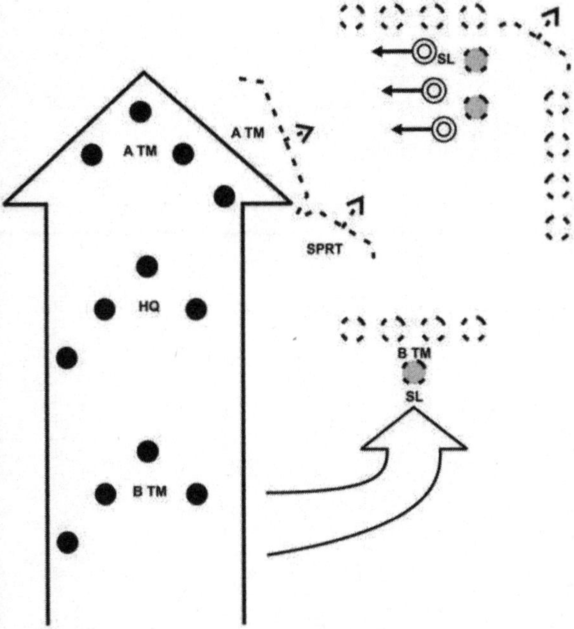

- React to contact.
- TM in the kill zone: (without orders)
  - Return fire.
  - Seek cover and concealment.
  - Suppress enemy call out 3 D's become overwatch.
- SL assess situation.
  - Determine COA (flank/attack).
- TM not in contact: (with SL)
  - Move along covered and concealed route.
  - Assault enemy on weak flank.
- Overwatch TM continue to suppress shift/cease fire as bounding team enters sector.
- Bounding team – continue to assault through enemy.
- SL – may request indirect fire.
- Consolidate and reorganize.

RAID BOARDS (LEFT)
GENERAL INFO

Raid References:
- RHB Chapter 5
- FM 3-21.8

Purposes:
1. Destroy
2. Liberate
3. Collect Intel

Planning Considerations:
- Mission
- Enemy
- Troops
- Terrain / OAKOC
- Time
- Civilians

STUDENT INPUT WRITTEN HERE

Raid:
A surprise attack on a fixed position or installation ending in a planned withdrawal.

Characteristics:
1. Surprise - Gain
2. Coordinated fires - Maintain
3. Violence of action - Retain
4. Planned withdrawal

INITIATIVE

RAID BOARDS (MIDDLE)
ACTIONS ON OBJECTIVE

Prep Clearing TM:
PL, RTO, WSL,
3 Ab's, SEC TM

ORP:
Leader's Recon
Prep MWE

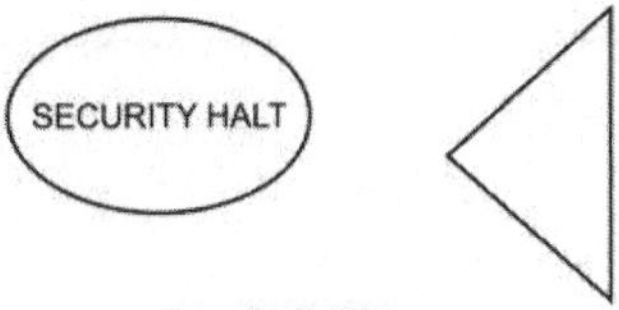

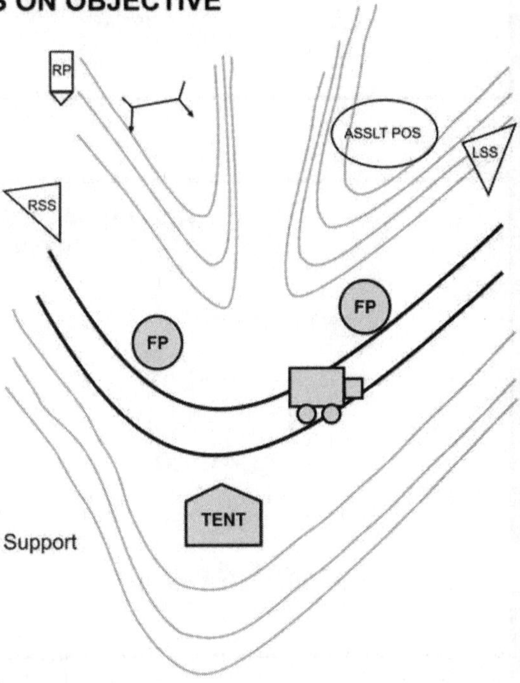

Leader's Recon
- Pinpoint OBJ
- Finalize ORP
- Determine Enemy Situation
- Emplace Surveillance
- Emplace Security
- Recon Positions and Routes
- ID Control Measures
- ID and Prioritize Targets
- Left and Right Limits, Aslt and Support
- Divide OBJ
- Signals
- LOA
- Plan to Secure OBJ
- Confirm Plan

RAID BOARDS (MIDDLE)
TASK ORGANIZATION

ELEMENT	WHO	WHY	WHAT
HQ	PL, APL RTO, FO, MEDIC	C2/FACILITATE SIGNALS, PEQ-2, STANO	1 x 119, ICOM, LITTER, PLUGGER
ASLT 1	SQUAD (+)(-)	DESTROY	ICOM, STANO, M18A1, AT4, SPECIAL TEAM KITS
ASLT 2	SQUAD (+)(-)	DESTROY	ICOM, STANO, M18A1, AT4, SPECIAL TEAM KITS
SUPPORT	WPNS SQD (+)(-)	SUPPRESS	3 x M240B COMPLETE, ICOM, PEQ-2, PVS-4, AT, STANO, SIGNALS, MAP
SECURITY	SQUAD (+)(-)	DELAY	ICOMs, M18A1, AT, STANO, MAP SIGNALS

RAID BOARDS (RIGHT)
SOP

CONTINGENCY

1. COMPROMISE
   a. ORP
   b. LEADER'S RECON
   c. OCCUPATION
2. MASS CAL
3. COUNTERATTACK

SOP

1. EPW SEARCH
2. AID LITTER
3. MEDEVAC
4. CCP
5. WITHDRAWAL PLAN

AMBUSH SOP (LEFT)

References:
- RHB Chapter 7
- FM 3-21.8

Ambush:
A surprise attack from a covered and concealed position on a moving or temporarily halted target.

Types:
- Point
- Area

Fundamentals:
1. Surprise - Gain
2. Fire superiority - Maintain
3. Violence of action - Retain

Purposes:
1. Disrupt/Destroy
2. Collect Intelligence
3. Block or Deny Access
4. Canalize

Categories:
Deliberate: Platoon has specific target at a predetermined time and location.

Hasty: Platoon makes visual contact with the enemy and has time to establish an ambush without being

Planning Considerations:
- Mission - Task/Purpose
- Enemy - MPCOA, MDCOA
- Terrain - Map Recon/Ldrs Recon
- Time - 1/3 - 2/3 Rule
- Troops - Task Org
- Civilians

AMBUSH BOARDS (MIDDLE)

HQ	PL, PL RTO, PSG, MEDIC, FO	COMMAND AND CONTROL	PRC-119s, ICOM, NVGs, BINO's, LITTER, PEQ-2, SIGNALS, MAP
SECURITY	MINIMUM OF 3-2 MAN TEAMS	EARLY WARNING, SEALS OFF OBJ, LFS, RFS, ORP SECURITY	COMMO, NVGs, CLAYMORES, AT WPNS, BINOs, SIGNALS, MAP
SUPPORT	WSL, 3 GUN TEAMS COMPLETE +/-	SUPPRESSIVE FIRE ON KILL ZONE, SEC ON OBJ DURING RECON/REORG	MGs, COMMO, NVGs, AT WPNS, BINOs, LITTER, PEQ-2, SIGNALS, MAP
ASSAULT	2 SQUADS +/-	MAIN EFFORT, BLOCK, DESTROY, CANALIZE, CAPTURE PIR	NVGs, BINOs, CLAYMORE, AT, FLEX CUFFS, POLELESS LITTER, MARKINGS, SIGNALS

LEADER'S RECONNAISSANCE

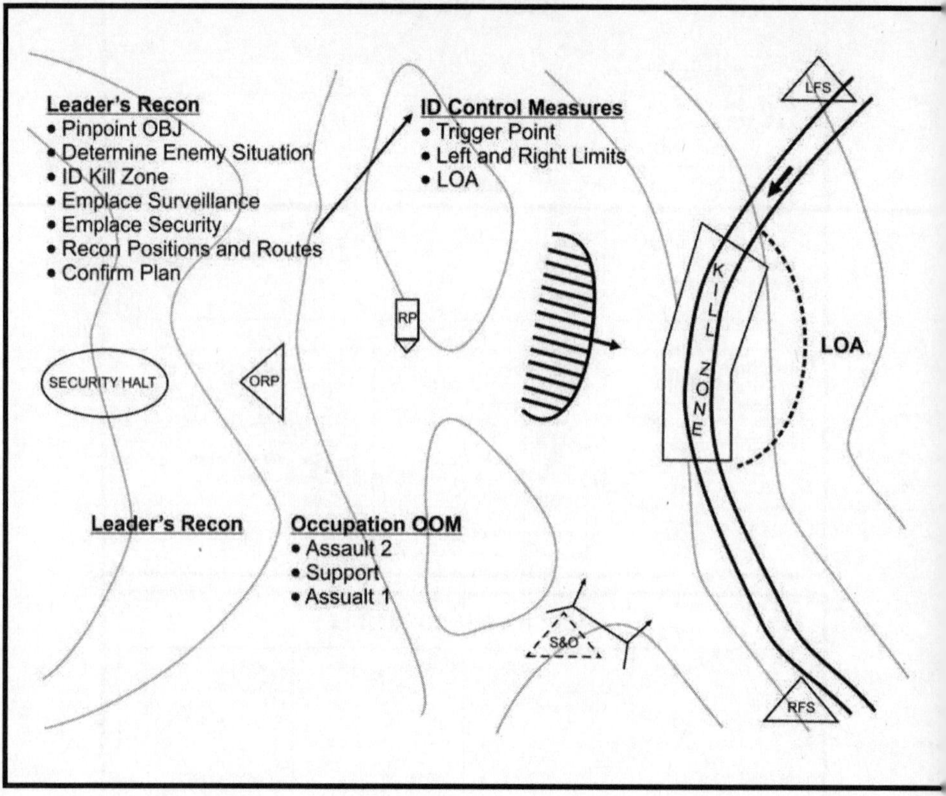

Appendix B
QUICK REFERENCE CARDS

CASUALTY FEEDER CARD

NAME ____________________ ROS # __________

COMPANY ________ ALLERGIES __________

LOCATION/EVENT ____________________________

TIME ______________ TIME ______________

TEMP ______________ TEMP ______________

PULSE ______________ PULSE ______________

RESP ______________ RESP ______________

B/P ______________ B/P ______________

MENTAL STATUS ____________________________

A & O x 1 2 3

NPA OPA ET Tube

SAO[2] ______________ GLUC ________________

IV _____ AC FOREARM HAND OTHER

1000-1000-500-500-500- NS RL D_5W

NOTES: ____________________________________

____

TROOP-LEADING PROCEDURES

1. Receive mission.
2. Issue warning order.
3. Make a tentative plan.
4. Start movement.
5. Reconnoiter.
6. Complete plan.
7. Issue plan.
8. Supervise.

FIRE REQUEST (FM 6-30)

1. Identification.
2. Warning order.
3. Target location.
4. Target description.
5. Method of engagement.
6. Method of control.

ESTIMATE OF SITUATION (FM 7-10)

1. Conduct a detailed mission analysis.
2. Analyze the situation and develop courses of action.
3. Analyze courses of action (wargame).
4. Compare courses of action.
5. Make a decision.

SPOT REPORT

1. Size.
2. Activity.
3. Location.
4. Unit/uniform.
5. Time.
6. Equipment.

SHELL REPORT (FM 6-121)

1. Observer identification.
2. Location (coded).
3. Azimuth to flash or sound.
4. Time (from and to).
5. Area shelled.
6. Nature of fire.
7. Type rounds received (artillery, mortars, etc.).
8. Damage (coded).

GTA 07-01-038
INFANTRY LEADERS' REFERENCE CARD
January 1995

HEADQUARTERS, DEPARTMENT OF THE ARMY
References: FM 3-3, FM 6-30, FM 6-121, FM 7-8, FM 7-10, FM 7-90
(Supersedes GTA 7-1-31)

DISTRIBUTION RESTRICTION: Distribution authorized to US Government agencies only to protect technical or operational information from automatic dissemination under the International Exchange Program or by other means. Other requests for this document will be referred to Commander, US Army Training Support Center, ATTN: ATIC-DMF, Ft Eustis, VA 23604-5166.
DESTRUCTION NOTICE: Destroy by any method that will prevent disclosure of contents or reconstruction of the document

DISTRIBUTION: US Army Training Support Centers (TSCs).

OPERATION ORDER (FM 7-10)

Task Organization
1. Situation.
   a. Enemy.
   b. Friendly.
   c. Attachments/detachments.
2. Mission
   - Who, what, where, when, and why.
   - Task and purpose.
3. Execution.
   a. Concept of operation.
      (1) Maneuver.
      (2) Fire support.
      (3) Engineer.
   b. Tasks to maneuver units
   c. Tasks to combat support units.
   d. Coordinating instructions.
4. Service support.
   a. General.
   b. Materiel and services.
   c. Casualty evacuation.
   d. Miscellaneous.
5. Command and signal.
   a. Command.
   b. Signal.

WEAPONS (FM 7-8)

TYPE	MAX EFF RANGE (m)
M16A2	580 (pt) 800 (area) 200 (mov)
M203	150(pt) 350 (area)
M249	600 (pt) 800 (area)
M136 (AT4)	300
M47 (Dragon)	1,000 (sta) 100 (mov)
MK19	1500 (pt) 2,212 (area)
M3 RAAWS	700 (sta) 60 (mov)
M60 MG	1,100 (600 grazing)
.50 Caliber MG	1,800 (1,000 grazing)
TOW	3,000 (ping purposes)
TOW 2	3,750
105-mm	11,500
105-mm Tank	*2 to 2.5 km
120-mm Tank	*2 to 2.5 km
25-mm BFV	2,200
155-mm M109A3	18,100
M198	24,000
8-in Howitzer	22,900

WEAPONS (MORTAR) HE ONLY (FM 7-90)

	MIN	MAX
60-mm	70m	3,500m
81-mm (M252)	80m	5,800m
81-mm (M29A1)	73m	4,790m
4.2-inch	770m	6,840m
120-mm	200m	7,200m

FPFs (FM 7-90)

GUNS	MORT	WIDTH		DEPTH
2	60-mm	60m	x	30m
4	81 (M252)	150m	x	50m
4	81 (M29A1)	140m	x	40m
6	4.2-in	240m	x	40m
6	120-mm	360m	x	60m
Plt	155-mm	200m	x	50m
Btry	155-mm	400m	x	50m

Optimum engagement ranges

MEDEVAC REQUEST	AIRCRAFT REQUEST
1. Requesting unit identification. 2. Location. 3. Number of patients by type (litter or ambulatory). 4. Type of injuries. 5. Special equipment needed. 6. Tactical situation.	1. Identification. 2. Precedence/priority. 3. Target description. 4. Target location. 5. Target time/date. 6. Desired ordnance/results. 7. Final control.

DELIBERATE ATTACK CONSIDERATIONS

1. Reconnoiter, pinpoint objective/enemy positions/obstacles.
2. Determine weak points; designate supporting positions.
3. Assign platoon/squad objectives—identify the decisive point.
4. Determine main attack, supporting attack, reserve.
5. Assign breach, support, assault missions.
6. Designate fire control measures.
7. Coordinate indirect/direct fires and CAS to time of attack.
8. Control measures during attack.
9. Secure (ground and air).
10. Consolidate and reorganize.

NBC 1 OBSERVER'S INITIAL OR FOLLOW-UP REPORT

Instructions

1. Line items D and H are mandatory for NBC 1 reports.
2. Line items A, E, G, I, K, L, M, S, Y, and ZA are optional for NBC 1 reports.
3. Line Items B, C, F, PAR, and PBR are reported if data is available.

Section I. Chemical or Biological Only

A. Strike serial number, if known (assigned by NBCE).
B. Position of observer.
C. Azimuth of attack from observer (state degrees or mils).
D. Date and time attack started (Zulu, local, or letter zone).
E. Time attack ended, if known.
F. Location of attack (UTM or place) (state actual or estimated).
G. Means of delivery, if known.
H. Type of agent and height of burst, if known.
I. Type and number of munitions or aircraft (state which).
K. Description of terrain (bare, scrubby vegetation, wooded, urban, or unknown).
S. Date and time contamination detected (Zulu, local, or letter zone).
Y. Representative downwind direction—4 digits (state degrees or mils), wind speed—3 digits (data kmph or knots).
ZA. Temperature (centigrade)—2 digits, cloud cover— 1 digit, significant weather phenomena—1 digit, air stability—1 digit.
ZB. Remarks

Section II. Nuclear Only

A. Strike serial number, if known (assigned by NBCE).
B. Position of observer.
C. Azimuth of attack from observer (state degrees or mils and grid or magnetic).
D. Date and time attack started (Zulu, local, or letter zone).
F. Location of attack (UTM or place) (state actual or estimated).
G. Means of delivery, if known.
H. Type of burst (state air, surface, or unknown).
J. Flash-to-bang time (seconds).
K. Crater diameter (meters), if known.
L. Cloud width at H +5 minutes (degrees or mils).
M. Cloud angle(top or bottom) or cloud height(top or bottom) at H +10 minutes (state degrees, mils, meters, or feet).
PAR. Location of radioactive cloud outline (UTM).
PAB. Downwind direction of radioactive cloud (state degrees or mils).
ZB. Remarks.

DEFENSE PLANNING CONSIDERATIONS

1. Establish security (OP/patrols, PWs, M8),
2. Position key weapons:
   a Coordinate with units on left and right
   b. Establish FPF or PDF for machine gun.
   c. Ensure mutual support between machine guns.
   d. Cover armor approaches with antiarmor systems.
   e. Establish fire control measures.
3. Prepare positions:
   a. Check sectors of fire.
   b. Check overhead cover and view positions from enemy vantage.
   c. Position in depth and achieve mutual support between positions.
   d. Select/prepare alternate and supplementary positions.
4. Integrate indirect fires, CAS, and obstacles with direct and indirect fire.
5. Check communications and establish emergency signals.
6. Designate ammunition, supply, PW, and casualty points.

IED / UXO

Procedures when IEDs are found

Security - Maintain 360 degrees. Scan close in and far out, up high and down low.

Always - Scan you immediate surroundings for more IEDs.

Move - Move away. Vary safe distances, but plan for 300m minimum safe distance and adapt to your METT-TC.

Attempt - To confirm suspected IEDs using optics while staying back as far as possible.

Cordon - Off the area. Direct people out of danger area. Do not allow anyone to enter except for EOD. Question, search, and detain suspects as defined by your Existing ROE.

Report - Your situation using the 9 line IED / UXO spot report format.

This could be your hand if you try to dispose of UXOs or IEDs. The enemy has developed anti-handling to catch you when you try defusing. Leave it to experts!

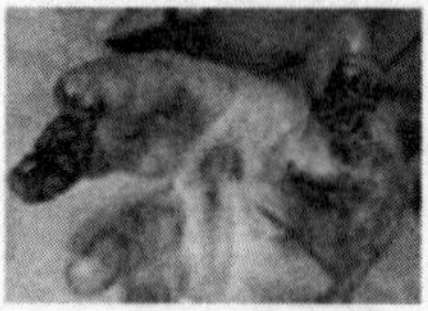

Call EOD
- Don't be a Hero!

IED / UXO Report

LINE 1. DATE-TIME-GROUP: When the item was discovered.

LINE 2. REPORT ACTIVITY AND LOCATION: Unit and grid location of the IED/UXO.

LINE 3. CONTACT METHOD: Radio frequency, call sign, POC, and telephone number.

LINE 4. TYPE OF ORDNANCE: Dropped, projected, placed, or thrown. Give the number of items, if more than one.

LINE 5. NBC CONTAMINATIONS: Be as specific as possible.

LINE 6. RESOURCES THREATENED: Equipment, facilities, or other assets that are threatened.

LINE 7. IMPACT ON MISSION: Short description of current tactical situation and how the IED/UXO affects the status of the mission.

LINE 8. PROTECTIVE MEASURES: Any measures taken to protect personnel and equipment.

LINE 9. RECOMMENDED PRIORITY: Immediate, Indirect, Minor, No Threat.

Priority

Immediate: Stops unit's maneuver and mission capability or threatens critical assets vital to the mission.

Indirect: Stops the unit's maneuver and mission capability or threatens critical assets important to the mission.

Minor: Reduces the unit's maneuver and mission capability or threatens non-critical assets of value.

No Threat: Has little or no effect on the unit's capabilities or assets.

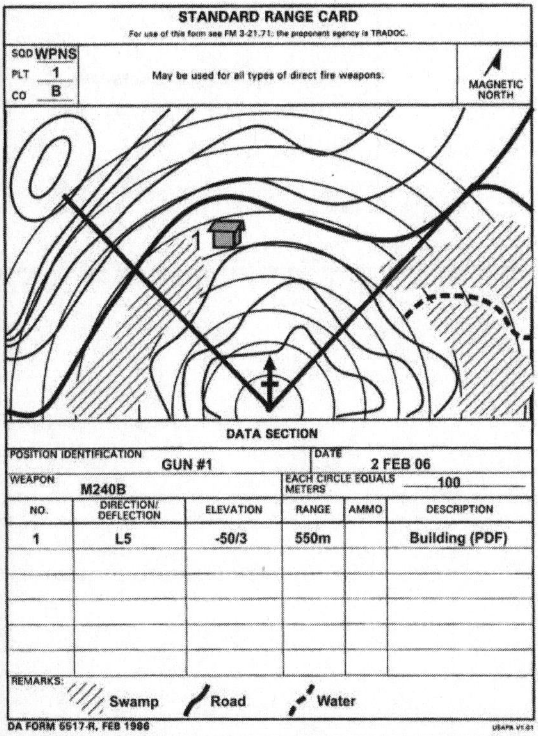
STANDARD RANGE CARD

For use of this form see FM 3-21.71; the proponent agency is TRADOC.

SQD WPNS
PLT 1
CO B

May be used for all types of direct fire weapons.

MAGNETIC NORTH

DATA SECTION

POSITION IDENTIFICATION: GUN #1
DATE: 2 FEB 06
WEAPON: M240B
EACH CIRCLE EQUALS 100 METERS

NO.	DIRECTION/ DEFLECTION	ELEVATION	RANGE	AMMO	DESCRIPTION
1	L5	-50/3	550m		Building (PDF)

REMARKS: Swamp Road Water

DA FORM 5517-R, FEB 1986

USAPA V1.01

PDF - Principal Direction of Fire

- First target is always the PDF

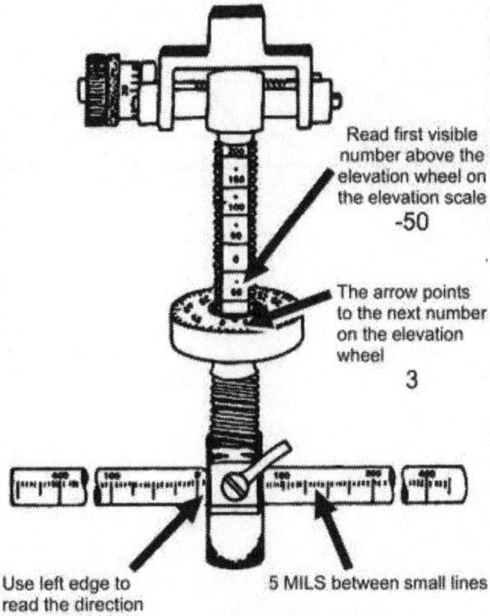

* The left or right direction is given for the direction the barrel points.

T&E Reading

L5 -50/3

STANDARD RANGE CARD

For use of this form see FM 3-21.71; the proponent agency is TRADOC.

SQD WPNS
PLT 1
CO B

May be used for all types of direct fire weapons.

MAGNETIC NORTH

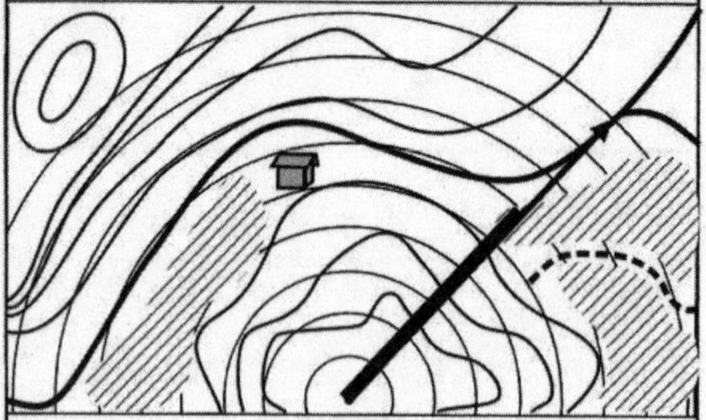

DATA SECTION

POSITION IDENTIFICATION: GUN #1 | DATE: 2 FEB 06

WEAPON: M240B | EACH CIRCLE EQUALS METERS: 100

NO.	DIRECTION/ DEFLECTION	ELEVATION	RANGE	AMMO	DESCRIPTION
1	R450	-50/10	600m		FPL

REMARKS: Swamp, Road, Water

DA FORM 5517-R, FEB 1986 USAPA V1.01

FPL - Final Protective Line

- FPL is always target #1
- FPL will always be metal-to-metal
- Represented by thick line
  - Break in thick line for dead space out to 600 meters
  - The gap is equal to the width of the dead space

FPL - Final Protective Line

STANDARD RANGE CARD

For use of this form see FM 3-21.71; the proponent agency is TRADOC.

SQD ______ PLT ______ CO ______	May be used for all types of direct fire weapons.	MAGNETIC NORTH

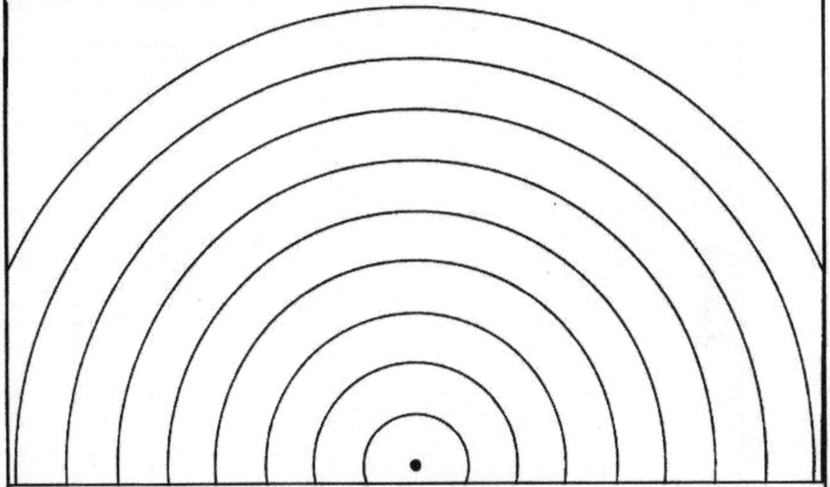

DATA SECTION

POSITION IDENTIFICATION	DATE
WEAPON	EACH CIRCLE EQUALS ______ METERS

NO.	DIRECTION/ DEFLECTION	ELEVATION	RANGE	AMMO	DESCRIPTION

REMARKS:

DA FORM 5517-R, FEB 1986 USAPA V1.01

OBSERVED FIRE REFERENCE CARD

GTA 7-1-32
JUNE 1987

GRID COORDINATE
(1) "___DE___AF/FFE, k"
(2) "GRID_______, k"
(3) (TGT DESC) "_______IN EFFECT, k"

(DIR SENT AFTER M.T.O.)

I. GRID:
A. DETERMINE A TWO CHARACTER, SIX DIGIT GRID FOR THE TARGET.
B. DETERMINE A GRID DIRECTION TO THE TARGET AND SEND AFTER THE CALL FOR FIRE AND BEFORE ANY SUBSEQUENT CORRECTIONS.

POLAR PLOT
(1) "___DE___AF/FFE, POLAR k"
(2) "DIR___DIST___UP/DOWN____, k"
(3) (TGT DESC) "_______IN EFFECT, k"

II. POLAR:
A. DETERMINE THE GRID DIRECTION TO THE TARGET.
B. DETERMINE THE DISTANCE FROM THE OBSERVER TO THE TARGET.
C. DETERMINE IF ANY SIGNIFICANT VERTICAL INTERVAL EXISTS.

SHIFT FROM A KNOWN POINT
(1) "___DE___AF/FFE, "SHIFT (TGT#/REG PT#), k"
(2) "DIR___, R/L___DD/DROP___UP/DOWN___, k"
(3) (TGT DESC) "_______IN EFFECT, k"

III.SHIFT:
A. DETERMINE THE GRID DIRECTION TO THE TARGET.
B. DETERMINE THE LATERAL SHIFT TO THE TARGET FROM THE KNOWN POINT.
W = Rm̸ (MIL RELATION FORMULA)
W = WIDTH OF LATERAL SHIFT (THE UNKNOWN)
R = DISTANCE TO THE KNOWN POINT DIVIDED BY 1000 AND ROUNDED TO ONE DECIMAL PLACE
m̸ = MEASURED ANGLE IN MILS FROM THE KNOWN POINT TO THE TARGET
C. DETERMINE THE RANGE SHIFT FROM THE KNOWN POINT TO THE TARGET

INCHES

OBSERVED FIRE REFERENCE CARD

FIST REPORT
OBS I.D.
DATE TIME GROUP
LOCATION ALT *
VISIBILITY
ENEMY ACTIVITY
FPF (GRIDS) *

INTELLIGENCE SPOT REPORT
IDENTIFICATION
S SIZE
A ACTIVITY
L LOCATION *
U UNIT *
T TIME
E EQUIPMENT

* ENCODE
FLASH-TO-BANG TIME
ELAPSED TIME BETWEEN IMPACT AND SOUND × 350 M (DISTANCE IN METERS)

MEASUREMENT OF ANGLES WITH THE HAND

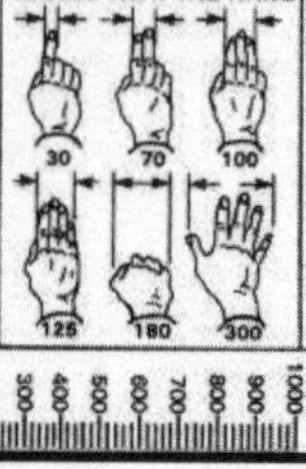

TGT LOCATION-SHIFT FROM KNOWN POINT
Complete Target Location
Direction 3570, Right 290, Add 500

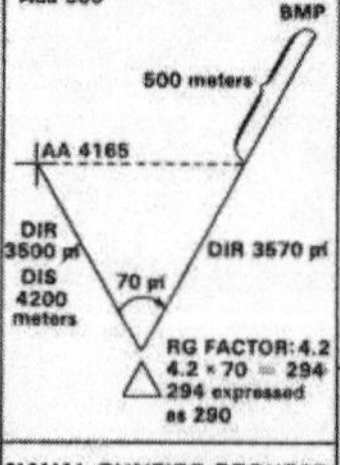

NAVAL GUNFIRE REQUEST

SEGMENT	DATA TRANSMITTED
Firing Alert	Spotter Identification Warning Order Target Number
	break and read back
Target Location	Grid, Polar, or Shift from a Known Point
	break and read back
Attack Data	Target Description Method of Engagement Method of Fire and Control

OT FACTOR
Distance to the target expressed in thousands
Example: 1800 OT Factor 2
2800 OT Factor 3
800 OT Factor 0.8
2500 OT Factor 2

ADJUSTMENT PROCEDURES
Deviation Correction (L R from the round to the target)
W R × m̸. R is the OT Factor, m̸ is mils from round to tgt

Range Adjustment (Add Drop from the round to the target)
Successive Bracket Technique
400
200
100
50 FIRE FOR EFFECT when you are within 50 meters

CAS REQUEST
OBS I D
WARNING ORDER (REQUEST CAS)
TGT LOCATION (GRID COORDINATES)
TGT DESC
-FRIENDLY LOCATION
-ADA THREAT
RESULTS DESIRED
DESIRED TIME ON TARGET (TOT)

ATTACK INFORMATION
DIRECTION IP TO TGT (DEGREES)
DISTANCE IP TO TGT (NAUT MILES)
TYPE MARKING ROUND
REIDENTIFY TGT (DESC)
REIDENTIFY ADA THREAT/RESTRICTIONS

1/25,000 METERS

1/50,000 METERS

HEADQUARTERS, DEPARTMENT OF THE ARMY

GPO: 1994 - 151-995

9 Line MEDEVAC

LINE ITEM	EXPLANATION
1. Location of Pickup Site.	Encrypt grid coordinates. When using *DRYAD Numeral Cipher, the same SET line* will be used to encrypt grid zone letters and coordinates. To preclude misunderstanding, a statement is made that grid zone letters are included in the message (unless unit SOP specifies its use at all times).
2. Radio Frequency, Call Sign, Suffix.	Encrypt the frequency of the radio at the pickup site, *not* a relay frequency. The call sign (and suffix if used) of person to be contacted at the pickup site may be transmitted in the clear.
3. No. of Patients by Precedence.	Report only applicable info & encrypt brevity codes. A = Urgent, B = Urgent-Surg, C = Priority, D = Routine, E = Convenience. (If 2 or more categories reported in same request, insert the word "*break*" between each category).
4. Spec Equipment.	Encrypt applicable brevity codes. A = None, B = Hoist, C = Extraction equipment, D = Ventilator.
5. No. of Patients by Type.	Report only applicable information and encrypt brevity code. If requesting MEDEVAC for both types, insert the word "*break*" between the litter entry and ambulatory entry: L + # of Pnt - Litter; A + # of Pnt - Ambul (sitting).
6. Security Pickup Site (Wartime).	N = No enemy troops in area, P = Possibly enemy troops in area (approach with caution), E = Enemy troops in area (approach with caution), X = Enemy troops in area (armed escort required).
6. Number and type of Wound, Injury, Illness (Peacetime).	Specific information regarding patient wounds by type (gunshot or shrapnel). Report serious bleeding, along with patient blood type, if known.
7. Method of Marking Pickup Site.	Encrypt the brevity codes. *A* = Panels, *B* = Pyrotechnic signal, *C* = Smoke Signal, *D* = None, *E* = Other.
8. Patient Nationality and Status.	Number of patients in each category need not be transmitted. Encrypt only applicable brevity codes. A = US military, B = US civilian, C = Non-US mil, D = Non-US civilian, E = EPW.
9. NBC Contamination (Wartime).	Include this line only when applicable. Encrypt the applicable brevity codes. N = nuclear, B = biological, C = chemical.
9. Terrain Description (Peacetime).	Include details of terrain features in and around proposed landing site. If possible, describe the relationship of site to a prominent terrain feature (lake, mountain, tower).

Reference: FM 8-10-6, *Medical Evacuation in a Theater of Operations, pages 7-7 through 7-9.*

GLOSSARY

1SG	first sergeant
5-C's	**Confirm**, **Clear**, **Call**, **Cordon**, **Control** (*reactions to contact with an IED*)
5-W's	**Who**, **What**, **Where**, **When**, **Why**?
5-S's	**Search**, **Silence**, **Segregate**, **Safeguard**, **Speed** to rear (*rules for handling prisoners*)
AA	avenue(s) of approach
AAR	after-action review
AATF	air assault task force
ABCDE	method of identification and response to life-threatening conditions: **Airway**, **Breathing**, **Circulation**, **Disability**, **Exposure**
ACE	ammunition, casualties, and equipment
ACL	allowable combat load
ACP	aerial checkpoint
ACU	Army combat uniform
ADA	air defense artillery
AG	assistant gunner
ATC	air traffic controller; a mechanical belay device that locks down on itself when tension is applied in opposite directions
ALT	alternate
AMC	air mission commander
ammo	ammunition
ANCD	automated net-control device
AO	area of operations
AOO	actions on the objective
AR	automatic rifleman
ABF	attack by fire
ATL	Alpha team leader
ATM	Alpha Team
bangalore torpedo	A manually emplaced, 1.5-meter long explosive-filled tube used to breach wire and detonate simple, pressure-activated antipersonnel mines. Ten tubes will clear a 1- by 15-meter lane.
belay	Any action taken to stop a climber's fall or to control the rate a load descends
binos	Binoculars
BMNT	begin morning nautical twilight
BN	battalion
body belay	belay that uses the belayer's body to apply friction by routing the rope around the his body
bow line on a coil	knot used to secure a climber to the end of a climbing rope

BP	battle position
British junction knot	knot used to join the ends of detonation cords from multiple charges to one initiation system
BTC	bridge team commander
BTL	Bravo Team leader
BTM	Bravo Team
C2	command and control
CAS	close air support
CASEVAC	casualty evacuation
CCA	close combat attack
CCIR	commander's critical information requirements
CCP	casualty collection point
CDR	commander
CDS	Camp Darby Special (*map*)
CLS	combat lifesaver
CO	company
COA	course of action
COMSEC	communications security
cordelette	short section of static rope or static cord. *Also called* **"sling rope"**
COTS	commercial off the shelf
CP	command post
CPR	cardiopulmonary resuscitation
CQC	close quarters combat
CS	combat support
CSS	*obsolete: now referred to as* sustainment
CTT	common task test
DAR	designated area of recovery
DOL	direction of landing
double Figure 8 knot	knot used to form a fixed loop in the end of the rope; loops are large enough to insert a carabiner
double overhand knot	knot used to secure the end of detonation cord
DP	duty position; decision point (depending on context)
DST	distance
DTG	date-time group
dynamic ropes	one of two classifications of kernmantle rope; used for climbing; *see also* **static rope**
DZ	drop zone
EA	engagement area
EEFI	essential elements of friendly information
EENT	end evening nautical twilight
End-of-the-Rope Clove Hitch	intermediate anchor knot that requires constant tension

End-of–the–Rope Prusik	knot used to attach a movable rope to a fixed rope; *see also* **Middle-of–the–Rope Prusik**
ENY	enemy
EPW	enemy prisoner of war
FA	field artillery
FDC	fire direction center
FFIR	friendly force information requirements
Figure 8 slip knot	knot used to form an adjustable bight in the middle of a rope
FLIR	forward looking infrared
FLOT	forward line of own troops
FO	forward observer
FPF	final protective fires
FRAGO	fragmentary order
FSC	fire support coordinator
FSO	fire support officer
GOTWA	**G** Where leader is Going
	O Others he is taking with him
	T Time he plans to go
	W What to do if the leader does not return in time
	A The unit's and the leader's actions on chance contact while the leader is gone
GPS	global positioning system
GRN	grenadier
HDG	heading
HE	high explosive
H-Hour	hit hour (the time the unit plans to accomplish the mission)
HI	high temperature (weather)
HPT	high-payoff target
HQ	headquarters
IAW	in accordance with
ICM	improved conventional munitions
ID	identification
IP	initial point
IR	information requirements
ERRP	en route to release point
JAAT	joint air attack team
JD	Julian date
KIA	killed in action
LAW	light antiarmor weapon
LBV	load-bearing vest
LD	line of departure

LDA	linear danger area
LO	low temperature (weather)
LOA	limit of advance
LOGSTAT	logistical status
LP	listening post
LZ	landing zone
material factor	the strength, hardness, and mass of the material to be demolished
mb	millibar (a metric unit used to measure air pressure)
MDI	modernized demolition initiator
ME	main effort
mechanical belay	a belay that uses mechanical devices to help the belayer control the rope, as in rappelling
MEDEVAC	medical evacuation
METL	mission-essential task list
METT-TC	mission, enemy, terrain (and weather), troops (and support) available, time available, and civil considerations
Middle– of–the–Rope Clove Hitch	knot that secures the middle of a rope to an anchor
Middle-of-the-Rope Prusik	knot that attaches a movable rope to a fixed rope, anywhere along the length of the fixed rope; *see also* **End-of–the–Rope Prusik**
MG	machine gunner
MOPP	mission-oriented protective posture
MR	moonrise
MRE	meal, ready to eat
MS	moonset
MSD	minimum safe distance
MSL	mean sea level
Munter hitch	commonly used belay that requires little equipment
NATO	North Atlantic Treaty Organization
NAV	navigation
NFA	no-fire area
NLT	no later than
NVD	night-vision device
NVS	night vision system
OAKOC	observation and fields of fire, avenues of approach, key terrain, observation, and cover and concealment
OBJ	objective
occlusive dressing	a dressing that seals a wound from air or bacteria
OD	olive drab
OOM	order of movement
OP	observation post

OPORD	operation order
OPSKED	operational schedule
ORP	objective rally point
OT	observer-target
PB	patrol base
PCC	precombat checks
PCI	precombat inspection
PDF	principal direction of fire
PI	probability of incapacitation
PIR	priority intelligence requirements
PL	platoon leader
PLD	probable line of deployment
PLOT-CR	purpose, location, observer, trigger, communication method, resources (a format for planning fire support)
PLT	platoon
POL	petroleum, oils, and lubricants
PRI	primary
protection	a piece of equipment, natural or artificial, that is used to construct an anchor
PSG	platoon sergeant
PSI	pounds per square inch
PZ	pickup zone
R	rifleman
R&S	reconnaissance and surveillance
RACO	rear area combat operations
RAP	rocket-assisted projectile
rappel seat	a rope harness used in rappelling and climbing
RED	risk-estimate distance
REQ	required
rerouted figure 8 knot	anchor knot that also attaches a climber to a climbing rope
RFA	restrictive fire area
RFL	restrictive fire line
RFLM	rifleman
RHB	Ranger Handbook
ROE	rules of engagement
round turn with two half hitches	a constant tension anchor knot
RP	release point
RTO	radio operator
S-2	intelligence staff officer
S-3	operations staff (and training) officer

SALUTE	**Size**, **Activity**, **Location**, **Unit/Uniform**, **Time**, and **Equipment**
SAW	squad automatic weapon
SBF	support by fire (position)
SDT	self-development test
SE	supporting effort
SEAD	suppression of enemy air defenses
SITREP	Situation Report
SITTEMP	situational template
SL	squad leader
sling rope	short section of static rope or static cord. Also called "cordelette"
SLLS	**Stop**, **Look**, **Listen**, **Smell**
SOC	succession of command
SOI	signal operating instructions
SOP	standing operating procedures
SP	start point
square knot	knot used to join two ropes of equal diameter; used to join the ends of the detonation cord to the explosive
SR	sunrise
SS	sunset
STANO	surveillance, target acquisition, and night observation
static ropes	one of two classifications of kernmantle rope; used where rope stretch is undesired, and when the rope is subjected to heavy static weight. *See also* dynamic rope
SURVIVAL	**S** Size up the situation, your surroundings, your physical condition, and your equipment. **U** Undue haste makes waste; don't be too eager to move. Plan your moves. **R** Remember where you are in relation to important friendly and locations and critical resources **V** Vanquish fear and panic. **I** Improvise. You can improve your situation. Use what you have. Use your Imagination. **V** Value your life. Remember your goal: to get out alive. Remain stubborn. Refuse to give in to problems and obstacles that face you. This will give you the mental and physical strength to endure. **A** Act like the natives; watch their daily routines. When, where, and how do they get food? Where they get water? **L** Live by your wits. Learn basic skills.
suspension traverse	used to move personnel and equipment over rivers, ravines, chasms, and up or down a vertical obstacle
TAC	tactical air controller
tamping factor	depends on the location and tamping of the charge
technical climbing	using safe and proper equipment and techniques to climb on a rock formation in parties of two or more
tensionless anchor	used to anchor rope on high-load installations such as bridging

TL	team leader
TLP	troop-leading procedures
TL	team leader
TOC	tactical operations center
triple roll knot	knot used to join branches of detonation cord
TTP	tactics, techniques, and procedures
uli knot	knot used to securely fasten detonation cord to explosive
VIXL	video image crosslink
WARNO	warning order
WBGT	wet bulb globe temperature
WFFs	warfighting functions (fire support, movement and maneuver, protection, command and control, and sustainment)
WIA	wounded in action
WSL	weapons squad leader
XO	executive officer

INDEX

- C's, 8–1
- abatis, 5–11 (illus)
- ABC's, 16–1
- aboveground fire, 14–22
- acclimatization, 16–9 (illus)
- accountability, 1–6
- ACE report, 1–4, 7–14, 7–17
- actions
  - at danger areas, 1–3
  - at halts, 6–5, 6–7
  - in the objective area, 1–3, 1–5
  - in the patrol base, 1–3, 1–5, 1–6
  - on the objective, 7–5, 7–7 (illus), 7–9, 7–12 (illus), 7–15, 7–17
- AH–1S Cobra, 15–15 (illus)
- AH–64A Apache, 15–16 (illus)
- aid and litter team, 7–2
- air
  - assault formations, 15–2 (illus)
  - movement
    - annex, 2–17
  - traffic controller, 9–15)
- aircraft
  - capabilities and limitations, 3–10
  - request, B–3 (illus)
- air tasking order special instructions (ATOSPINS), 14–1
- airway management, 16–1
- alert plan, 7–19
- alternate routes, 7–3
- alternating bounds, 6–2
- ambush, 7–10, 7–11, 7–12
  - boards, A-17 (illus)
  - deliberate, 7-15 (illus)
  - formations, 7-11 (illus)
  - SOP, A-16 (illus)
  - types, 7-10
- ammunition, 2-10
  - planning, 10-11
- anchoring traverse rope, 9-20 (illus)
- anchors, 9-5
- annexes, 2-17
- antennas, 4-6, 4-7, 4-8
  - base, 4-9 (illus)
  - field expedient, 4-8 (illus), 4-9, 4-10 (illus), 4-11 (illus)
  - length considerations, 4-12
  - whip, 4-7
  - wire, 4-7
- Apache , AH-64A, 15-15, 15-16 (illus)
- approach
  - march, 7-23, 7-24
  - to building or breach point, 12-7
- area
  - ambush, 7-10, 7-12
  - reconnaissance, 7-5 through 7-9
- Army aviation coordination checklist, 2-26
- artificial
  - anchors, 9-1
  - protection, 9-4 (illus), 9-6 (illus)
- assault
  - element, 7-2, 7-10, 7-15, 7-16, 7-17
- assumption of command, 1-8
- ATOSPINS, 14-1
- attachments and detachments, 2-12
- attack helicopters, 15-15
- automated net control device (ANCD), 4-1, 4-5
- avenues of approach, 2-3, 12-5
- aviation operations, 15-1

B

- backbrief, 2-4
- battle drills, 8-1
- beaching a boat, 13-4
- beaten zone, 10-2
- begin morning nautical twilight (BMNT), 2-11, 2-23
- Be, Know, Do, 1-1 (illus)
- belays, 9-13
- Big Dipper, 14-6
- bites, 16-6
- Blackhawk , UH-60A, 15-18
- blasting caps, 5-3
- bleeding, 16-2
- boats, crew positions on, 13-6 (illus)
- body
  - belay, 9-13
  - rappel, 9-22 (illus)
- bounding overwatch, 6-2, 6-3 (illus), 6-4 (illus)
- box method, 7-9 (illus)
- breach.
  - mined wire obstacle, 8–25
  - point, approach to, 12–7
- breaching
  - charge, 5–7
  - material factors, 5–10 through 5–14
  - radius (R), 5–11 (illus)
- break contact, 8–6
- breathing, 16–2
- bridges, 9–3, 9–17, 13–1
- british junction, 5–7 (illus)
- buddy teams, 7–2
- burns, 16–3
- burst of fire, 10–2
- butchering large game, 14–17 (illus)

C

- call for fire, 3–5, 3–8 (illus)
- camming device, 9–4 (illus)
- camouflage, 14–2
- capsize drill, 13–8
- captured documents, 2–5
- carabiner(s), 9–2, 9–24 (illus), 9–25 (illus)
- care under fire, 16–1
- cargo helicopters, CH–47B Chinook, 15–19 (illus)
- carry technique, 12–6
- Cassiopeia, 14–6 (illus)
- casualty
  - collection point (CCP), 1–3
  - criterion, 3–3
  - evacuation (CASEVAC), 2–13
  - feeder card, B–1 (illus)
  - report, 1–2
- cavity, 5–5
- Cayuse , OH–6A, 15–14 (illus)
- CH–47B Chinook, 15–19 (illus)
- chain gun, 30 mm, 3–10
- charge
  - bare, 5–8 (illus)
  - breaching, 5–8, 5–9, 5–10 (illus)
  - material factor (K), 5–10 (illus)
  - expedient, 5–4 through 5–6 (illus)
  - internal or external, 5–13 (illus), 5–14 (illus)
  - ring, 5–11 (illus)
  - size, cutting timber, 5–11 (illus)
  - timber cutting, 5–12 (illus)
- checklist coordination
  - adjacent unit, 2–16
  - aviation, 2–27, 2–28,
  - fire support. 2–15
  - forward unit, 2–25

intelligence, 2–24
rehearsal area, 2–26
chemlights, 12–9
chillblain, 16–5
Chinook, CH–47B, 15–19 (illus)
circulation, 16–1
civilians, 12–5
classes of fire, 10–3 through 10–7 (illus)
clear a room, 8–18
climbing commands, 9–15 (illus)
close
air support, 3–8, 3–9 (illus), 3–10 (illus)
combat attack (CCA), 3–10, 3–11
quarters combat, 12–6
Clove Hitch, 9–9, 9–10
Coal Sac, 14–7 (illus)
Cobra , AH–1S, 15–15
cold injuries, 16–5 (illus)
cobra head, 4–10 (illus)
combat
intelligence, 2–5
patrols, 7–2, 7–9
commando crawl, 9–17, 9–18 (illus)
common sense, 7–1
communication, 4–1, 14–2
desert, 15–11
compass man, 7–2
complete plan, 2–5
concrete, breaching charges for, 5–9 (illus)
conduct a detailed mission analysis, 2–2
cone of fire, 10–2
constant tension anchor knot, 9–8 (illus)
constellations, 14–6 (illus), 14–7 (illus)
contingency plan, 7–4
converging routes method, 7–9 (illus)
conversion factors, materials other than concrete, 5–10 (illus)
convoy operations, 11–1
coordinated fires, 7–15
cordelette, *See sling rope and static rope*, 9–3
cords, 9–3
correction of errors, 3–7
course of action, 2–2, 2–3
cover and concealment, 2–3, 12–5
coxswain, duties, 13–6
crossing a small open area, A–10 (illus)
crossing site, 13–4
current, 13–9

D

DA Form 1155, 1–2
DA Form 1156, 1–2
DA Form 5517–R, B–7 (illus)
Dakota fire hole, 14–22 (illus)
danger
area, 1–3, 6–7 through 6–10, A–8, A–9
dead
space, 1–4, 10–3
water, 13–8
deadfall trap, 14–14 (illus)
debarkation, 13–6, 13–8 (illus)
defense, 10–9
in urban operations, 12–2
definitions, 7–1, 10–2
dehydration, 16–5
delay, 3–1
demolition, 5–1
explosives, 5–2
knots, 5–7
team, 7–2
density altitude, mountain, 15–12
descender
carabiner wrap, 9–23 (illus), 9–24(illus), 9–25 (illus)
figure 8, 9–22, 9–23 (illus)
desert, 14–8, 15–11
detonation (firing) systems, 5–4
detonation point, 5–5
diamond formation, 15–3 (illus)
direct fire, 8–1, 8–2, 8–5
dismounted mobility, 9–1
distribution of fire, 10–10
documents, captured, 2–6
drag noose snare, 14–13 (illus)
duties, responsibilities, actions, 1–2 through 1–8
dynamic ropes, 9–3

E

edibility test, 14–9
embarkation, 13–6
end evening nautical twilight (EENT), 7–22
end–fed antenna, 4–8
enfilade fire, 10–5
en route recorder, 7–2
enter
and clear a room, 8–18
a trench to secure a foothold, 8–21
environmental injuries, 16–4 (illus)
equipment, 4–1, 7–22
mountain, 15–17
navigation, 13–9
personnel and, 13–5
poncho raft, 13–4
rescue, 9–2
rope bridge, 9–17
suspension traverse, 9–20
wet crossing (One Rope Bridge), 13–1
essential elements of friendly information (EEFI), 2–13
estimate process, 2–3
evacuation team, 9–1
evasion, 14–1
expedient
antenna, 4–8
explosives, 5–4 throug 5–6 (illus)
exploitation, 12–2
explosives
demolition, 5–2 (illus)
expedient, 5–4 through 5–5 (illus)
extremities, 16–2

F

factor K, 5–10 (illus)
False Cross, 14–7 (illus)
fan method, 7–9 (illus)
field
artillery, 3–2 (illus)
sanitation, 6–6
field–expedient (FE)
omnidirectional antennas, 4–8
figure 8
descender, 9–22, 9–23 (illus)
double, 9–11
rerouted, 9–11 (illus)
slip knot, 9–12 (illus)
final
protective fire (FPF), 10–2
protective line (FPL), 10–2, B–6 (illus)
fire
aboveground, 14–22
classes, 10–3 through 10–7 (illus)
commands, 10–11
control, 10–10
Dakota fire hole, 14–22
fixed, 10–7
laying methods, 14–23 (illus)
lean–to, 14–21 (illus)

site preparation, 14–21
superiority, 8–15
support, 1–8, 3–1, 3–5, 6–4
coordination checklists, 2–25
overlay, 3–3, 3–4 (illus), 3–5
traversing and searching, 10–7
first aid, 16–1
fishing, 14–10 (illus)
five phases of truck movement, 11–1
fixed
fire, 10–7
protection, 9–3 (illus)
flank security, 7–12
flanking fire, 10–5
food
animals, processing, 14–15 through 14–18 (illus)
insects, 14–10
meat, smoking, 14–19 (illus)
plants, 14–9, 14–10
snakes, 14–15, 14–16 (illus)
foot care, 16–7
force protection, 2–13
forced stop, 11–8 (illus), 11–9 (illus)
formations, 6–1, A–6
air assault, 15–2 (illus)
and orders of movement, A–6 (illus)
boat, 13–9, 13–10 (illus)
landing, 15–12
formula
antenna size, 4–12 (illus)
charge size, 5–11 (illus)
conversion factors, 5–10 (illus)
fallen tree obstacle, 5–12 (illus)
internal charge, 5–13
safe distances for large charges, 5–8
size of charge to breach concrete, masonry, rock, 5–9 (illus)
forward observer (FO), 1–8 (illus)
fragmentary order (FRAGO), 7–14, 7–15, 7–17
free gun, 10–7
frequency ranges, 4–4
frontal fire, 10–5
frostbite, 16–5
full spectrum operations, 12–1
fundamentals
of movement, 6–4
of reconnaissance, 7–5
fuse igniter, M81, 5–3

G

game, 14–15 through 14–18 (illus)
gill net, 14–10 (illus)
grapeshot charge, 5–6 (illus)
graphic training aid (GTA), 07–01–038, B–2 (illus)
grenades, 8–18
ground
movement plan, 11–2
slope, 15–9, 15–10 (illus)
tactical plan, 11–2, 15–1

H

halts, 6–7
heat injuries, 16–4
heavy formation, 15–3 (illus), 15–4 (illus)
helicopters, 15–13 (illus) through 15–19 (illus)
attack, 15–15 (illus)
cargo, 15–19 (illus)
observation, 15–13 (illus)
utility, 15–17 (illus)
Hellfire missile, 3–10
hide site, 14–2
high–carry technique, 12–6
hole–up area, 14–2
hydration, 16–9 (illus)
hypothermia, 16–5

I

IED/UXO report, B–4 (illus)
immediate assault, 8–1
implied task, 2–2, 7–3
improvised
explosive device (IED), 8–1
shaped charge, 5–4, 5–5 (illus)
indirect fire, 3–1, 8–28
Infantry Leader's Reference Card, B–2 (illus)
inflation of watercraft, 13–5
initial
evasion point, 14–1
planning and coordination, 7–1
initiating (priming) systems, 5–3
injuries, 16–4 through 16–6 (illus)
inspection, 2–5, 13–1
insulator, 4–7
intelligence, 2–5, 2–24
interdiction, 3–1
intermediate anchor knot Clove Hitch, 9–8, 9–9 (illus)
internal charge, 5–13, 5–14
Iroquois, UH–1H, 15–17 (illus)
island, 13–8
issue
a warning order, 2–6
the complete order, 2–10

K

key terrain, 2–4, 12–5
Kiowa, OH–48D, 15–13 (illus)
knock out bunker, 8–15
knots
demolition, 5–7
double figure 8, 9–11 (illus)
double overhand, 5–7 (illus)
End–of–the–Clove Hitch, 9–8, 9–9 (illus)
End–of–the–Rope Prusik, 9–12 (illus)
figure 8 slip, 9–12 (illus)
Middle–of–the–Rope Clove Hitch, 9–9 (illus)
Middle–of–the–Rope Prusik, 9–13 (illus)
rappel seat, 9–10
rerouted figure 8, 9–11 (illus)
round turn, 9–8 (illus)
square, 9–8 (illus)
uli, 5–7 (illus)

L

laid rope, 9–3
landing formations, 15–9
landing
plan, 15–1
sites, 13–7, 15–11, 15–12
zones, 15–1
landings, desert, 15–11
large open danger area, A–9 (illus)
laying a fire, 14–21, 14–22 (illus), 14–23 (illus)
lazy W, 14–4
lead team, 6–6
leader's reconnaissance, 7–4, A–18 (illus)
leadership, 1–1, 7–1
lean–to, 14–19, 14–20 (illus), 14–21 (illus)
lifesaving steps, 16–1

lift or shift fires, 7–13
liftoffs, desert, 15–11
limit, 3–1
line of sight, 4–4, 10–2
linear
ambush, 7–10
danger area, 6–7, 6–8 (illus), A–8 (illus)
linkup, 7–18, A–7 (illus)
litter, 16–7
live, virtual, and constructive training, 12–3
loading
plan, 11–2
procedure, 9–2
sequence, UH–60, 15–7 (illus)
locking down a room, 12–7
locking snare loop, 14–13, 14–14 (illus)
logroll method, 9–2
long
count, 13–6 (illus), 13–7
halt, 6–7
–range surveillance, 7–5, 12–5
low–carry technique, 12–6
L–shaped ambush, 7–10

M

M81 fuse igniter, 5–3
machine guns, 10–1 (illus)
M2, 10–8, 10–10
M240B, 10–1, 10–8
M249, 10–1, 10–8, 10–10
MK 19, 10–8, 10–10
effective ranges, 10–1 (illus)
ordinate, 10–2, 10–3 (illus)
main body, 6–5
make a tentative plan, 2–2
malfunction, weapon, 12–6
maneuver, 3–3, 7–9
element, 10–9
mark, 3–6
matches, 5–3
material factor K, 5–10 (illus)
meat, smoking, 14–19 (illus)
mechanical belay, 9–14 (illus)
medic, 1–7(illus), 6–6
medical evacuation (MEDEVAC),
nine–line report, B–9 (illus)
plan, 6–6
request, B–3 (illus)
memory aids, GOTWA, 7–4 (illus)
mess plan, 7–22
message to observer, 3–7
method
of engagement, 3–6
of fire, 10–11
of fire and control, 3–6
METT–TC, 6–5, 12–4
military mountaineering, 9–1
minimum safe distances, 5–8 (illus)
mission, 2–1
accomplishment, 6–5
analysis, 2–2, 11–1
–essential tasks, 2–2
grid, 3–8 (illus)
preparation, 7–23
receive the, 2–1
restarted, 2–2
MK 19, 10–8, 10–10
monkey crawl, 9–19 (illus)
mortars, 3–2 (illus), 3–3 (illus)
mountain
aircraft operations, 15–1
density altitude, 15–12
landing sites, 15–12
navigation, 15–12
site assessment, 15–13
winds, 15–12
mountaineering
equipment, 9–3
training, 9–1
movement, 6–1
during limited visibility conditions, 6–6
formations, 6–1
plan, 2–12, 11–2, 15–1
techniques, 6–1
to contact, 7–23, 12–2
munter hitch, 9–14 (illus)

N

NATO standard markings, 12–8
natural anchors, 9–5
navigation, 13–9, 14–3
navigator–observer method, 13–9
night vision devices, 6–4, 6–7
nine–line MEDEVAC report, B–9 (illus)
North star, 14–6

O

objective
actions on the, 7–5, 7–6, 7–7 (illus)
area, 1–3, 1–5
rally point (ORP), 7–4, 7–19, 7–20 (illus)
oblique fire, 10–5 (illus)
observation helicopters, 15–13 (illus), 15–14 (illus)
observed fire reference card, B–8 (illus)
obstacles, 2–3, 5–12, 12–5, 15–1
occupation, controlled, 10–9
offense, 10–8, 12–1
offensive considerations, 2–3 (illus)
OH–48D Kiowa, 15–13 (illus)
OH–6A Cayuse, 15–14 (illus)
operation
order (OPORD), 2–10, 2–11 through 2–14 (illus)
pickup zone, 15–6
signals, 15–9
operations, 2–1
convoy, 11–1
types, , 2–30
orders. *See also* fragmentary, operation, and warning orders
annexes, 2–17
of movement, A–6 (illus)
standing, Rogers' Rangers, inside front cover
overlay, fire support, 3–3 through 3–5 (illus)
overwatch, 6–2, 6–3 (illus), 6–4 (illus)

P

packing list, 2–9 (illus)
paintball guns, 12–8
passive patrol base, 7–21
passwords, 7–3
patrol, 7–1
base, 7–21, 7–22, 7–23 (illus)
personal hygiene, 7–22
pickup zones, 15–1, 15–6 (illus)
pitons, 9–5 (illus)
plan, 7–3
alert, 7–19
communications, 6–5
completion, 7–3
contingency, 7–4
loading, 11–2
medical evacuation (MEDEVAC), 6–6
mess, 7–22
movement, 11–2
tactical, 11–2, 15–1
withdrawal, 7–22
planning, 7–1
ammunition, 10–11

completion, 7–3
considerations, 7–20, 14–1
initial, 7–3,
planning considerations, 14–1
plant edibility, 14–9
plants, poisonous, 14–9, 16–7
platoon
bounding overwatch, 6–2, 6–3 (illus) 6–4 (illus)
leader, 1–2 (illus), 6–6
ORP, 7–4, 7–19, 7–20 (illus)
sergeant, 1–2 (illus), 1–3, 1–4 (illus),
platter charge, 5–5, 5–6 (illus)
point ambush, 7–10, 7–12
pointer stars, 14–6 (illus)
poisonous plants, 16–7 (illus)
polar, 3–8 (illus)
Polaris, 14–6 (illus)
poncho
lean–to, 14–21 (illus)
raft, 13–4
preparation
actions during, 1–6
mission, 7–22
site, 14–21
primary
routes, 7–3
principal direction of fire, 10–2, B–5 (illus)
principles, 7–1
of leadership, 1–1
priorities of work, 7–21
priority intelligence requirements (PIR), 2–13
prisoners, 2–6
probable line of deployment (PLD), 7–13
processing fish or game, 14–15
protection, 9–4 (illus)
Prusik knots, 9–13 (illus)
purpose, location, observer, trigger, communication, resources (PLOT–CR), 3–5 (illus)
pursuit, 12–2

R

radiating element, 4–10, 4–12 (illus)
radio, 4–1
AN/PRC 119F, 4–4
frequencies, 4–4 (illus)
operator, 1–7
transmitter, 4–7
raid, 7–15
assessment, 7–17 (illus)
boards, A–12 (illus) through A—15, (illus), A–19 (illus)
rally point, 6–7, 7–4, 7–19
rappel
seat, 9–10 (illus), 9–18, 9–22 (illus)
site selection, 9–25
rappelling, 9–22 (illus)
rate of fire, 10–1 (illus)
reach, 13–9
react to
ambush, 8–9, 8–12, 11–6 (illus), 11–7 (illus)
contact, 8–1, A–2 (illus)
far ambush, A–4 (illus)
indirect fire, 8–28, A–1 (illus)
near ambush, A–3 (illus)
reconnaissance, 7–1, 7–5 through 7–9
method comparison, 7–9
recovery operations, 11–11 (illus)
rehearsal, 2–4, 12–6
area, 2–26
coordination checklist, 2–26
stream crossing, 13–1
release point, 6–5
rendezvous point, 6–5
repairs, 4–6
reports, 2–5
rerouted figure 8 knot, 9–11 (illus)
rescue equipment, 9–2
respect
to ground, 10–3, 10–4 (illus)
to gun, 10–7 (illus)
to target, 10–4, 10–5, 10–6 (illus)
responsibilities, 1–2, 6–6
rest/sleep plan, 7–22
restated mission, 2–2
resupply, 7–22
reverse planning, air assaults, 15–1
ring charge, 5–12 (illus)
risk, 3–2
estimate distances (RED), 3–3 (illus)
river, 13–8
Rogers' Rangers, inside front cover
rope
bridge, 9–17, 13–1
installations, 9–15, 9–16
ropes, 9–3
round–turn with two half hitches knot, 9–8(illus)
route of march, 6–5
routes, 7–3, 7–4 (illus) 14–1
rubber boats, 13–6
ruler, B–8 (illus)
rules of engagement, 2–13, 8–18
running passwords, 7–3

S

safe distances, minimum, 5–8 (illus)
safety, 5–3, 15–9
SALUTE format, 2–5
sandbar, 13–9
sanitation, 6–6
search, 7–23
and attack, 7–23
and rescue numeric encryption grid (SARNEG), 14–2
team, 7–2
searching fire, 10–7
seat–hip rappel, 9–10, 9–18 (illus)
security, 1–4, 6–4, 6–6, 6–7, 7–1, 9–1, 12–4
continuous, 7–22
element, 7–3, 7–5, 7–9, 7–14, 7–16
measures, 7–5, 7–21
squad, 6–6
semipermanent protection, 9–5 (illus)
shadow tip, 14–3, 14–4 (illus)
shelters, 14–19, 14–20, 14–21 (illus)
shift from a known point, 3–6
shock, 16–1
tube, 5–3
short
count, 13–6
halt, 6–7
–range surveillance, 7–8 (illus)
signals, 15–9
simplicity, 12–4
site
assessment, mountain, 15–13
selection, 7–18, 7–20, 9–25
situation.report (SITREP), 1–2

size, activity, location, unit/uniform, time, equipment (SALUTE) report, 2–5 (illus)
skinning and butchering game, 14–15 (illus) through 14–18
sleep plan, 7–22
slide method, 9–2, 16–8
sling rope, *See cordalette and static rope*, 9–3
slings, 9–4
slit trench, 7–22
slough, 13–8
smoking meat, 14–19 (illus)
snake bite, 16–6 (illus)
snakes
- edible, 14–15, 14–16 (illus)
- venomous, 14–16

snares, 14–11
- drag noose, 14–13 (illus)
- treadle, 14–12 (illus)

snow blindness, 16–5
Soldier's Load, 7–23
Southern Cross, 14–7 (illus)
spacing stick, 4–11 (illus), 4–12 (illus)
spears and fish hooks, 14–11 (illus)
special teams, 7–14
specified task, 2–2
speed principle, 12–4
spider bite, 16–6
spoilage, 14–15
spray paint, 12–8
spring–loaded camming device, 9–4 (illus)
squad
- attack, A–11 (illus)
- bounding overwatch, 6–2, 6–3 (illus), 6–4 (illus)
- leader (SL), 1–4 (illus), 6–1
- ORP, 7–20 (illus)

square knot, 9–9 (illus)
stability, 12–2
staggered trail formation, 15–2, 15–5 (illus)
staging plan, 11–1, 15–1
stance, 12–6
standard range card, B–7 (illus)
standards of movement, 6–4
standing operating procedure (SOP), 1–2,
- ambush, A–16
- Nato markings,12–8
- squad ammunition, 2–10

Standing Orders, Rogers' Rangers, inside front cover
standoff, 5–5
static rope, *See cordalette and sling rope*, 9–3
stealth, 6–4
stop, look, listen, smell (SLLS), 6–7
stream crossing annex, 2–21
stretcher, 9–2, 16–8
subordinate guidance, 2–9 (illus)
successive bounds, 6–2, 6–3 (illus)
supervise and refine, 2–4
support–by–fire position, 10–9
suppress, obscure, secure, reduce, assault (SOSRA), 8–21
suppression, 8–8 (illus)
surface conditions, 15–1, 15–9
surprise, 12–4
surveillance, 7–5
- team, 7–2, `12–5

survival, 14–3, 14–8, 14–9 (illus)
suspension traverse, 9–20 (illus), 9–21 (illus), 9–22 (illus)
swinging traverse, 10–7

T

tactical
- cross–loading, 11–2
- loading sequence, UH–60, 15–8 (illus)
- marches, 6–5
- unloading sequence, UH–60, 15–8 (illus)

tamping factor (C), 5–11 (illus)
target
- description, 3–8
- location, 3–5
- overlays, 3–3

task
- organization, 7–1, 9–1, 11–2, 12–1,
- standards, 7–5, 7–24

tasks, 7–3
- to maneuver units, 7–9
- to subordinate units, 2–6

team leader (TL), 1–6 (illus)
temperate zones, 14–5
tensionless anchor, 9–5, 9–7 (illus)
tentative
- ORP, 7–19, 7–20 (illus)
- plan, 2–2

tepee, fire, 14–19 (illus)
terrain model, 2–30, 2–31 (illus)
- –dimensional battlefield, 6–5
- –point anchor, 9–6 (illus)

timber–cutting charges (external), 5–11 thru 5–14 (illus)
time, 12–5
- schedule, 2–9, 2–13 (illus)

TNT (P), 5–11 (illus), 5–12 (illus), 5–14 (illus)
tools of the tactician, 2–1 (illus)
tracer burnout, 10–1 (illus)
traditional protection, 9–4 (illus)
trail team, 6–6
trajectory, 10–2, 10–3 (illus
transmissions, 3–8 (illus), 15–11 (illus)
transport–tightening system, 9–17 (illus)
traps and snares, 14–11, through 14–14 (illus), 14–16 (illus)
traveling, 6–1
- overwatch, 6–2

traversing and searching fire, 10–7
treadle snare, 14–12 (illus)
triple roll knot, 5–7 (illus)
trip–string deadfall trap, 14–14 (illus)
troop–leading procedures (TLP), 2–1 (illus)
troubleshooting, 4–6
truck annex, 2–22

U

UH–1H Iroquois, 15–17 (illus)
UH–60A Blackhawk, 15–18 (illus)
uli knot, 5–7 (illus)
unit tasks, 2–2
unloading sequence, UH–60, 15–7 (illus)
urban
- operations, 12–1
- warriors, 12–3

utility helicopters, 15–17, 15–18 (illus)

V

vee formation, 6–1 (illus)
vehicle recovery, 11–11 (illus)
venomous snakes, 14–16
vertical hauling line, 9–16 (illus)

violence of action, 7–15, 12–4
visual signals, 7–11, 15–9

W

warfighting functions, 2–12 (illus)
war–game, 2–4
warning
  order (WARNO), 2–2, 2–6, (illus)
  shock tube, 5–1
wasp or bee sting, 16–6
watch method, 14–5 (illus)
water
  consumption, 16–9, 16–10 (illus)
  discipline, 6–6, 14–8
  purification, 14–7, 14–8
  supply/resupply, 7–22
  survival still, 14–9 (illus)
waterborne operations, 13–1
watercraft, other, 13–5
weapons
  maintenance, 7–22
  squad leader, 1–5 (illus), 10–11
  squad TTP, 12–8
webbing, 9–4
wedge formation, 6–1 (illus)
wet crossing, 13–1
whip antenna, 4–7
winds, 15–11, 15–12
wire antenna, 4–7
withdrawal, 7–22, 10–9
wolf tail, 12–9
work, rest, and water, 16–9, 16–10 (illus)
zone reconnaissance, 7–5, 7–8, 7–9 (illus)